# Separations of *f* Elements

# Separations of *f* Elements

Edited by

Kenneth L. Nash

*Argonne National Laboratory*
*Argonne, Illinois*

and

Gregory R. Choppin

*Florida State University*
*Tallahassee, Florida*

Springer Science+Business Media, LLC

Library of Congress Cataloging-in-Publication Data

On file

Proceedings of an American Chemical Society Symposium on *f* Elements Separations, held March 13–17, 1994, in San Diego, California

ISBN 978-1-4899-1408-8 ISBN 978-1-4899-1406-4 (eBook)
DOI 10.1007/978-1-4899-1406-4


Originally published by Plenum Press, New York in 1995
Softcover reprint of the hardcover 1st edition 1995

10 9 8 7 6 5 4 3 2 1

## PREFACE

The symposium which provided the incentive for this volume was conducted in San Diego, California as a part of the 207$^{th}$ National Meeting of the American Chemical Society, March 13-17, 1994. It was conceived partly to continue an informal decennial sequence of symposia dedicated to the topic of f element separations. A lot has changed in the world of f elements over the last ten years, precipitating a change in emphasis which should be evident to most practitioners in the field. Production and reprocessing of nuclear fuels are no longer the principal drivers of f element separation technology. Separations technology for environment restoration, waste disposal, and the preparation of high purity lanthanides are now the defining parameters in this important field. These imperatives are reflected in the contributions to this volume. The symposium itself must be considered a success, as the attendance at all sessions was above expectations, despite the fact that it was conducted on the last two days of a large five day meeting. Our thanks to the speakers for their quality presentations, and to the audience who persevered to the end of a long meeting and against the temptation of the excellent weather of San Diego in the springtime. A complete list of symposium participants is given in Appendix 1.

Preparation of this volume has been a relatively painless undertaking, largely as a result of the high quality of the submitted papers. The participating authors have tried mightily to meet the short deadlines imposed both by us and by the publisher. We are proud to have completed this volume in less than one year, and we must give credit again to the contributing authors and to the publisher. Our thanks to Jan Nolan and Jennifer Edwards for their editorial support, which included faxing, phoning, mailing, and manuscript preparation. We trust and believe that the papers contained in this book represent a worthwhile and useful contribution to the scientific and technological literature on the subject of f element separation science. We are also looking forward to participating in the next f element separation symposium in 2004-5.

Kenneth L. Nash
Group Leader,
Heavy Element Coordination Chemistry
Chemistry Division
Argonne National Laboratory

Gregory R. Choppin
R. O. Lawton Professor of Chemistry
Department of Chemistry
Florida State University

# CONTENTS

Introduction: A Brief History of f Elements Separations .................... 1

In the Beginning, There Was the Actinide Hypothesis .................... 9

New Methods of Transplutonium Elements Isolation, Purification, and Separation from Rare Earth Elements and Selected Fission Products .................... 11
B. F. Myasoedov

Extraction of Eu(III) and Am(III) with Thio and Amide Extractants .................... 31
G. R. Choppin, J. Yao, R. M. Wharf

Selective Extraction of Trivalent Actinides from Lanthanides with Dithiophosphinic Acids and Tributylphosphate .................... 43
G. D. Jarvinen, R. E. Barrans, Jr., N. C. Schroeder, K. L. Wade, M. M. Jones, B. F. Smith, J. L. Mills, G. Howard, H. Freiser, S. Muralidharan,

Design of Ligands for f Element Separations .................... 63
R. T. Paine

Synthesis and Evaluation of Polyhydroxamate Chelators for Selective Actinide Ion Sequestration .................... 77
A. S. Gopalan, V. J. Huber, N. Koshti, N. Jacobs, O. Zincirioglu, P. H. Smith, G. D. Jarvinen

Synthesis, Characterization, and Actinide Extraction Behavior of Bridge-Modified Carbamoylmethyl-Phosphonates and Phosphine Oxides .................... 99
B. M. Rapko

New Water-Soluble Phosphonate and Polycarboxylate Complexants for Enhanced f Elements Separations .................... 125
K. L. Nash, P. G. Rickert, E. P. Lessmann, M. D. Mendoza, J. F. Feil, J. C. Sullivan

Separation of Americium from Europium by Solvent Extraction from Aqueous Phosphonate Media ... 143
D. D. Ensor, K. L. Nash

Europium(III) Interaction with a Water Soluble Complexant: From Speciation to Photodestruction ... 153
J. V. Beitz

Use of Selective Inorganic Ion Exchangers for Separation of Rare Earths ... 165
A. Clearfield, R. A. Cahill, S. B. Wright, P. C. Bellinghausen, B. Shpeizer

Factors Influencing the Efficiencies of Multistage Separations of Lanthanides ... 177
S. Muralidharan, G. Ma, H. Freiser

Removal of Plutonium and Americium from Hydrochloric Acid Waste Streams Using Extraction Chromatography ... 199
L. D. Schulte, J. R. FitzPatrick, B. T. Martinez, R. R. Salazar, B. S. Schake

Sorption Behavior of Uranium onto Reillex™-HPQ Anion Exchange Resin from Nitric and Hydrochloric Acid Solutions ... 209
K. D. Abney, A. B. Pinkerton, R. C. Staroski, N. C. Schroeder, K. R. Ashley, J. M. Adams, J. R. Ball

Issues in the Scale-up of Aqueous Biphasic Extraction of Uranium from Contaminated Soils ... 225
D. J. Chaiko, R. Mensah-Biney

Plutonium and Americium Separation Using Organophosphorus Extractant Adsorbed onto Ferromagnetic Particles ... 241
L. Nunez, G.F. Vandegrift

Magnetic Separation for Environmental Remediation ... 257
A. R. Schake, L. R. Avens, L. A. Worl, K. J., de Aguero, D. D. Padilla, F. C. Prenger, W. F. Stewart, D. D. Hill, T. L. Tolt,

Appendix 1. List of Participants in the Symposium ... 271

Index ... 273

# INTRODUCTION: A BRIEF HISTORY OF *f* ELEMENT SEPARATIONS

Kenneth L. Nash, Argonne National Laboratory

Gregory R. Choppin, Florida State University

In 1787, C. A. Arrhenius found a black mineral near Ytterby, north of Stockholm. He named it yttria. Subsequent studies of yttria led to the isolation and identification of yttrium (1794). By 1907 all the lanthanides except element 61 had been discovered. The final lanthanide was produced in nuclear fission and identified in 1947. A major problem over the intervening 160 years was the extreme difficulty of separating these elements of almost identical chemical properties. However, the development of separation procedures based on cation exchange resins using aminopolycarboxylate eluent solutions resulted in isolation of macro quantities of high purity lanthanides. The subsequent development of separation methods of the 4f elements has continued to be an exciting and successful area of separation science since the 1950's.

The transuranium (5f) elements are not naturally occurring. They were produced in the last 50 years as a result of the research of nuclear scientists who predicted how they could be synthesized and developed the methods to do so as well as to isolate and purify the few atoms made in the discovery experiments. Since their discovery, these elements have been made in amounts which range from tons (Pu) to micrograms (Es) as a result of nuclear power and weapons operations. Some small amounts have been introduced into the environment as a result of such operations. Highly specific, rapid separations as well as remote automated procedures continue to be developed to meet the needs in 5f element science and technology.

In the reminiscences of Glenn Seaborg at the symposium commemorating the 40$^{th}$ anniversary of the first isolation of measurable amounts of Pu, he recounted the early development of laboratory scale separation processes for plutonium isolation. The first weighable amounts of Pu were isolated as minor products in large quantities of cyclotron irradiated uranium. The process began with the observation that "reduced 94" could be precipitated with fluoride while "oxidized 94" remained in solution. Next came the introduction of a coprecipitation process based on $BiPO_4$ developed by S. G. Thompson in Seaborg's laboratory at the University of Chicago in 1942 . This bismuth phosphate - oxidation/reduction process became the first production scale method for isolation of Pu

*Separations of f Elements*, Edited by K.L. Nash
and G.R. Choppin, Plenum Press, New York, 1995

from uranium and fission products at the Hanford extraction plant. The process was efficient enough to satisfy the demands of wartime production of Pu.

From an engineering perspective, processes which require liquid-solid separations present significant challenges. Design and operation of processes based on liquid-liquid separations methods offer many practical advantages. This led to the development of first the REDOX process (using methyl isobutyl ketone (MIBK - 1) as an extractant and subsequently the Plutonium Uranium Recovery by EXtraction (PUREX) process. The latter method is based on the extraction of nitrate salts of actinides by solutions of tributyl phosphate (TBP - 2). These processes greatly accelerated the production rate for plutonium

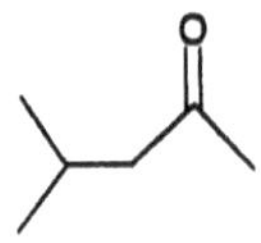
(1)

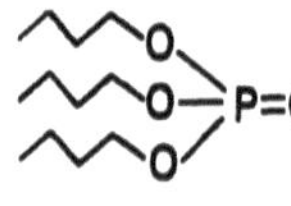
(2)

from irradiated fuel. The latter is still the standard method of reprocessing of nuclear fuels worldwide. The introduction of ion exchange resins permitted more specialized smaller scale separations methods to be developed.

In several laboratories around the world, research in the 50's, 60's, and 70's continued to extend the actinide series towards the final element of the series (Z = 103). It was quickly learned that the rich redox chemistry of the light actinides, which was central to most of the successful separations of these elements, did not persist beyond americium. The transamericium elements in aqueous solutions behaved chemically more like the 4f analog lanthanides that the light members of the series. This circumstance created difficult separations problems which could not be ignored. Because the transplutonium elements were prepared by nucleosynthetic methods, the neutron (or charged particle) capture reactions were always accompanied by some fission. Fission yields for lanthanides are generally high and the synthesis therefore yielded dissolver solutions which contained not only small amounts of the target transamericium elements but also significant concentrations of lanthanides. The lanthanides in the fuel also created problems in actinide production by neutron irradiation since these metals are very efficient neutron poisons which reduce the efficiency of the reactor operations used to produce more transamericium elements.

Two difficult separations problems resulted from this circumstance. First it became necessary to develop a method to separate trivalent lanthanides from the transamericium elements because of their efficiency in capturing neutrons. The solution chemistry of the lanthanides and trivalent actinides are not just similar, they are very nearly identical in "normal" aqueous solutions. Second, in order to elucidate the basic chemistry of the transamericium elements, pure samples of each of the new elements were required. Because multiple oxidation states are essentially no longer available, subtle differences between either the chemistries of lanthanides and trivalent actinides or between individual members of the series had to be exploited. The precipitation/fractional crystallization methods which had been employed for lanthanide production required thousands of crystallization cycles for isolation of pure lanthanides. Such methods would hardly be acceptable for the separation challenges presented by the chemistries and minute quantities available for transplutonium element production. These methods were, in fact, unable to keep up with the commercial demands for lanthanides.

With the development of first inorganic ion exchangers (synthetic zeolites) and later organic polymeric resins, separations of large amounts of lanthanides became possible. Such materials per se generally exhibit only marginal separation ability for the lanthanides based on the small, regular changes in the cation radius. However, when these materials are employed in a multiplate system combined with a strong complexant in the eluting solution, efficient separations based on differences in the stability of the aqueous complexes can be designed. Commercial ion exchange procedures were successfully adapted for the laboratory based separation of the transamericium elements.

Beginning with the work of Street and Seaborg in 1950 and followed by Diamond, Street and Seaborg in 1954, methods for the separation of trivalent actinides from lanthanides by ion exchange techniques were developed. These authors found that below 6 M HCl, trivalent actinides and lanthanides eluted from a cation exchange column together (Figure 1). Above 6 M HCl, actinides were eluted from the column while the lanthanides

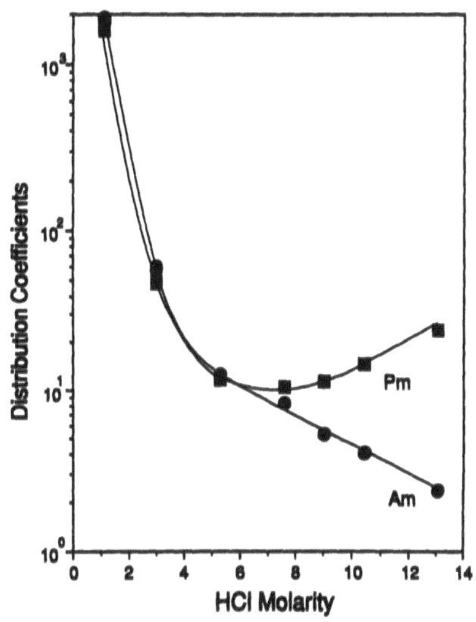


**Figure 1.** Lanthanide-actinide separations by cation exchange from HCl solutions (1-3).

were retained. Anion exchange employing either lithium chloride or ammonium thiocyanate (Figure 2) were reported as efficient methods for the separation of lanthanides from trivalent actinides. Separations by both methods were the result of enhanced interaction strength of actinide cations with soft-donor ligands relative to those of lanthanides. Such differences are only manifested in concentrated solutions where the cation solvation effects, which mask differences between the series, are suppressed.

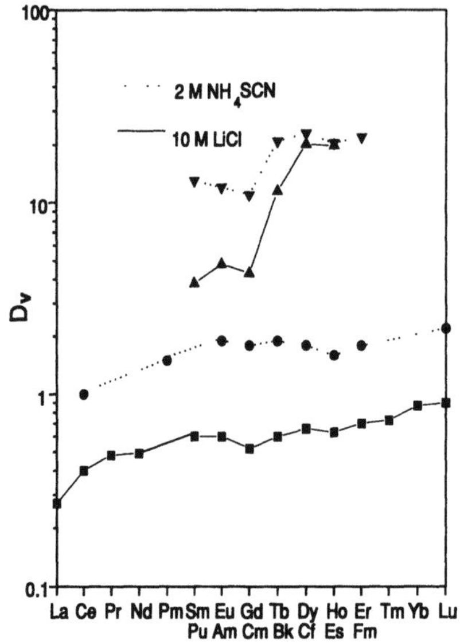


**Figure 2.** Anion exchange separations of lanthanides and actinides using $NH_4SCN$ or (H,Li)Cl solutions (4-6).

Separations of individual members of the two groups is an even more difficult task, in many respects, than group separations. The first successful separation of individual trivalent actinides by ion exchange was reported by Thompson and coworkers in 1950 and 1954. They used either ammonium citrate or ammonium lactate at pH 3-5 to elute actinides from Dowex 50. The technique was significantly improved by the introduction of α-hydroxyisobutyrate as the eluant in 1956 (Figure 3).

It is remarkable from the perspective of 50 years later that the insights of the original f element separation scientists were so good as to have designed chemical processes that are often still the best method for accomplishing the isolation of these elements. For example, chloride based anion exchange for lanthanide-actinide separation is used today in the TRU facility at Oak Ridge National Laboratory for separation of transplutonium elements from lanthanides (Figure 4). The separation of individual transplutonium actinides is accomplished at this facility still by cation exchange using α-hydroxyisobutyrate. As noted above, the PUREX process, introduced in the late forties, is still the method of choice for plutonium processing.

Of course, the "pioneers of f element separations" did not exhaust all of mankind's good ideas in f element separations. Several additional noteworthy developments have appeared on the f elements separation scene in more recent years. One of the current principal methods for isolation of individual lanthanides commercially is solvent extraction with bis(2-ethylhexyl)phosphoric acid (HDEHP), a reagent first reported in 1958 (Figure 5). Adjacent lanthanide separation factors are on average nearly twice as high for HDEHP separations than α-hydroxyisobutyrate cation exchange.

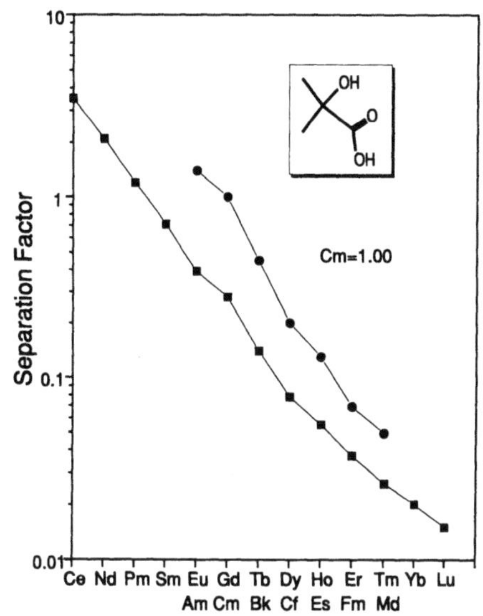


**Figure 3.** Separation of individual trivalent lanthanides or actinides by cation exchange using α-hydroxyisobutyrate (shown inset) as the eluting agent (7-10).

A particularly noteworthy example of a process for group separations is the Trivalent Actinide/Lanthanide Separations by Phosphorus Extractants from Aqueous Komplexes (TALSPEAK) process from Oak Ridge National Laboratory. This process combines features of the liquid cation exchanger extraction by HDEHP with cation dehydration provided by lactic acid and selectivity between the groups provided by a soft-donor chelating agent in the aqueous phase. Typically, the latter role is filled by diethylenetriamine-N,N,N',N",N"-pentaacetic acid, which complexes the trivalent actinides more strongly than the lanthanides. Other combinations of reagents have been examined, most of which provide similar behavior. The minimum group separation factor is about 10 (Figure 6).

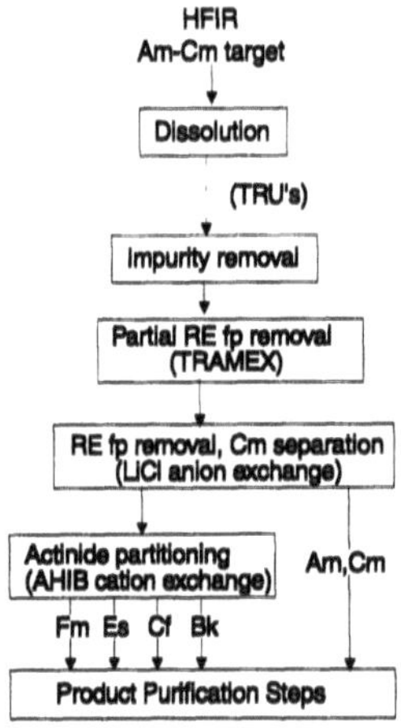


**Transuranium Processing Plant (TRU) and High Flux Isotope Reactor (HFIR) at ORNL produce transuranium elements for research**

**Production 1967-1979**

| | |
|---|---|
| $^{242}Pu$ | 141 g |
| $^{243}Am$ | 252 g |
| $^{244}Cm$ | 1852 g |
| $^{249}Bk$ | 456 mg |
| $^{252}Cf$ | 4131 mg |
| $^{253}Es$ | 18.4 mg |
| $^{257}Fm$ | 10 pg |

**Figure 4.** Basic flow scheme and past production efficiency for transuranium elements at the High Flux Isotope Reactor (11).

When the emphasis in actinide processing shifted from production to waste treatment, it became clear that, as good as the PUREX process was for plutonium recovery, it was

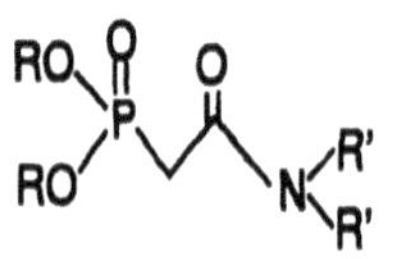

(3)

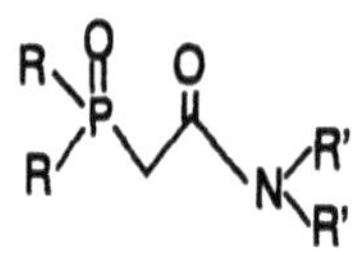

(4)

clearly inadequate for extraction of trivalent actinides. The need for a complementary process capable of extracting all actinides led to the development of the TRansUranium EXtraction (TRUEX) process, developed primarily at Argonne National Laboratory. The application of the carbamoylmethylphosphonates (CMP's -3) and, more significantly, carbamoylmethylphosphine oxides (CMPO's -4), created the concept of total actinide recovery from nitric acid solutions. The promise of the combined application of PUREX and TRUEX is a potential orders of magnitude reduction in the volume of TRU wastes requiring deep geological disposal. Methods based on molten salt technologies or liquid metal recycling concepts (like the Integral Fast Reactor) have also been extensively examined.

But there is one major difference between the f element separation scientists of today and those of yesterday. In the beginning the goal was simple and straightforward - to tap an unbelievably powerful energy source first to end a war and, subsequently, to keep the

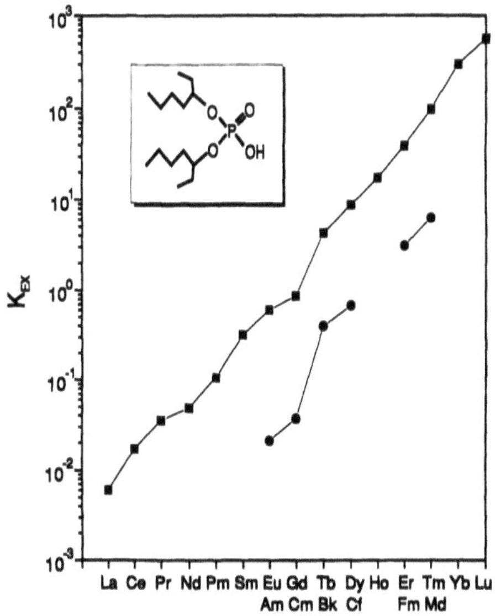


**Figure 5.** Separation efficiency for trivalent actinides and lanthanides by solvent extraction with bis(2-ethylhexyl)phosphoric acid (shown inset) (12).

peace. "Harnessing the atom" for peaceful purposes drove developments through the 60's and 70's. Today the production of transuranium elements is at a standstill because the world has changed. The principal driving force for continued development of new separations methods for f elements is no longer production but environment restoration. Evolution of technology toward "greener solutions" has challenged today's separation scientist in ways never imagined by previous generations. Many of the contributions to this volume reflect the changing priorities of modern life in f elements separations.

This volume marks an informal 10 year anniversary since the previous symposium dedicated to f elements separations, just as that symposium occurred 10 years after the one before. Those of us involved in f element separation science are confident that there will be adequate interest and sufficient continued development in this field to justify another symposium on f elements separations 10 years from now. We remain optimistic about the future for f elements separation science despite concerns that nuclear technology is too dangerous for man to pursue further. We believe that nuclear power is destined to play an important role in maintaining and improving the quality of life in the next century and beyond. We also believe the lanthanide elements will continue to find expanding and valuable applications in the future. The spirit of adventure and the excitement of discovery have always driven the human spirit to new heights. May it ever be so!

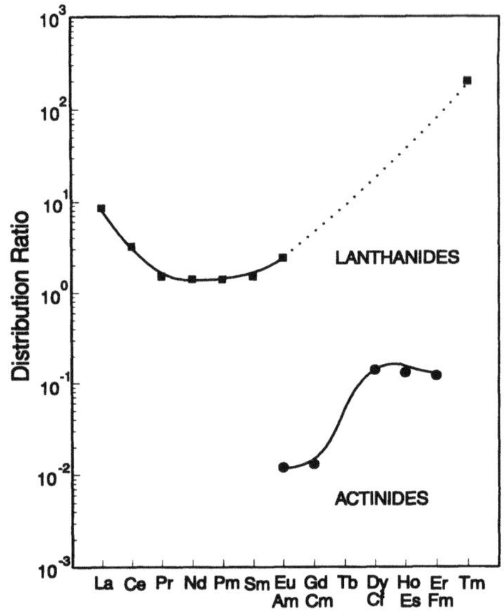


**Figure 6.** Trivalent Actinide/Lanthanide Separation by Phosphorus Reagent Extraction from Aqueous Komplexes (TALSPEAK) process (13).

## BIBLIOGRAPHY

1. K. Street, Jr. and G. T. Seaborg, *J. Am. Chem. Soc.* 72:2790 (1950).
2. R. M. Diamond, K. Street, Jr., and G. T. Seaborg, *J. Am. Chem. Soc* 76:1461 (1954).
3. J. P. Surls and G. R. Choppin, *J. Inorg. Nucl. Chem.* 4:62 (1957).
4. S. G. Thompson, B. G. Harvey, G. R. Choppin, and G. T. Seaborg, *J. Am. Chem. Soc* 76:6229 (1954).
5. J. P. Surls, G. R. Choppin, *J. Am. Chem. Soc* 79:855 (1957).
6. E. K. Hulet, R. G. Gutmacher, and M. S. Coops *J. Inorg. Nucl. Chem.* 17:350 (1961).
7. A. Giorso, B. G. Harvey, G. R. Choppin, S. G. Thompson, and G. T. Seaborg, *Phys. Rev* 98:1518 (1955).
8. G. R. Choppin, B. G. Harvey, and S. G. Thompson, *J. Inorg. Nucl. Chem.* 2:66 (1956).
9. G. R. Choppin and R. J. Silva, *J. Inorg. Nucl. Chem.* 3:153 (1956).
10. H. L. Smith and D. C. Hoffman *J. Inorg. Nucl. Chem.* 3:243 (1956).
11. L. J. King, J. E. Bigelow, and E. D. Collins in "Actinides Separations" ACS Symposium Series Vol. 117 (Eds. J. D. Navratil and W. W. Schulz) American Chemical Society, Washington, D.C. (1981) p133.
12. D. F. Peppard, G. W. Mason, J. L. Maier, and W. J. Driscoll, *J. Inorg. Nucl. Chem.* 4:334 (1957).
13. B. Weaver and F. A. Kappelmann, J. Inorg. Nucl. Chem. 30:263 (1968).

In the beginning,
there was the Actinide Hypothesis ...

And Professor Seaborg said to "These new elements are not like the Old Ones (d-transition elements) but like the Young Ones (Lanthanides)." And the Doubters said "Prove it!". But these elements were nowhere to be found. So Professor Seaborg said "We will revive the alchemists' quest and bring the new elements back." And he gathered the modern Alchemists (whose language was Physics) and the Sorcerers (who spoke Chemistry), and began pursuit of the Grail. And the Alchemists and Sorcerers combined their arts and soon restored the New Elements to the earth. But the New Elements were very much like the Young Ones and parted from them only with much encouragement from the Sorcerers. Thus the art and science of f-element separation was born.

Ken Nash
March, 1993

# NEW METHODS OF TRANSPLUTONIUM ELEMENTS ISOLATION, PURIFICATION AND SEPARATION FROM RARE EARTH ELEMENTS AND SELECTED FISSION PRODUCTS

B. F. Myasoedov and M. K. Chmutova

V. I. Vernadsky Institute of Geochemistry and
Analytical Chemistry
Russian Academy of Sciences
Moscow 117975, Russia

## INTRODUCTION

It is well known that the production of transplutonium (TPE) is associated with the processing of highly radioactive materials containing a mixture of several elements formed in nuclear reactors as a result of neutron irradiation of plutonium and heavier elements. High degrees of purification from radioactive isotopes and inert admixtures are necessary when isolating TPE from various natural objects and highly radioactive materials, including technological solutions for their subsequent determination. The choice of methods of TPE isolation and separation is determined by the chemical composition of the analyzed solution and by the purpose of the work. For group removal of actinides from the solutions of different composition including highly radioactive wastes (HRW) from nuclear fuel reprocessing we will consider the application of bidentate neutral organophosphorus compounds, extraction from alkaline solutions and the use of two phase water-poly(ethylene glycol)-salt systems. The counter-current chromatography (CCC) is more effective for the group separation of actinides and lanthanides. The best separation of transplutonium elements (TPE) is obtained by using methods based on the various oxidation states of the separated elements and by membrane extraction.

## METHODS OF ACTINIDE ISOLATION AND CONCENTRATION

The quantitative isolation and concentration of actinides are important for many cases, viz. for their production, the isolation of actinides from HRW from nuclear fuel reprocessing, when determining them in different substances and first of all in objects of environmental media.

### Extraction of Actinides by BNOC

The radioactive wastes remaining after spent fuel reprocessing are usually water solutions containing large concentrations of strong acids (often nitric acid) and salts. Different extractants can be used for deep purification of such high radioactive wastes from long-lived radionuclides (LRN). Recent investigations evidence that bidentate neutral organophosphorus compounds (BNOC) are the most effective extractant for this purpose.

Two types of BNOC have been studied: tetraalkyl(aryl)alkylene(arylene) diphosphine dioxides (DO)[1]: R'R"PO-X-PO(R"')$_2$ and dialkyl(diaryl)(dialkylcarbamoyl-methyl) phosphine oxides (CMPO, or "carbamoyls")[2, 3]: R'R"PO-X-CO-N(R"')$_2$, where R', R", R"', X - alkyl or aryl.

Wide possibilities of varying structures of these compounds enabled to change both extractive power and selectivity of these reagents together with their compatibility with solvents. Thus, the nature of the substituents at phosphorus atom significantly influences extraction capacity of reagent. As seen from Fig. 1 the aryl-substituted reagents have the highest extraction capacity with respect to Am(III) as well as to U(VI) and Pu(IV).

The change of substituent nature at phosphorus atoms influ ences not only reagent extraction capacity but also its solubility in both phases. Thus, tolyl-substituted reagents are much more compatible with a wide range of diluents than phenyl-substituted ones. Aryl-substituted reagents are less soluble in water and in $HNO_3$ than alkyl-substituted ones.

The nature of substituents at nitrogen atoms in CMPO mainly influences their solubility in organic and aqueous phases, i.e. butyl-substituted reagents are much more compatible with diluents than ethyl-substituted and more soluble in $HNO_3$ than octyl-substituted. The change of bridge structure between functional groups influences reagent extrac tion capacity. Thus, the reagents with methylene bridge between functional groups have the highest extraction capacity. The increase of linear bridge length results in a decrease of reagent extraction capacity.

When studying possibilities to isolate TPE from various media unusually high efficiency of extraction from perchloric acid media has been established by aryl-substituted CMPO[4]. The efficiency of extraction is so high that addition of very small amounts of perchloric acid to solutions of other acids or complexing agents results in a formidable increase in the efficiencies of extraction of TPE, lanthanides, U(VI), Pu(IV) and makes it possible to concentrate these elements from various solutions. Figure 2 represents the effect of perchloric acid on Am(III) extraction from $HNO_3$ and U(VI) extraction from $H_3PO_4$

solutions by CMPO. It is shown that Am extraction is significantly increased. The addition of small amount of $HClO_4$ allows to extract U(VI) from phosphoric acid quantitatively.

Together with many advantages DO and CMPO have substantial drawbacks that one can not overcome even by changing their structures, e.g. they are practically insoluble in aliphatic diluents. One means to eliminate these drawbacks is to add a modifier such as tributyl phosphate (TBP) to solutions of bidentate reagents, that are well compatible with many organic solvents. Addition of TBP not only results in eliminating the third phase but also provides a non-additive increase in distribution coefficients ($K_d$) of TPE (synergistic effect). The addition of TBP allowed application of alyphatic hydrocarbons widely used in radiochemical industries as solvents for CMPO5,[5,6].

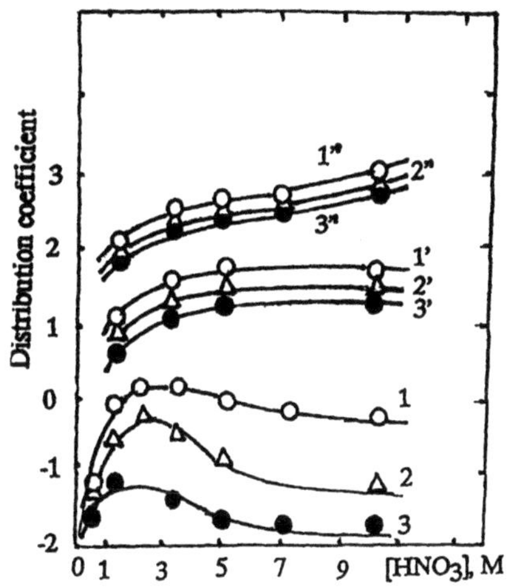


**Figure 1**. Extraction of Am(III) (1-3), U(VI) (1'-3') and Pu(IV) (1"-3") by 0.05 M dichloroethane solutions of CMPO from $HNO_3$ solutions. 1, 1', 1": R'= R" = phenyl, R" '= ethyl; 2, 2', 2": R'= phenyl, R" = butyl, R"'= ethyl; 3, 3', 3": R' = R" = butyl, R" '= ethyl.

One of the alternatives was also the study of principally new diluent classes such as, for example, fluoroethers and fluoroderivatives of some other compounds. These diluents have a good compatibility with reagents and their complexes, high density and boiling temperature (130°). They are poorly soluble in water and have high chemical and radiation stability[7].

One of the most important requirement to reagents for TPE extraction is their radiation stability. It was shown that the extraction properties of phenyl-substituted DO and CMPO do not change even under high gamma- and alpha-radiation[8].

BNOC solutions in solvating diluent ("Fluoropol-732") were suggested for isolating actinides from strongly acidic solutions of HRW. Such systems have the advantage of possibility to use readily available efficient reagents "carbamoyls" and to avoid the addition of solvating TBP. The extraction of TPE and accompanied elements from the model acidic HRW solutions (2.5 M $HNO_3$) by CMPO ($Ph_2Bu_2$, i.e. R'=R"=phenyl, R'''= butyl) in

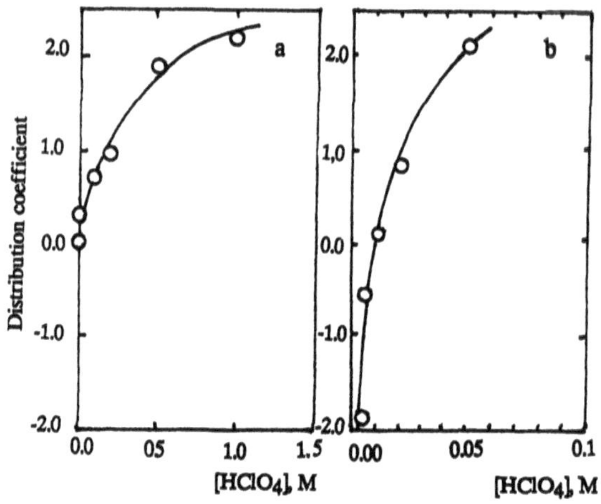


**Figure 2.** Effect of $HClO_4$ on extraction of Am(III) from 3 M $HNO_3$ (a) and U(VI) from 6 M $H_3PO_4$ (b) by CMPO in dichloroethane. a: 0.05 M solution, R'= R"= phenyl, R'''= ethyl; b: 0.1 M solution, R'= R"= butyl, R'''= ethyl.

"Fluoropol-732" was studied. It was shown that lanthanides and Zr(IV) are extracted as effectively as TPE; Mo(VI) and Fe(III) are extracted to a less extent than TPE while alkaline and alkaline-earth metals and Cr, Co, Ni are not extracted at all. Mo(VI), Fe(III), Zr(IV) can be separated from TPE and lanthanides by multiple stripping of organic phase by acetohydroxamic acid (AHA).

The technological scheme of actinide and lanthanide isolation using 18 - step set of centrifuge extractors with Ph2Bu2 in "Fluoropol-732" and with AHA as stripping reagent has been studied (Fig. 3). Using a model HRW solution containing more than 13 g/l of lanthanides and actinides and up to 5 M $HNO_3$ a possibility has been shown to isolate more than 99.5% lanthanides and actinides as well as to purify them from Fe, Zr and Mo (coefficient of purifying >50) and to transfer them to a weakly acidic extract with concentrating by 4-6 times[9].

The back-extract of TPE and lanthanides contains as low as 0.02 - 0.04 M $HNO_3$ that makes it possible to choose any other water system for group separation of these elements.

## Extraction of Actinides and Lanthanides from Alkaline Solution

The application of alkaline solutions is worth-while for the TPE group isolation and concentration. The extraction of metals from such solutions had not been studied until recently. Our investigations have shown that TPE and many other elements can be extracted from alkaline solutions in the presence of complex-forming reagents by extractants of various classes: quaternary ammonium bases, amines, alkylpyrocatechols (DOP), alkyl-derivatives of amine alcohols (AA) and ß-diketones[10]. Some of the extractants studied, especially alkylpyrocatechols and α-oxy-5-alkylbenzyldiethanolamine are characterized by a high extraction capacity in relation to TPE, since with their help it is possible to isolate effectively those elements from the alkaline and carbonate solutions with a concentration of the order of 5-6 M (Fig. 4).

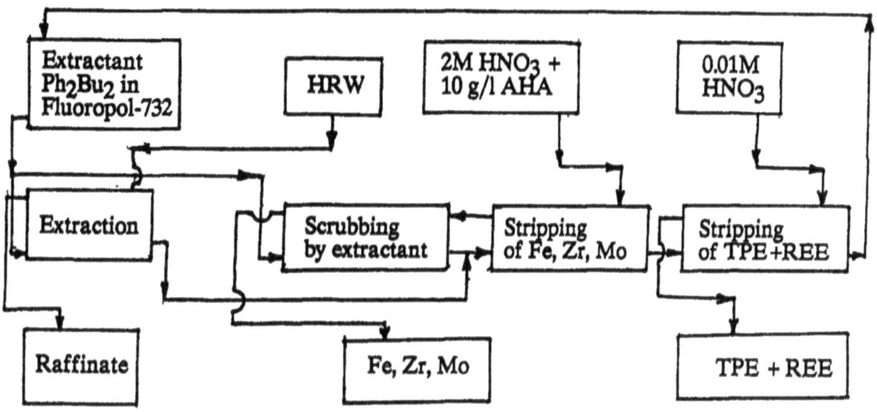


**Figure 3.** Scheme of isolation of TPE and lanthanides from HRW by 0.1 M solution of diphenyl(di-butylcarbamoylmethyl) phosphine oxide in "Fluoropol-732".

The efficiency of the extraction isolation of TPE depends on the nature of the complex-forming reagents capable of keeping elements in alkaline solutions in a soluble form. Depending on the conditions, the elements may be extracted from alkaline and carbonate solutions in the form of ion associates, the anionic part of which contains either hydroxocomplexes of the corresponding metals, or their compounds with the complex forming ligand, while the cationic part is the extractant. They can also be extracted in the form of the coordinationally saturated and hydrolyzed ß-diketonates.

Methods of extraction and extraction chromatography in alkaline and carbonate solutions are used to solve a number of practical problems. First, under certain conditions with the help of all the extractants studied, one can quantitatively isolate and concentrate the elements capable of existing in alkaline solutions in the form of various complex compounds. Secondly, the group concentration of tervalent TPE is achieved in these systems with simultaneous isolation from the elements in other oxidation states as well as

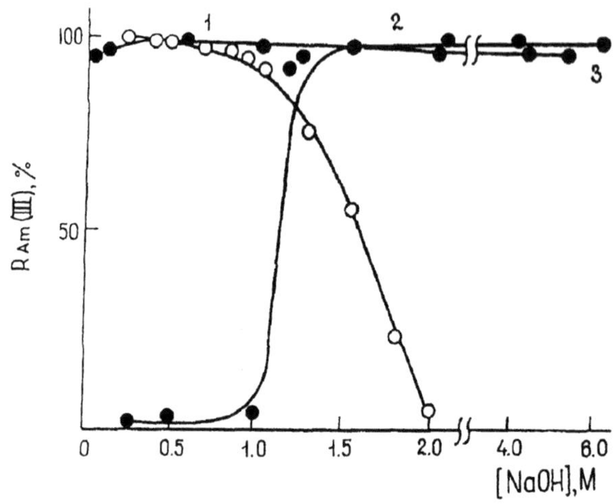


**Figure 4.** Extraction of Am(III) by 0.2 M Aliquat•336•OH (1), 0.04 M DOP in toluene (2) and 0.1 M AA in hexane (3) in the presence of 0.025 M tartaric acid as a function of the NaOH concentration.

from iron (separation factor $Am^{III}/U^{VI} = 2.10^2$; $Pu^{IV}/Am^{III} = 1.10^2$; $Ce^{IV}/Am^{III} = 3.10^2$; $Fe^{III}/Am^{III} = 1.10^3$; $Am^{III}/Cs^{I}=2.10^3$).

## Extraction of Actinides in Two-Phase Water-Poly (Ethylene Glycol)-Salt Systems

Aliquid-liquid extraction system based on water-soluble polymers such as poly (ethylene glycol) (PEG) has been suggested for concentration , separation and extraction of actinides from various salt solutions including HRW[11]. These systems are of interest from a practical point of view, since they contain no organic solvents, which are typically volatile, explosive, or toxic, and makes it possible to use well-known water soluble reagents. Extraction of actinides from sulfate-, carbonate-, phosphate-, rodanide- and nitrate solutions in the presence of different complexing agents has been investigated. Conditions have been found for quantitative group extraction on the TPE and for separation of these elements from U, Th and lanthanides. Figure 5 shows the extraction of the complexes of actinides and lanthanides with (curve 1) and without (curve 2) Arsenazo III as a function of equilibrium salt-phase pH. The distribution of elements in the absence of Arsenazo III has been less than 10% into the PEG-enriched phase. The quantitative extraction of actinides takes place only in pres ence of Arsenazo III which correlates with metals complexing with the reagent. The distribution coefficients of elements increase with increasing Arsenazo III concentration.

A high degree of separation of trivalent actinides from neptunium (V) in the system PEG-potassium phosphotungstate (PW) has made it possible to develop a method for separation of $Am^{243}$ from the daughter $Np^{239}$ and to work out an isotope generator of $Np^{239}$. In Fig. 6 are shown γ-spectra of initial americium solution and a solution ammonium sulfate and PEG after extraction with PW.

From γ-spectra analysis it follows that it is only americium is extracted into the PEG phase while all of the neptunium remains in the salt phase. Neptunium-239 accumulated in the PEG phase may be washed out from the PEG solution by any salt solution that makes up a heterogeneous system with the PEG and no complexing agent.

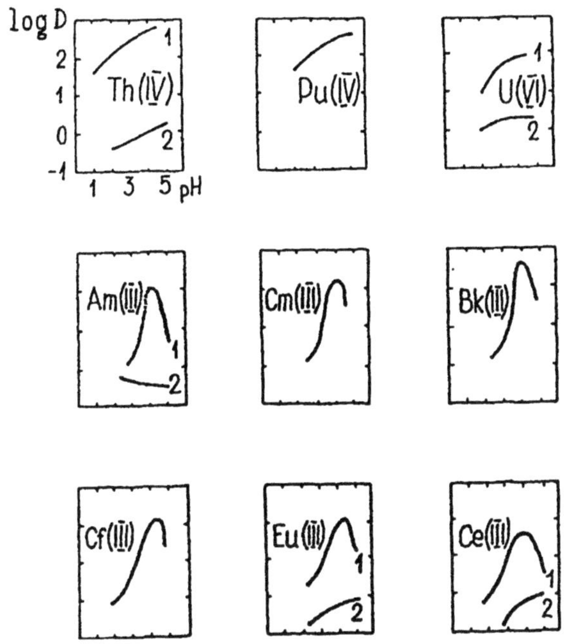


**Figure 5.** Dependence of distribution coefficients of different metals on pH in the system PEG - $(NH_4)_2SO_4$ - $H_2O$. (1) - with Arsenazo III, (2) - without Arsenazo III.

## SEPARATION OF ACTINIDES AND LANTHANIDES BY PARTITION COUNTER-CURRENT CHROMATOGRAPHY

Partition counter-current chromatography (CCC), which may also be called a liquid chromatography with free stationary phase, is a relatively new method of separation. This method suggested by american scientist Ito is based on the retention of stationary organic phase in a rotating column under the action of centrifugal forces while the mobile aqueous phase is being continuously pumped through. Substances are separated due to difference in their distribution coefficients when the mixture to be separated moves with the flow of the aqueous phase through the column.

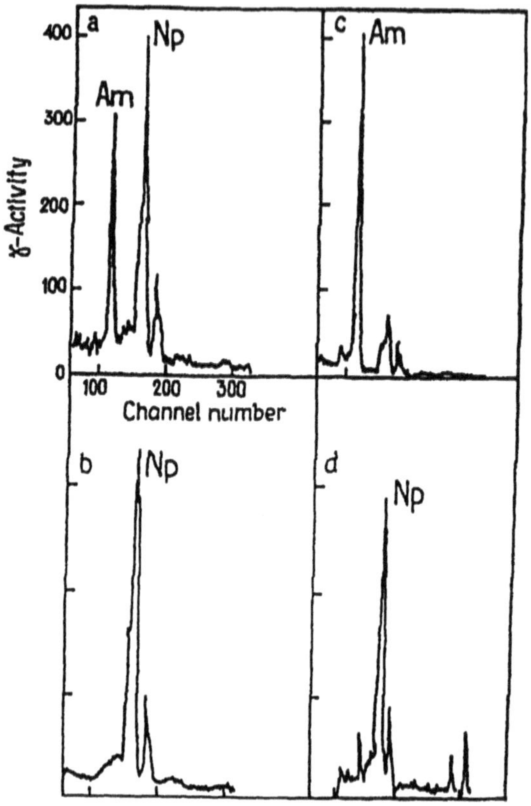


**Figure 6.** Separation of Am-243 and Np-239 in the system PEG - (NH4)2SO4 - H2O - PW. ϒ-spectra: americium initial solution (a), salt phase after extraction (b). PEG phase after extraction (c), ammonium sulfate solution with accumulated Np-239 back extracted from the PEG phase.

The CCC method is similar with the method of extraction chromatography and with counter-current extraction. A feature of the CCC as a chromatographic method is the absence of sorbent or solid support for retaining the stationary phase. This feature defines the main advantages of the method: 1. The absence of losses or denaturation of the substances being separated due to interaction with solid support, 2. Variety of the two-phase liquid systems which may be used and hence tasks accomplished by using this method, 3. Easy change from one system to another, 4. Absence of the problem of the column regeneration, and 5. High preparation capacity.

The main difference of CCC from solvent extraction consist in the possibility of obtaining separation efficiency of several thousand theoretical plates which is very difficult to realize by the use of traditional extraction apparatus, particularly when working with small volumes of solutions.

A few devices providing the retention of the stationary phase in the field of mass forces in the absence of solid support have been suggested. The most successful is a planetary centrifuge.

Our investigations were made on a device consisting of a planetary centrifuge with a vertical column drum, peristaltic pump and a fraction collector. The column rotates around its axis and the same time revolves around central axis of the device. The basic element of the column is a teflon tube: a column wound on a rigid core (spiral column). The rotation and revolution speens were equal (350 r/min). The planetary centrifuge model had the following design parameters: revolution radius R=140 mm, rotation radius r=50 mm. The column was made of a teflon tube with an inner diameter of 1.5 mm and a wall thickness 0.75 mm. The total inner capacity of the column was 20 ml.

The amount of the stationary organic phase in the system is characterized by the retention factor ($K_r$) equal to the ratio of the organic phase volume to the total capacity of the column. The $K_r$ value depends on parameters of the planetary centrifuge, on specific process conditions, particularly on the rotation and revolution speed of the column, speed of the aqueous phase pumping through and on the properties of the two-phase system, of which the wettability of the materials of the column with the organic phase is of importance.

The separation of actinides and rare earth elements along with commonly used techniques can be carried out by CCC. A number of BNOC have been tested as extractants. These reagents are not selected and cannot be used for the separation of TPE from trivalent REE. There are two possibilities to increase the separation factors for the elements in BNOC-based systems: 1. Improvement of reagent selectivity by changing the reagent structure, mainly by introduction of "hard" bridge fragments (arresting the arrangement of the donor atoms) to the extractant molecule; 2. Addition of selective complex-forming agents to the aqueous phase. However, neither the first nor the second means enable one stage separation of TPE from REE as they have very similar properties. Multistage extraction separation is required to separate TPE and REE. For example, the method has been suggested of group separation of TPE from weighable amounts of REE in tetraphenylmethylenediphosphine oxide-$CHCl_3$-$NH_4SCN$-HCl sys tem[12]. Figure 7 illustrates the effect of reagent concentration on the element separation by CCC. It is seen that the higher TPMDPD concentrations the better separation is achieved. Addition of such complex-forming agents as hydroxyethylidenediphosphonic acid (HEDPA) to the aqueous phase enables the elution of elements with smaller volume of the eluent. The separation time and the eluent volume can be reduced if step elution is used. Figure 8 illustrates the separation of TPE from 10 mg of REE. First a REE fraction containing 95.4% of REE and no TPE is eluted with $NH_4SCN$-HCl solution, than 97.5% TPE (without REE) are stripped out with hydroxyethylidene diphosphonic acid (HEDPA) solution in water. The separation takes 20 minutes.

The following investigations have been based on the use of the extractant more selective with respect to TPE then the above reagent. The extraction capacity and selectivity of new reagents-2,4,6,-tris[diarylphosphoryl]-1,3,5 triazines (Tol-triazine) have been investigated. In these reagents an aromatic fragment serves as "hard" bridge connecting the P=O-groups. This fragment limits the conformational flexibility and brings about an

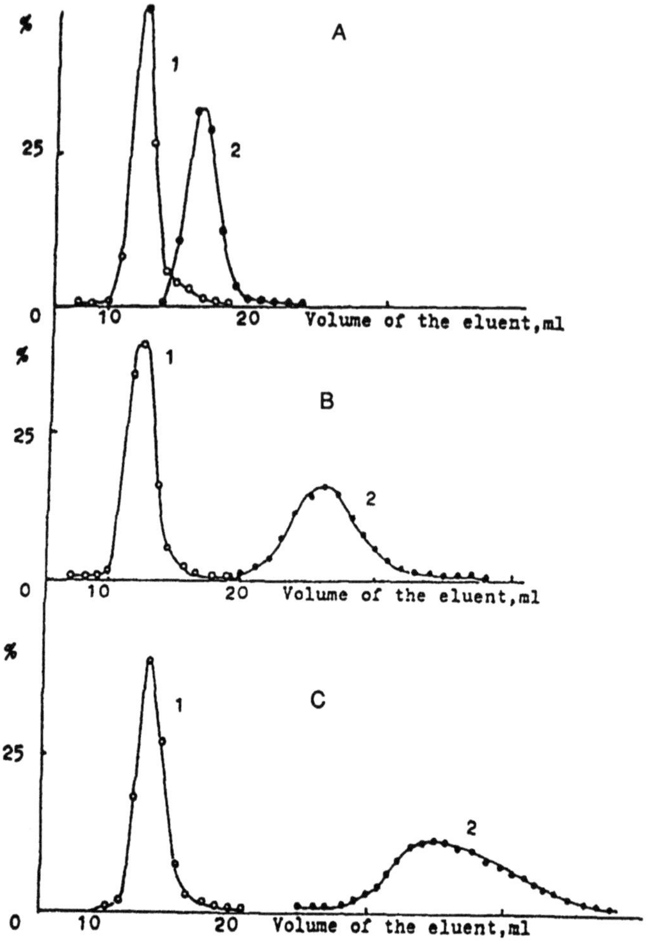


**Figure 7.** Influence of TPMDPD concentration in $CHCl_3$ on the separation of Eu(III) (1), and Am(III) (2). Stationary phase: A. 0.002; B. 0.0025; C. 0.003 M TPMDPD in $CHCl_3$. Mobile phase: 0.5 M $NH_4SCN$ - 1 M HCl.

increase in the selectivity toward certain ions. Figure 9 illustrates the separation of TPE from 20 mg of stable REE. The groups have been completely separated: each of the two fractions contained more than 99% of pure sample. The separation took 18-20 min. The separation of TPE and REE in nitric acid is more difficult problem because it is difficult to choose an extraction system with the moderate $HNO_3$ concentration in which the distribution coefficients for all TPE would be higher than that for all REE. In the system with Tol-triazine distribution coefficients for all TPE are higher than that for all REE. It is believed that TPE can be completely separated from heavy REE. Figure 10 illustrates the attempts to separate Am from Eu and Ce which confirm this as sumption.

The factor of group separation can be increased by use of complexones capable of selective complex formation with REE in acidic solutions.

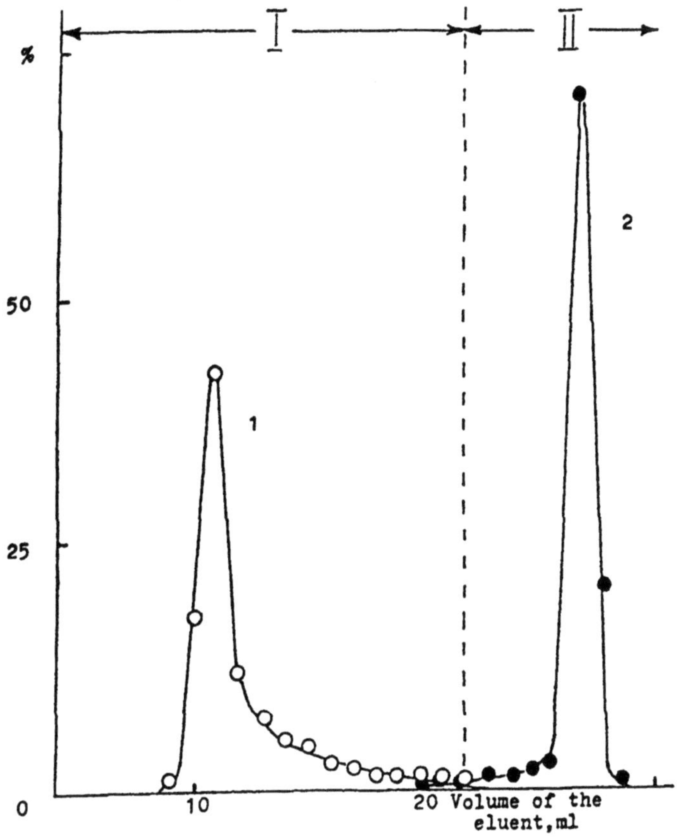


Figure 8. Separation of the sum of REE (1) and Am(III) (2). Stationary phase: 0.003 M TPMDPD in $CHCl_3$. Mobile phases: I. 0.5 M $NH_4SCN$ - 1 M HCl; II. 0.025 HEDPA in water; 10 mg REE in sample.

## METHODS OF TRANSPLUTONIUM ELEMENTS SEPARATION

Modern methods of transplutonium element separation are usually based on small differences in the properties of these in the oxidation state +3 which is most stable. The best separation of TPE is obtained by using methods based on the various oxidation states of the separated elements. Almost all the transplutonium elements (except Lr) can exist in solution in "unusual" oxidation states, from +1 up to +7. Systematic in vestigation of TPE properties in solution has resulted in the creation of quite a large number of practically convenient methods for the preparation and stabilization of $Am^{IV}$ and $Bk^{IV}$, $Am^{V}$ and $Am^{VI}$. This has expanded their application in analytical practice and technology, and primarily in the extraction methods of isolation and separation of those elements.

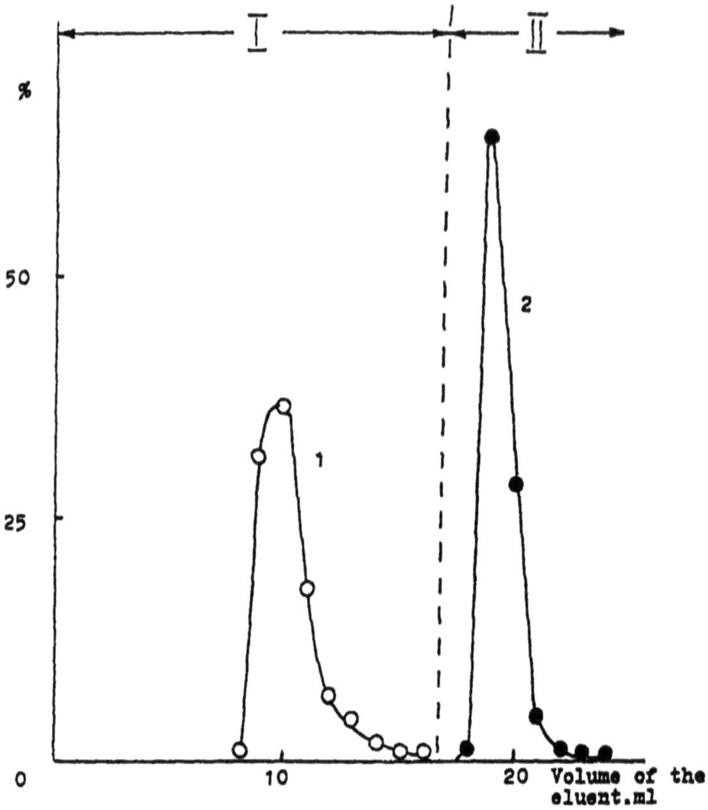


**Figure 9.** Separation of the sum of REE (1) and Am(III) (2). Stationary phase: 0.02 M Tol-triazine in $CHCl_3$. Mobile phases: I. 0.5 M $NH_4SCN$ - 1 M HCl; II. 0.025 HEDPA in water; 20 mg REE in sample.

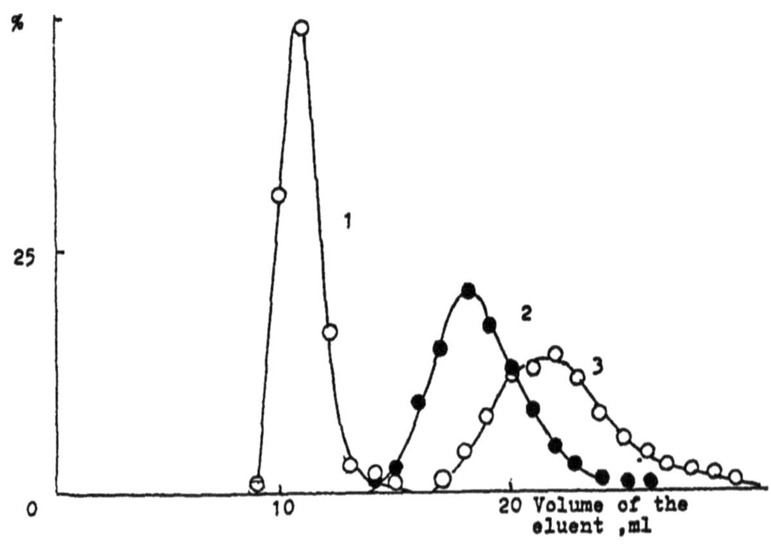


**Figure 10.** Separation of Eu(III) (1), Am(III) (2) and Ce(III) (3). Stationary phase: 0.07 M Tol-triazine in $CHCl_3$. Mobile phases: 3 M $HNO_3$.

The presented results show, that CCC is the promising method for separation of elements of similar properties.

The developed procedures can be applied for the analytical and preparative purposes. Development of a device of greater size will make it possible to separate TPE and REE on an industrial scale.

## Separation of Americium, Curium and Berkelium

Until recently $Am^{IV}$ was considered to exist in concentrated solutions of fluoride ions only. Presently a number of methods of americium stabilization in the tetravalent state are available (Table 1).

**Table 1.** Methods of production and stabilization of AmIV in solutions

| Medium | Oxidation method | Conditions of oxidation | Year |
|---|---|---|---|
| Aqueous solutions of phosphoric acid | Electrochemical: $E_a$ = 1.75 V | 10-12 M $H_3PO_4$<br>Oxidation time 1-1.5 hrs, t = 20 °C | 1973 |
| Acetonitrilic solutions of phosphoric acid | Electrochemical: $E_a$ = 2.07 V | 0.5-2 M $H_3PO_4$<br>Oxidation time 1 hr, t = 20 °C | 1983 |
| Aqueous solutions potassium of phosphotungstate | Electrochemical: $E_a$ = 1.77 V | 0.5-3.5 M $H_2SO_4$, $10^{-2}$ M $K_{10}P_2W_{17}O_{61}$<br>Oxidation time 0.5-1 hr, t = 25 °C | 1983 |
| | Chemical:<br>0.05 M $(NH_4)_2S_2O_8$,<br>0.01 M $Ag^+$ | 0.1-3 M $H_2SO_4$, $HNO_3$<br>$5.10^{-3}$ M $K_{10}P_2W_{17}O_{61}$, t = 20 °C | 1985 |
| Aqueous solutions of potassium hydrocarbonate | Electrochemical: $E_a$ = 1.0 V | 3 M $KHCO_3$, pH = 9, t = 25 °C<br>Oxidation time 0.5-1 hr | 1986 |
| Aqueous solutions of linear condensed phosphates | Electrochemical: $E_a$ = 1.8 V | 0.1-1 M $Na_4P_2O_7$ or $Na_5P_3O_{10}$, pH < 2,<br>Oxidation time 1 hr, t = 25 °C | 1987 |
| Aqueous solutions of ring condensed phosphates | Chemical:<br>0.1 M $Na_2S_2O_8$ | 0.5 M $Na_3P_3O_9$ or $Na_6P_2O_{18}$, pH = 1-2<br>Oxidation time 10 min, t = 90 °C | 1988 |

Americium (IV) is stabilized in the presence of phosphotungstate to permit its quantitative extraction from solutions of rather concentrated sulfuric and nitric acids by solutions of dioctylamine in dichloroethane. Under these conditions, extraction of americium (III) did not exceed 5% and a high degree of americium (both micro- and macroamounts) separation from curium and other trivalent elements was achieved.

A number of extraction methods of americium separation from other TPE and fission products based on the use of $Am^V$ and $Am^{VI}$, have been developed. Such methods are especially promising to isolate americium (V) traces for subsequent de termination. The conditions for quantitative extraction of $Am^V$ with thenoyltrifluoroacetone (TTA) 1-phenyl-3-methyl-4-benzoylpyrazolone-5 (PMBP), di(2-ethyl hexyl)phosphoric acid (HDEHP) and ammonium pyrrolidinedithiocarbamate (PDTC) have been found. Table 2 shows that $Am^V$ is extracted well by different extractants at pH 5 while the extraction of tervalent actinide ions may be suppressed by complexing agents, such as acetate and potassium phosphotungstate. AmV (both micro and macro amounts) can be extracted by HDEHP from acetate buffer at pH 4-5 with a distribution coefficient as high as 30. The separation factor of Am from Cm by extraction of $Am^V$ with PMBP or HDEHP is $(3\text{-}6)\bullet 10^3$. Pentavalent americium can be separated from other actinides and lanthanides by extraction with mixtures of picrolonic acid (PA) and sulphoxides in methyl isobutyl ketone from nitric acid since $Am^V$ is extracted significantly under these conditions.

**Table 2.** Conditions for $Am^V$ separation from transplutonium elements[2].

| Aqueous phase composition | Organic phase composition | Distribution ratio $Am^V$ | Distribution ratio $Me^{III}$ | Separation factor, $Am^V/Me^{III}$ |
|---|---|---|---|---|
| 0.01 M $NH_4NO_3$, pH 5 | 0.06 M $NH_4$PDTC in isopentanolethanol mixture | 30 | 0.02 | 1.5x103 |
| 0.1 M $NH_4NO_3$, pH 5 (acetate buffer 0.001 M PW*) | 0.05 M PMBP in isopentanol | 12.7 | 0.002 | $6.4\times10^3$ |
| 0.1 M $NH_4NO_3$, pH 5 (Acetate buffer 0.001 M PW) | 0.5 M HDEHP in octane | 30 | 0.01 | $3.0\times10^3$ |
| 0.1 M $HClO_4$ + + 0.01 M $H_3PO_4$ | 0.05 M PMBP + 0.025M TOPO in cyclohexane | 0.02 | 900 | $4.5\times10^4$** |
| 0.1 M $HN0_3$ | 0.16M PA in methyl isobutyl ketone | 0.26 | 194 | $7.3\times10^2$ |

*PW=potassium phosphotungstate; **Me(III)/Am(V)

Finding conditions for oxidation and stabilization of TPE in different oxidation states gave the basis to develop a number of isolation methods of TPE from solutions of complex compositions, such as the effective methods of separation of weighable amounts of Am and Cm. Americium preliminarily oxidized to the hexavalent state reduces upon extraction to the pentavalent state and stays in the aqueous phase. A mixture of extractants (such as TTA and TOPO as synergist) is used to increase isolation factor of curium. Under these conditions separation factor of Am and Cm reaches $10^3$ in a single extraction cycle. Americium sample of about 1 g with curium content of $10^{-6}$ % has been obtained by this method.

Trivalent curium can be separated from americium by extraction after electrochemical oxidation of the latter to the hexavalent state in sodium pyrophosphate solution with pH=10. The coefficient of curium separation from americium during one extraction cycle was about $10^3$ (Table 3).

**Table 3.** Separation of Am and Cm from 0.1 M pyrophosphate solution at pH=10 by extraction with 0.1 M solution of PMBP in chloroform.

| Element | Taken | | Found in aqueous phase | | Found in organic phase | | |
|---|---|---|---|---|---|---|---|
| | Oxidation state | mcg | Oxidation state | mcg | Oxidation state | mcg | % |
| Am | VI | 241.2 | V | 227.0 | III | 14.2 | 5.8 |
| Cm | III | 25 | III | 0.04 | III | 25 | 98.4 |

A number of extraction methods of $Bk^{IV}$ have been developed for selective isolation of berkelium. TOPO and TBPO solutions extract $Bk^{IV}$ quantitatively from 1-12M nitric acid and TBP solutions extract it from 8-12 M nitric acid. The berkelium (IV) can be efficiently separated from trivalent americium and curium, rare earths and some fission products using TOPO and TBP as extractants.

The use of the high molecular-weight amines to extract berkelium is most promising. By extraction of berkelium with 30% Aliquat solution in carbon tetrachloride from 10-12M nitric acid after oxidation with dichromate, a separation from the other TPE and the lanthanides is achieved. In the presence of heteropoly acid anions, $Bk^{IV}$ is quantitatively extracted from nitric acid and sulfuric acid by primary, secondary or tertiary amines and quaternary ammonium bases. Table 4 shows the distribution coefficients of some elements and separation factors of $Bk^{IV}$ from other metals in extraction with amines from nitric acid.

## Membrane Separation of Actinides from the Solutions of Inorganic Complex-Forming Agents

Membrane extraction processes of inorganic substances attract lately more and more attention of specialists. Membrane extraction is considered to be quite perspective method in comparison with its nearest analogue-liquid extraction owing to higher degree of element isolation from very diluted solutions and, in some cases, more effective their separation.

The kinetics of Np(VI), Pu(IV), Am(IV), Am(III), Cm(III) and Cf(III) transfer through solid supported liquid membranes (SLM) with HDEHP, TOA, DOA, TOPO, TBP, ALQ-336, Primene-JMT and DOP solutions and emulsion membranes has been studied. Permeability coefficients of some actinide compounds through different types of membranes are calculated. The optimal conditions of TUE separation have been chosen on the base of data obtained. It has been shown, that the most effective separation is attained

**Table 4.** Distribution coefficients (D) and separation factors (S) of berkelium (IV) and several metals during the extraction with amines from 1 M nitric acid containing $4\text{x}10^{-5}$ M $K_{10}P_2W_{17}O_{61}$ [13].

| Metal | Extractant | | | | | |
|---|---|---|---|---|---|---|
| | 3% decylamine in chloroform* | | TOA in carbon tetrachloride | | 5% Aliquates-33 in carbon tetrachloride | |
| | D | S | D | S | D | S |
| $^{249}$Bk | 305.5** | | 23.4 | 30.9 | | |
| $^{241}$Am | 0.014 | 2.2x104 | 0.0045 | 5.2x103 | 0.0035 | 8.8x103 |
| $^{243}$Cm | 0.017 | 1.8x104 | 0.014 | 1.7x103 | 0.0045 | 6.9x103 |
| $^{144}$Ce | 49 | 6.2 | 4.5 | 5.3 | 15.4 | 2.0 |
| $^{152-154}$Eu | 0.006 | 5.1x104 | 0.006 | 3.9x103 | 0.0015 | 2.1x104 |
| $^{137}$Cs | 0.007 | 4.4x104 | 0.0016 | 1.5x104 | 0.0033 | 9.4x103 |
| $^{106}$Ru | 0.055 | 5.6x103 | 0.044 | 534 | 0.064 | 483 |
| $^{90}$Sr | 0.020 | 1.5x104 | 0.003 | 7.8x103 | 0.0025 | 1.2x104 |

* Extraction from 6M HNO3; ** 250Bk was used.

when using the method of electrochemical and chemical stabilization of oxidation state of elements removed through the membrane. There are three cases when membrane extraction is preferable to liquid/liquid extractions:

1. Elements are quite stable in the given conditions, but the selective extractant does not have sufficient capacity. For example, the technique of Am(IV) separation from TPE and other fission elements by extraction with DOA from sulfuric acid solutions, containing $K_{10}P_2W_{17}O_{61}$ (PW) is well known. The low solubility of DOA salts even in a polar organic solvent is a substantial disadvantage of the methods. The application of membrane emulsion extraction allows considerably to increase the degree of Am(IV) selective isolation (Fig.11). Nitric acid destroys the phosphotungstate complex of Am(IV) allowing its subsequent reduction to Am(III) in receiving phase for back extraction in this system.

2. Ions of isolated elements have the tendency to the reactions of disproportionation. The extraction purification of plutonium from microimpurities of Np, or vice versa-neptunium from microimpurities of Pu, which is complicated by interaction of different ionic forms in solution, can serve as an example. However, using electrochemical stabilization and constant removing of impurities through the membrane, they manage to purify microcomponent with high degree of purification (Fig. 12).

3. Isolated element is unstable in needed oxidation state or reduces by extractant. Separation of Am(VI) and Cm(III) is the most important example for practice. Am(VI) reduces to Am(V) practically by all extractants. But, nevertheless, at simultaneous electrochemical oxidation the part of Am(VI) manages to diffuse through thin membrane

(Fig. 13), transmembrane flow being approximately one-tenth of calculated one under condition of Am(VI) stability. Thus, combining electrochemical stabilization and membrane extraction they can isolate americium even when liquid extraction can not be applied.

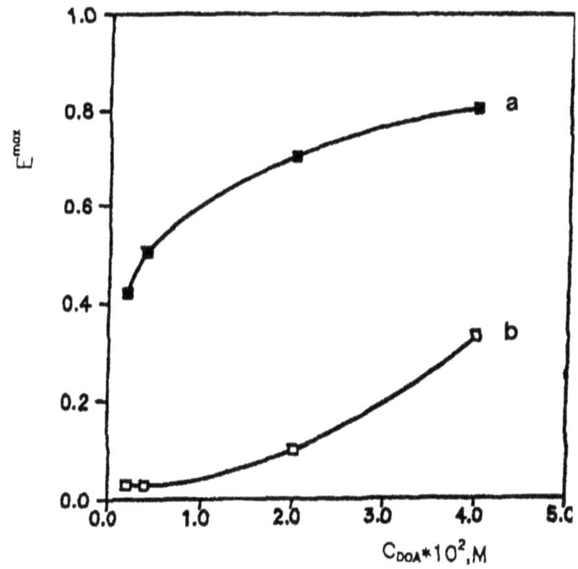


**Figure 11.** Am(IV) extraction from $10^{-3}$ PW ($H_2SO_4$ = 1.5 M) solutions: a - DOA, $CCl_4$ - 3 M $HNO_3$ emulsion; b - DOA solution $CCl_4$.

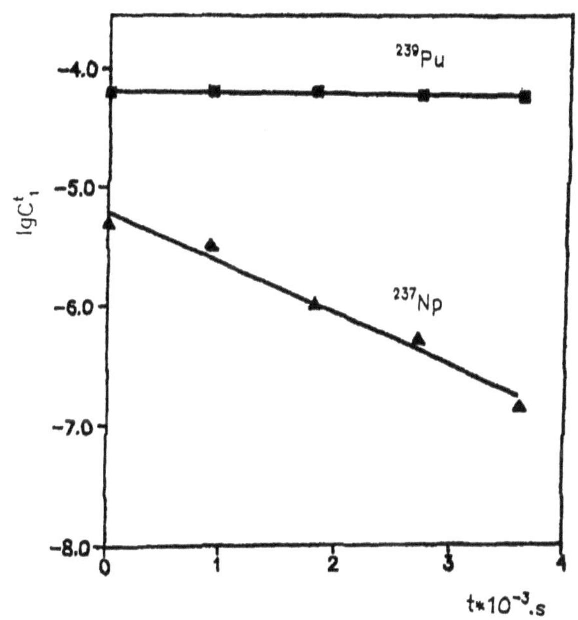


**Figure 12.** Semilogarithmic plot of Np(VI) and Pu(IV) concentration vs time, through a HDEHP = 0.1 M SLM. Feed: $[HNO_3]$ = 1 M, [PW] = $10^{-4}$ M; Strip: $[(NH_4)_2CO_3]$ = 1 M.

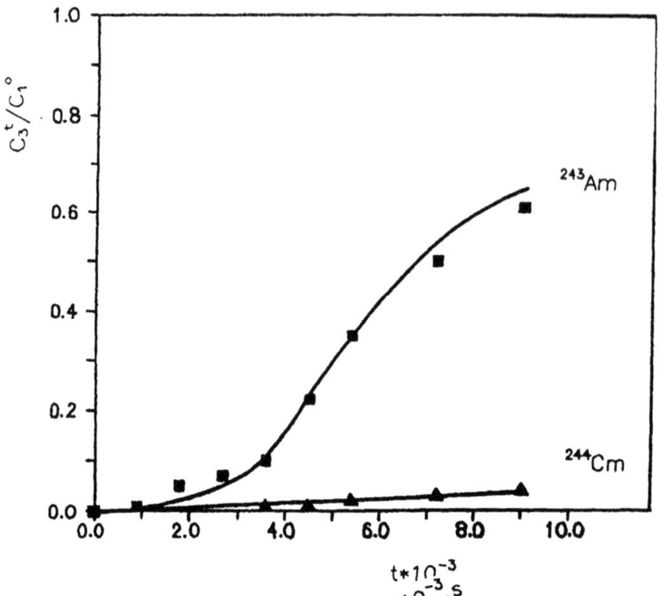


**Figure 13.** Semilogarithmic plot of Am and Cm conc... ...ume, through HDEHP = 0.1 M SLM. Feed: $[HNO_3]$ = 1 M, $[H_3PO_4]$ = $10^{-4}$ M; Strip: $[HNO_3]$=1 M, $[H_2O_2]$.

## REFERENCES


1. M.K. Chmutova, N.E. Kochetkova, and B.F. Myasoedov, Polydentate neutral organophosphorus compounds as extractants of transplutonium elements, *J. Inorg. and Nucl. Chem.* 42:897 (1989).
2. M.K. Chmutova, N.E. Kochetkova, O.E. Koiro, B.F. Myasoedov, T.Ya. Medved', N.P. Nesterova, and M.I. Kabachnik, Extraction of transplutonium elements with diphenyl(alkyl)(dialkylcarbamoylmethyl)phosphine oxides, *J. of Radioanalyt. Chemistry* 80:63 (1983).
3. B.F. Myasoedov, M.K. Chmutova, N.E. Kochetkova, O.E. Koiro, G.A. Pribylova, N.P. Nesterova, T.Ya. Medved', and M.I. Kabachnik, Effect of the structure of dialkyl(aryl)(dialkylcarbamoylmethyl)phosphine oxides on their extraction capacity and selectivity, *Solv. Extr. and Ion Exch.* 4:61 (1986).
4. M.K. Chmutova, M.N. Litvina, N.P. Nesterova, B.F. Myasoedov, and M.I. Kabachnik, Extraction of Am(III), Eu(III) and U(VI) from perchloric acid and mixtures of acids by dialkyl(aryl)(diethylcarbamoylmethyl)phosphine oxides, *Solv. Extr. Ion Exch.* 10:439 (1992).
5. M.K. Chmutova, G.A. Pribylova, N.P. Nesterova, B.F. Myasoedov, and M.I. Kabachnik, The extraction of americium (III) from nitric acid with the mixtures of neutral phosphorus organic reagents, *Radiokhimiya.* 31:73 (1989).

6. G.A. Pribylova, M.K. Chmutova, N.P. Nesterova, B.F. Myasoedov, and M.I. Kabachnik, Extraction of americium(III) by solutions of diaryl(dialkylcarbamoyl-methyl) phosphine oxides in aliphatic diluents, *Radiokhimiya.* 33:70 (1991).
7. G.A. Pribylova, M.K. Chmutova, V.I. Babain, and A.Yu. Shadrin, Extraction of Am(III) by di alkyl(diaryl)dialkylcarbamoylmethylphosphine oxides solutions in mixed and fluorocontaining solvents, *in*: "Proceedings of the International Conference ACTINIDES'89, Nauka, Moscow (1989).
8. N.E. Kochetkova, M.K. Chmutova, I.A. Lebedev, and B.F. Myasoedov, Investigation of radiation stability of tetraphenylmethylene diphosphine dioxide, *Radiokhimiya.* 23:420 (1981).
9. B.F. Myasoedov, M.K. Chmutova, I.V. Smirnov, and A.Yu. Shadrin, Isolation of actinides and separation from rare earth elements and selected fission products by bidentate neutral organophosphorus compounds, *in*: Proc. Int. Conf. and Technology Exposition on Future Nuclear Systems: Emerging Fuel Cycles and Waste Disposal Options (GLOBAL - 93), Publ. Amer. Nucl. Soc., Washington (1993).
10. Z.K. Karalova, B.F. Myasoedov, T.I. Bukina, and E.A. Lavrinovich, Extraction and separation of actinides and lanthanides from alkaline and carbonate solutions, *Solv.Extr. on Exch.* 6:11O9 (1988).
11. N.P. Molochnikova, V.M. Shkinev, and B.F. Myasoedov, Two-phase aqueous systems based on poly(ethylene glycol) for extraction separation of actinides in various media, *Solv. Extr. Ion Exch.* 10:697 (1992).
12. M.K. Chmutova, T.A. Maryutina, B.Ya. Spivakov, and B.F. Myasoedov, Separation of americium(III) and europium(III) in the systems with neutral bidentate organophosphorus compounds using countercurrent chromatography, *Radiokhimiya* 34:56 (1992).

# SOLVENT EXTRACTION OF EU(III) AND AM(III) WITH THIO AND AMIDE EXTRACTANTS

Jing Yao, Rosalind M. Wharf and Gregory R. Choppin

Department of Chemistry
Florida State University
Tallahassee, FL 32306-3006 USA

## ABSTRACT

Extraction studies of trivalent f-elements are reported for two thio donor ligands and for a diamide. The systems studied were tributylphosphate (TBP) + thiothenoyltrifluoracetone (HSTTA), triisobutyl phosphine sulfide (TiBPS) + thenoyltrifluoracetone (HTTA) and N,N'-dimethyl-N,N'-dihexyl-3-oxapentanediamide + HTTA. HSTTA and TiBPS were found to be much weaker extractants than the oxo analogs HTTA and TBP. Moreover, these soft donor ligands show little enhancement in Eu/An separation. A number of species were observed in the diamide extraction. The largest separation factor ($\alpha$ = 10) was obtained for the extraction of the $M(TTA)(DMDHOPDA)(ClO_4)_2$ species.

## INTRODUCTION

Separations of the trivalent lanthanide and actinide elements are difficult as both series are hard acid cations which have slowly varying ionic radii of similar range across each series. As a result, the bonding is strongly ionic and their stereochemical behaviors are very similar. Group separations have been achieved in concentrated solutions using ion exchange and solvent extraction techniques. A novel approach has attempted to exploit an expected small increase in covalency in bonding of the actinides by use of complexants with soft donor (e.g., N, S) atoms [1]. Some success has been achieved

*Separations of f Elements*, Edited by K.L. Nash
and G.R. Choppin, Plenum Press, New York, 1995

with a thiopyrazolone extractant in which one oxygen donor has been substituted by a sulfur atom [2]. In this study we have followed that approach by using thiothenoyltrifluoroacetone, HSTTA, as the extractant in the presence of tributyl phosphate, TBP. We compare these results with those from an analogous system in which we replace the oxygen donor of the synergist with a sulfur atom (e.g., triisobutylphosphine sulfide, TiBPA) with thenoyltrifluoracetone, HTTA, as the anionic extractant [3].

Another approach to separation of Ln/An trivalent cations has been pioneered by Musikas and colleagues [4], who studied N,N'-tetralkyl-1,3-propane diamides. Reported advantages of these diamide extractants are their good solubility in aliphatic diluents, lack of third phase formation problems, low aqueous solubility, and non-deleterious radiolytic and hydrolytic degradation products. The ether-dicarboxylic-diamides [5] offer the same advantages while serving as open chain analogs of the crown ethers. We have chosen, N,N'-dimethyl-N,N'-dihexyl-3-oxapentane-diamide, DMDHOPDA, for further study as a separating agent for trivalent lanthanides and actinides. The hexyl chains ensure hydrophobicity while the methyl groups offer minimal steric hindrance. The amides are strongly basic while the ether oxygen could be an extra donor site. We have also investigated the extractive properties of N,N,N',N'-3-oxapentanediamide, TEOPDA. However, it was found to have poorer organic phase solubility and less suitable steric properties as a chelating extractant and, hence, was studied less extensively. The structures of the extractants are shown in Figure 1.

## EXPERIMENTAL

**Chemicals:** Thiothenoyltrifluoroacetone was synthesized by a slight modification of a published method [6]. The red precipitate of HSTTA was collected by vacuum filtration, washed with water and recrystallized. The recrystallized solid had a melting point of 73° C.

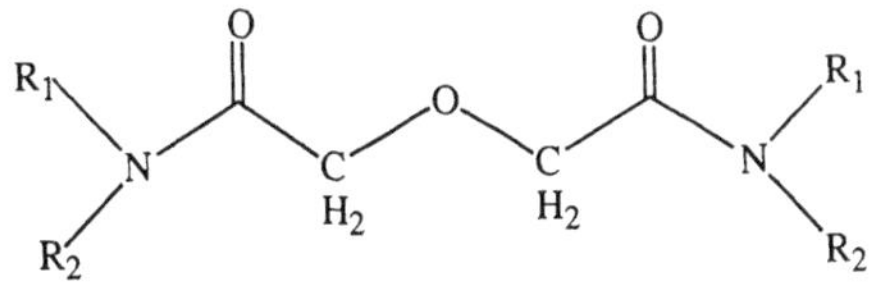

$R_1 = R_2 = CH_2CH_3$; N,N'-tetraethyl-3-oxapentanediamide (TEOPDA)
$R_1 = C_6H_{11}$, $R_2 = CH_3$; N,N'-dimethyl-N,N'-dihexyl-3-oxapentanediamide (DMDHOPDA)

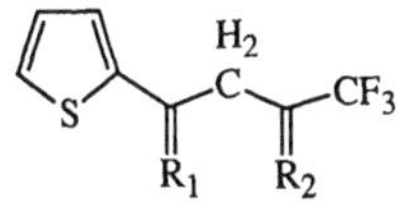

$R_1 = R_2 = O$; Thenoyltrifluoroacetone (HTTA)
$R_1 = O$, $R_2 = S$; Thiothenoyltrifluoroacetone (HSTTA)

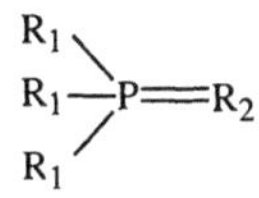

$R_1 = Bu$, $R_2 = O$; Tri-n-butylphosphate (TBP)
$R_1 = i\text{-}Bu$, $R_2 = S$; Tri-iso-butylphosphine sulfide (TBPS)

**Figure 1.** Structures of the extraction ligands.

The DMDHOPDA was synthesized by first converting diglycolic acid, DGA, to digylcolyl chloride refluxing a solution of DGA in thionyl chloride for 2.5 hours. A yellow, clear liquid remained after distilling off the thionyl chloride. N-methyl-N-hexylamine was mixed with benzene and frozen to avoid an explosion. This frozen mixture was added to the liquid diglycol chloride and refluxed at 90°C for 10 hours. After cooling for 3 hours, the liquid phase was filtered under vacuum, then washed successively with HCl, 1 M NaOH and water. The residual benzene was removed by rotary evaporation. A yellow, clear liquid remained that was confirmed to be DMDHOPDA by $^1$H-NMR, IR and elemental analyses. The latter indicated the purity to be ≥98%. Details of the synthesis and the analyses are given in reference 7.

The TEOPDA was synthesized and analyzed in a similar manner and the purity determined to be ≥97%.

Tributylphosphate, TBP, was purified by addition of anhydrous $Na_2CO_3$ followed by four washings of the TBP with distilled water before vacuum drying. Thenoyltrifluoroacetone, HTTA, (Aldrich) was purified by sublimation under vacuum at 42 °C.

Acetic and chloroacetic acids (Baker) were used as buffers at total concentrations of 0.007 - 0.01 M. The ionic strength was adjusted with $NaClO_4$ to 0.10 M. The organic solvents, cyclohexane and toluene, were reagent grade and were used as received. All other reagents were spectral or reagent grade.

**Radioassay:** $^{152,154}Eu$ and $^{241}Am$ (Oak Ridge National Laboratory), were assayed by counting their gamma radiations on a well type NaI(Tl) counter and on a Micromedic Isoflex Gamma Counter.

**Extraction:** Both aqueous and organic solutions were "pre-equilibrated" by contact with aliquots of the opposite phase in the absence of Am and Eu. Extractions were carried out in 20 ml borosilicate glass scintillation vials with polyethylene screw caps. Samples of either 3.00 or 5.00 ml of the aqueous phase, spiked with tracer, were mixed with an equal volume of organic solution containing the appropriate concentration of extractant(s). The mixtures were equilibrated for 2 - 24 hours by rotation in a thermostated bath at 25 ± 1 °C. The equilibrated samples were centrifuged, the phases separated and 0.50 or 1.00 ml aliquots of each phase removed for radioassay. By measuring samples withdrawn at different times, it was shown that the extraction equilibrium was established within two hours in all systems.

**Luminescence:** The experimental equipment and procedure has been described earlier [8].

## RESULTS

The extraction of a metal ion by an anionic extractant can be represented by:

$$M^{n+}_{(a)} + nHE_{(a)} = ME_{n(o)} + nH^{+}_{(a)} \tag{1}$$

In the presence of a neutral donor, the equation becomes:

$$M^{n+}_{(a)} + nHE_{(o)} + mS_{(o)} = ME_nS_{m(o)} + nH^{+}_{(a)} \tag{2}$$

$$K_{n,m} = [ME_nS_m][H^+]^n/[M^{n+}][HE]^n[S]^m$$

In our systems E=TTA and STTA while S=TBP, TIBPS and DMDHOPDA. The equilibrium constant of reaction 1 is denoted by $K_{n,o}$ and that of reaction 2 by $K_{n,m}$.

The distribution coefficient D is defined as:

$$D = \Sigma M_{(o)}/\Sigma M_{(a)} = \frac{\text{(cpm M per ml in organic phase)}}{\text{(cpm M per ml in aqueous phase)}} \tag{3}$$

From eq. (1) and (2), we see that the slope of a plot of log D vs. pH would indicate the value of n and that of log D vs. log [S] at constant pH the value of m. Typical plots are shown in Figure 2 for the HSTTA system. The dependency of D on variation of the concentration of pH and TBP was also determined. The slope values and the species inferred from them are presented in Table 1.

**Table 1.** Results of Extraction Studies of HSTTA and TiBPS Systems. T = 25 °C; I = 0.1 M ($NaClO_4$).

| | | | |
|---|---|---|---|
| | A. HSTTA[a] | | |
| | HSTTA | pH | |
| Dependency: | 2.8 | 2.1 | |
| Species: | M(STTA)$_2$(X)(HSTTA) (X = $ClO_4^-$ or $OAc^-$) | | |
| | B. HSTTA + TBP[a] | | |
| | HSTTA | pH | TBP |
| Dependency: | 2.8 | 2.7 | 1.8 |
| Species: | M(STTA)$_3$(TBP)<br>M(STTA)$_3$(TBP)$_2$ | | |
| | C. HTTA + TiBPS[a] | | |
| | HTTA | pH | TiBPS |
| Dependency: | 1.8 | 2.4 | 1.0 |
| Species: | M(TTA)$_2$(X)(TiBPS) (X = $ClO_4^-$ or $OAc^-$) | | |

[a]The values of Am and Eu agreed within the uncertainty of ±0.1 for these values.

**Table 2.** Extraction Constants for the HSTTA + TBP and TTA + TiBPS Systems. T = 25.0 °C; I = 0.1 M ($NaClO_4$).

| | Cation | Solvent | $\log K_{2,1}$ | $\log K_{3,1}$ | $\log K_{3,2}$ |
|---|---|---|---|---|---|
| | | | A. HSTTA + TBP | | |
| *i.* | Eu(III) | Cyclohexane | --- | -4.85±0.02 | -0.94±0.02 |
| | | Toluene | --- | -7.55±0.13 | -4.87±0.03 |
| *ii.* | Am(III) | Cyclohexane | --- | -4.71±0.10 | -1.11±0.01 |
| | | Toluene | --- | -7.37±0.11 | -4.74±0.03 |
| | | | B. HTTA + TiBPS | | |
| *i.* | Eu(III) | Cyclohexane | -2.35±0.01 | -7.80±0.08 | --- |
| *ii.* | Am(III) | Cyclohexane | -2.58±0.01 | -8.34±0.13 | --- |

Although our data were not sufficiently reliable to calculate a value of $\log K_{3,0}$ for the HSTTA system, Nag and Chandhury have reported a value of -13.40 for the Nd-HSTTA system in cyclohexane and -12.82 in benzene [9]. These authors also reported $\log K_{3,1}$ = 6.36 and $\log K_{3,2}$ = -2.96 for Nd-HSTTA-TBS in benzene. Our values of $K_{n,m}$ were obtained from a regression analysis of the extraction data as [$H^+$], [HE] and

[S] were varied. They are reported in Table 2 for the HSTTA + TBP system. The results of the earlier study [3] of the Eu/Am extraction by HTTA and TIBPS are included in Table 2 for comparison.

Figure 3 is a plot for Am (III) extraction from log D vs. log [DMDHOPDA]. The data were separated into three regions for analysis (all at [HTTA] = 0.015 M) with log concentration values of DMDHOPDA, from -5.00 to -3.60 (pH 3.05), from -3.60 to -2.20 (pH 3.5) and from -3.60 to -2.20 (pH 2.7). The resultant slopes, reflecting the ratio of M:DMDHOPDA, are listed in Table 3. All extractions with DMDHOPDA used toluene as the organic solvent.

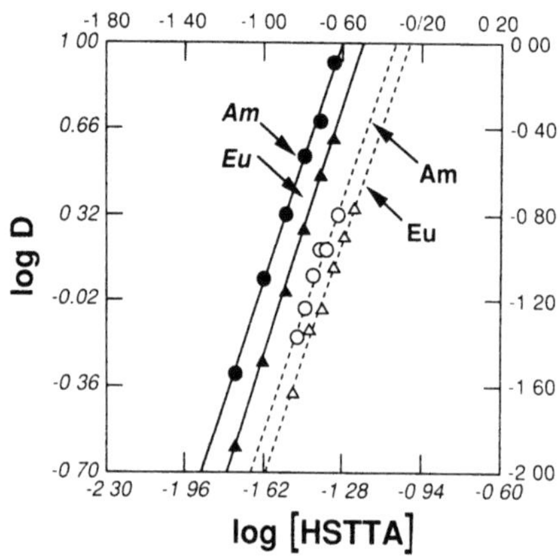


**Figure 2.** Variation of log D with log [HSTTA] for Am and Cm in toluene at pH ca. 4.7. Data shown by solid symbols have [TBP] = 0.01M while the open symbols are in the absence of TBP.

Figures 4 and 5 show the variation of log D with pH and HTTA concentration, respectively. In both figures, the results of varying the concentration of DMDHOPDA concentration is also shown. The results of the analysis of the slopes of these plots are also presented in Table 3. From the species postulated from the dependencies, it is possible to write the probable reactions as:

$$M^{3+}_{(a)} + 3HTTA_{(o)} + A_{(o)} = M(TTA)_3(A)_{(o)} + 3H^{+}_{(a)},\ K_{3,1} \quad (5)$$

$$M^{3+}_{(a)} + 2HTTA_{(o)} + 2A_{(o)} + X^{-}_{(a)} = M(TTA)_2(A)_2(X)_{(o)} + 2H^{+}_{(a)},\ K_{2,2} \quad (6)$$

$$M^{3+}_{(a)} + HTTA_{(o)} + 3A_{(o)} + 2X^{-}_{(a)} = M(TTA)(A)_3(X)_{2(o)} + H^{+}_{(a)},\ K_{1,3} \quad (7)$$

where X = $ClO_4^-$ and A = DMDHOPDA.

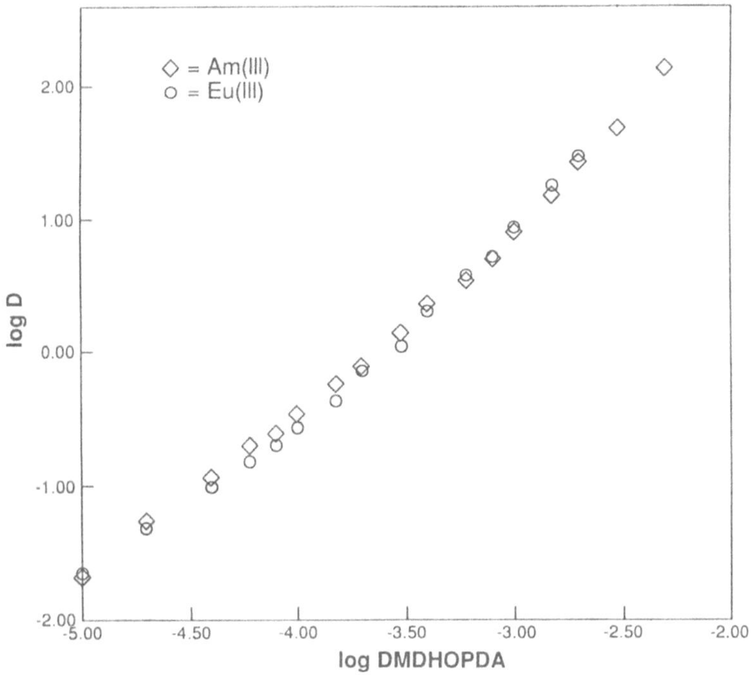


**Figure 3.** Variation of log D with log [DMDHOPDA] in toluence for pH = 3.0, [HTTA] = 0.015 M.

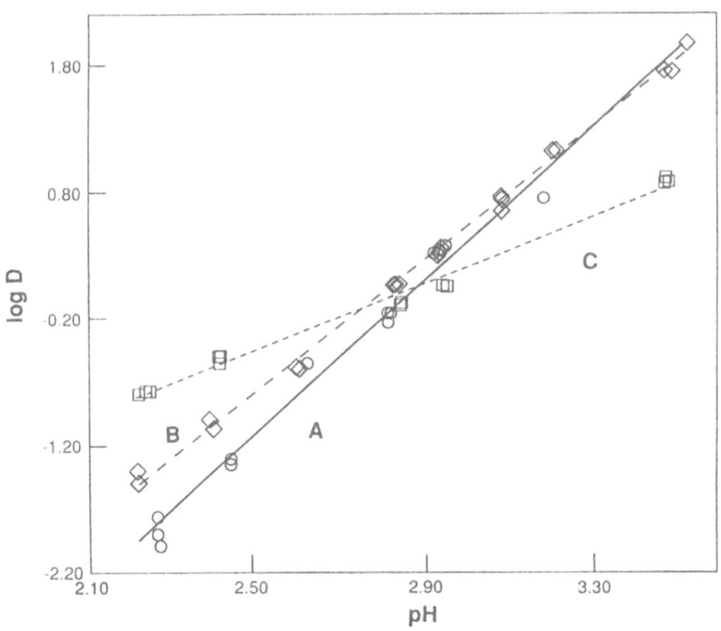


**Figure 4.** Dependence of log D on pH: A, [DMDHOPDA] = $10^{-5}$ M, [HTTA] = 0.06 M; B, [DMDHOPDA] = $10^{-4}$ M, [HTTA] = 0.03 M; C, [DMDHOPDA] = $10^{-3}$ M, [HTTA] = 0.004 M.

The distribution ration can be represented by:

$$D = \frac{[MA(TTA)_3 + [M(TTA)_2(A)_2(X) + [M(TTA)(A)_3(X)_2]}{[M]}$$

$$= K_{3,1}\frac{[H]^3}{[HTTA]^3}[A] + K_{2,2}\frac{[H]^2[X]}{[HTTA]^2}[A]^2 + K_{1,3}\frac{[H]^2[X]^2}{[HTTA]}[A]^3 \qquad (8)$$

The values of $K_{n,m}$ were obtained by regression analysis of appropriate sets of data in which one of the terms in equation 8 was dominant. The results are listed in Table 4.

**Table 3.** Results of Extraction Data for the Am + HTTA + DMDHOPDA System. T = 25 °C; I = 0.1 M ($NaClO_4$)

| | HTTA | pH | DMDHOPDA |
|---|---|---|---|
| | **A. HTTA** | | |
| Dependency: | 3.0±0.1 | | |
| Species: | $M(TTA)_3$ | | |
| | **B. HTTA + DMDHOPDA** | | |
| Dependency: | 1.2±0.1 | 1.3±0.1 | 1.1±0.1 |
| | 2.2±0.1 | 2.0±0.1 | 2.2±0.1 |
| | 3.1±0.2 | 2.9±0.2 | 2.9±0.1 |
| Species: | $M(TTA)_3(A)$<br>$M(TTA)_2(A)_2(X)$<br>$M(TTA)(A)_3(X)_2$<br>(A = DMDHOPDA, X = $ClO_4^-$) | | |
| | **C. DMDHOPDA** | | |
| Dependency: | --- | --- | 3.8±0.2 |
| Species: | $M(A)_4(X)_3$ | | |

The hydration of the Eu(III) in the organic phase was evaluated from the fluorescent lifetimes. The measurements were made for extraction from $H_2O$ and $D_2O$ solutions, and the hydration numbers obtained as described in reference 8. For ratios of [HTTA]:[DMDHOPDA] from 1 to 500, covering the range of species proposed in Tables 3 b, c and 4, the hydration number ranged from 0 to 0.2 with an uncertainty of ± 0.5, indicating no residual water in the primary coordination sphere of Eu(III).

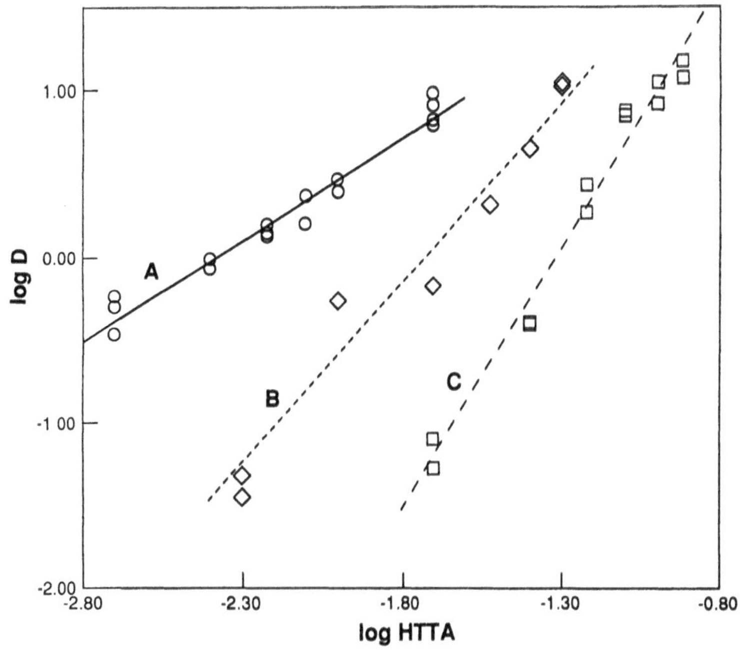


**Figure 5.** Dependence of log D on log [HTTA]: A, [DMDHOPDA] = $10^{-3}$ M; B, [DMDHOPDA] = $10^{-4}$ M; C, [DMDHOPDA] = $10^{-5}$ M.

**Table 4.** Extraction Constants of the HTTA + DMDHOPDA System. T = 25 °C; I = 0.1 M ($NaClO_4$).

| Species[a] | Constant | Eu(III) | Am(III) |
|---|---|---|---|
| $M(TTA)_3$ | log $K_{3,0}$ | -6.9±0.2 | -6.6±0.2 |
| $M(A)_4(X)_3$ | log $K_{0,4}$ | 10.27±0.12 | 9.93±0.07 |
| $M(TTA)_3(A)$ | log $K_{3,1}$ | -0.11±0.01 | -0.11±0.01 |
| $M(TTA)_2(A)_2(X)$ | log $K_{2,2}$ | 5.40±0.01 | 5.18±0.01 |
| $M(TTA)(A)_3(X)_2$ | log $K_{1,3}$ | 10.7±0.05 | 9.71±0.04 |

[a]A = DMDHOPDA; X = $ClO_4^-$.

## DISCUSSION

Our studies confirm that HSTTA is a much weaker extractant than its dioxo analog HTTA. An earlier study of Nd(III) extraction with HSTTA and TBP in benzene reported values of log $K_{3,0}$ = -12.82, log $K_{3,1}$ = -6.36 and log $K_{2,3}$ = -2.96. The latter two values fall between our values for cyclohexane and toluene. HSTTA is a stronger acid (pKa = 5.53) than HTTA (pKa = 6.63); however, this difference does not seem large enough to account for the difference in log $K_{n,m}$ values, suggesting weaker interaction with the S donor than with the O donor. Further there was no enhancement in the Am(III)/Eu(III) separation with HSTTA + TBP compared to that for HTTA + TBP and, hence no indication of improved separations with the thio system.

The extraction constants for the TiBPS system are also smaller than for the analogous TBP system. This again reflects the weakness of the metal-S donor interaction. The separation factor in the extraction by TiBPS of Am(III) and Eu(III) was not sufficient to indicate a use for these systems in separations of trivalent actinides and lanthanides.

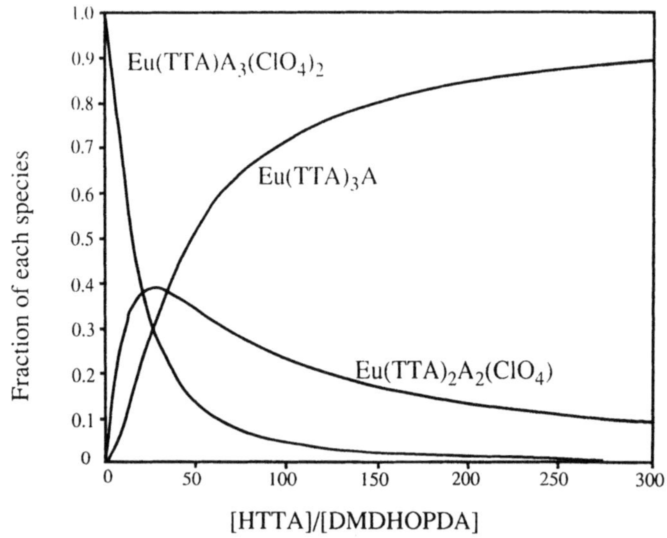


**Figure 6.** Speciation diagram of extracted species of Eu + TTA + DMDHOPDA from $NaClO_4$ solution.

The speciation in the DMDHOPDA + HTTA system, using the extraction constants in Table 4, is shown in Figure 6. The fluorescence data indicates that these species are anhydrous. Assuming both TTA and DMDHOPDA act as bidentate ligands, the coordination number of 8 can be assigned to Eu(III) and Am(III). This assumes the perchlorate anion forms an outer sphere ion pair; e.g., $[M(TTA)_2(A)_2^+]$--$[ClO_4^-]$ and $[M(TTA)(A)_3^{+2}]$--$[ClO_4^-]_2$. This assessment of the coordination number also omits metal-ether oxygen interaction. The ether oxygen is very weakly basic so any interaction would be very weak unless it were sterically promoted. Inclusion of metal-ether oxygen interaction would result in coordination numbers of 9, 10, 11 for the species in Table 3.

The separation factors for these species based on $\alpha_{Am}^{Eu} = K_{n,m}(Eu)/K_{n,m}(Am)$, are

| | |
|---|---|
| $M(TTA)_3(A)$ | $\alpha = 1.0$ |
| $M(TTA)_2(A)_2(X)$ | $\alpha = 1.7$ |
| $M(TTA)(A)_3(X)_2$ | $\alpha = 10.0$ |

Comparison of these values of α with the speciation plots in Figure 6 shows that separation is greater when the [DMDHOPDA] to [HTTA] ratio is high. Also, the separation factor for extraction by HTTA alone reflects enhanced extraction of Am relative to Eu whereas in HTTA + DMDHOPDA extraction, Eu is preferred. We did observe synergism, suggesting that a combination of DMDHOPDA with other synergistic extractants could lead to improved separation factors and, perhaps, result in a useful separation system.

Extraction constants have been reported for N,N'-dimethyl-N,N'-dioctyl-malonamide, DMDOMA, from nitrate solutions and involved the species $M(DMDOMA)_4(NO_3)_3$ [10]. These constants had values (log K) of 2.36 (Eu) and 2.42 (Am), almost eight orders of magnitude lower than our values for $M(DMDHOPDA)_4(ClO_4)_3$ extraction. Presumably this reflects the enhanced hydrophobicity of the ether diamides, although some of this difference is due to the use of $ClO_4^-$ rather than $NO_3^-$ as the anion. The ether group can increase the basicity of the amide donors as well as increase the flexibility of the ligand.

The HTTA + TEOPDA had a lower extration than the analgous DMDHOPM for comparable conditions (ca. $\geq 10^2$ lower D value). The variation of log D with log[TEOPDA] indicated the extracted species was $M(TTA)_3,(A)$, where the HTTA:TEOPDA ratio was ca. 600. Also, the data indicated a limiting solubility of ca. $10^{-4}$M for TEOPDA.

**Acknowledgement**

This research was supported by a grant from the Division of Chemical Sciences, OBES-USDOE.

## REFERENCES

1. G. R. Choppin, *Radiochim. Acta*, 32:43 (1983).
2. D. D. Ensor, G. D. Jarvinen and B. F. Smith, *Solv. Extr. Ion Exch.*, 6:439 (1988).
3. R. M. Wharf and G. R. Choppin, *Solv. Extr. Ion Exch.*, 8:615 (1990).
4. C. Musikas and H. Hubert, *Solv. Extr. Ion Exch.*, 5:151 (1987); ibid. 877 (1987).
5. H. Stephan, K. Gloe, J. Beger and P. Muhl, *Solv. Extr. Ion Exch.*, 9:435 (1991).

6. S. H. Chaston, S. E. Livingstone, T. N. Lockyer, V. A. Pickles, and J. S. Shannon, *Aust. J. chem.*, 18:673 (1965).

7. J. Yao, "Synergistic Extraction of Eu(III) and Am(III) by Thenoyltrifluoracetone and N,N'-Dimethyl-N,N'-dihexyl-3-Oxapentanediamide," M. S. Thesis, 1994, The Florida State University.

9. K. Nag and M. Chandhury, *J. Inorg. Nucl. Chem.*, 38:309 (1976); ibid., 39:1213 (1977).

10. N. Condamines and C. Musikas, *Solv. Extr. Ion Exch.*, 10:69 (1992).

# SELECTIVE EXTRACTION OF TRIVALENT ACTINIDES FROM LANTHANIDES WITH DITHIOPHOSPHINIC ACIDS AND TRIBUTYLPHOSPHATE

Gordon D. Jarvinen,[1] Richard E. Barrans Jr.,[1] Norman C. Schroeder,[1] Karen L. Wade,[1] Marianne M. Jones,[1] Barbara F. Smith,[1] Jerry L. Mills,[2] Gregory Howard,[2] Henry Freiser,[3] and S. Muralidharan[3]

[1]Los Alamos National Laboratory
Los Alamos, New Mexico

[2]Texas Tech University
Lubbock, Texas

[3]University of Arizona
Tucson, Arizona

## ABSTRACT

A variety of chemical systems have been developed to separate trivalent actinides from lanthanides based on the slightly stronger complexation of the trivalent actinides with ligands that contain soft donor atoms. The greater stability of the actinide complexes in these systems has often been attributed to a slightly greater covalent bonding component for the actinide ions relative to the lanthanide ions. We have investigated several synergistic extraction systems that use ligands with a combination of oxygen and sulfur donor atoms to achieve a good group separation of the trivalent actinides and lanthanides. For example, the combination of dicyclohexyldithiophosphinic acid and tributylphosphate has shown separation factors of up to 800 for americium over europium in a single extraction stage. Such systems could find application in advanced partitioning schemes for spent nuclear fuel and nuclear waste.

## INTRODUCTION

The separation of the trivalent actinides from the lanthanides has provided a notable challenge to chemists during the past half century because of the many similarities in the chemistry of these metal ions. A variety of chemical systems have been developed to accomplish this separation based on the slightly stronger complexation of the trivalent actinides with ligands that contain "soft" donor atoms, such as sulfur and nitrogen. The greater stability of the actinide complexes in these systems has often been attributed, beginning with a paper by Diamond, Street, and Seaborg,[1] to a slightly greater covalent bonding component for the actinide ions relative to the lanthanide ions. There is a considerable amount of experimental data that is consistent with this hypothesis, but thus far, attempts to measure a difference in the enthalpy of binding of trivalent actinide and lanthanide ions to ligands containing soft donor atoms have not clearly shown a stronger actinide bond. These concepts are discussed in more detail in a recent review on developments in trivalent f-element separations by Nash.[2]

Musikas and coworkers[3] have evaluated some extraction systems containing sulfur donor ligands to determine their ability to discriminate between the trivalent actinides and lanthanides. The synergistic system di-2-ethylhexyldithiophosphoric acid /tributylphosphate (HDEHDTP/TBP) gave excellent separation factors of up to 100 at certain concentrations of aqueous acid, HDEHDTP, and tributylphosphate (TBP). Such selectivity is not observed for the oxygen-donor analogue di-2-ethylhexylphosphoric acid. The TBP synergist increased the extraction coefficient to useful values as HDEHDTP alone is a poor extractant relative to its oxygen analog. The synergistic system also showed increased selectivity for americium over europium relative to HDEHDTP or TBP alone. One disadvantage of these systems is the relatively low stability of the dithiophosphoric acid derivatives to hydrolysis. The hydrolysis products are stronger extractants of both americium and europium, but they show low selectivity.

Extraction studies reported in the literature indicated that dithiophosphinic acid derivatives have improved stability relative to the dithiophosphoric acid systems. Tjioe *et al.*[4] reported on the extraction of Cd(II) from "black" phosphoric acid at 90°C with diphenyl- and dicyclohexyldithiophosphinic acids. These compounds showed considerable stability under these rather severe conditions, with the dicyclohexyl derivative exhibiting the greatest resistance to oxidation and hydrolysis. Rickelton[5] also described the separation of Cd(II) from phosphoric acid solutions using Cyanex™ 301, a dithiophosphinic acid derivative. In this paper, we report on the initial results of testing three different dithiophosphinic acid/synergist systems for their ability to separate trivalent actinides from trivalent lanthanides.

## EXPERIMENTAL

### Instrumentation

Radioactivity measurements were carried out with NaI well-type counters (Packard Models 500C, 5000, or MINAXI g Auto-Gamma® 5000). Other instrumentation included NMR (Varian Model Gemini 200), FT-IR (Mattson Model Galaxy 5020), UV-VIS diode array (Hewlett-Packard Model 8451A), GC-FID (Hewlett-Packard Model 5710A, DB-1 column, 12 m, 0.2 mm bore, 0.33 μm coating), and pH meter (Fisher Model 610A ). Elemental analyses were performed by Galbraith Laboratories.

### Materials

Dicyclohexyldithiophosphinic acid (HDCHDTP) was prepared at Texas Tech University using the method of Rauhut et al.[6]. Bis(2,4,4-trimethylpentyl)dithiophosphinic acid (Cyanex™ 301) was prepared by American Cyanamid and was purified at the University of Arizona. Diphenyldithiophosphinic acid (HDPhDTP) was purchased from Alfa/Aesar and was recrystallized from hexane. Other chemicals such as tributylphosphate (TBP), $NaNO_3$, $HNO_3$, toluene, and sulfanilic acid (4-aminobenzenesulfonic acid, $pK_a$ = 3.23) were Baker Analyzed reagents and were used as received. The TBP was checked by thin layer chromatography (TLC) for acidic impurities. All water used in this work was deionized to a resistance of 18 MW by a Millipore MILLI-Q™ system.

### Ligand Synthesis/Purification

**Dicyclohexyldithiophosphinic Acid (HDCHDTP).** HDCHDTP was synthesized according to the procedure of Rauhut and coworkers.[6]

**Bis(2,4,4-trimethylpentyl)dithiophosphinic acid (Cyanex™ 301).** Commercial, crude Cyanex™ 301 (40 g, 0.12 mol), which had a green color, was dissolved in 200 mL of analytical-grade hexane. It was washed once with 200 mL of 6 M $H_2SO_4$. The green color in the hexane phase became light. The hexane solution was then equilibrated with 170 g (0.6 mol) of $ZnSO_4$ in 400 mL of 0.05 M $H_2SO_4$ for 15 min. This procedure was repeated with fresh $ZnSO_4$ solution. A yellow zinc complex was obtained in the hexane phase. This phase was then dried over 25 g of anhydrous $Na_2SO_4$ for 4 hr. The resulting hexane solution was pale yellow, in contrast to the darker fresh extract. The hexane was removed by rotary evaporation to obtain a pale yellow, viscous product. This product was washed three times with 150 mL of methanol; the methanol was decanted after phase separation. This step yielded a milky white product. The traces of methanol were removed by rotary evaporation in a 38°C water bath to yield a zinc-Cyanex™ 301 chelate as an almost clear viscous product. (CAUTION: higher temperatures lead to the decomposition of the zinc chelate as evidenced by the darkening of the color to deep yellow.) The material was stored as the zinc complex because it is more stable in this form. Stripping of zinc was effected by redissolving the zinc-Cyanex™ 301 chelate in 100 mL of hexane and equilibrating with 100 mL of 4 M HCl a total of three times. The resulting hexane solution was dried over anhydrous $Na_2SO_4$, after which it was rotary evaporated at room temperature to remove the hexane. The viscous liquid obtained after this procedure was placed in a refridgerator where it crystallized after a couple of days. This solid was removed from the remaining liquid by filtration and was air-dried for 2 to 3 hr. It had a melting point of 34°C.

**Diphenyldithiophosphinic Acid (HDPhDTP).** The material, as received from Alfa/Aesar, required purification before use. The procedure that follows was adapted from Higgins et al.[7] The greenish solid (20.7 g) was dissolved in 225 mL of distilled water and 12.5 mL of 50% NaOH. The solution was separated from considerable undissolved solid by filtration through Whatman #2 filter paper. The solution was washed with 2 x 50 mL of toluene in a separatory funnel. To the slightly yellowish solution, 100 mL of 2.4 M HCl was added dropwise over 75 min with cooling in an ice water bath and rapid stirring. A milky mixture and a white solid consisting of small crystals resulted. The crystals were collected on a glass frit, washed three times with 50 mL of cold water, and dried under vacuum. The yield was 11.2 g, and the melting point was 56.8 to 57.8°C.

**Ligand Evaluation for Actinide/Lanthanide Separations**

**Distribution Determinations.** Europium-152 (SRM #4370) was obtained from the National Institute of Standards and Technology (NIST); it has a carrier europium concentration of 0.002 M. Americium-241 was obtained from Los Alamos in-house supplies. We prepared separate aqueous radioisotope stock solutions of $^{241}$Am(III) and $^{152}$Eu(III) by fuming the individual isotopes to dryness several times in $HNO_3$ and redissolving the tracers in dilute $HNO_3$ (pH ~3.5). Aqueous solutions for extraction contacts were buffered with 0.01 M sulfanilic acid, adjusted to a pH between 3.00 and 3.60 with 50% NaOH, and set at an ionic strength of 0.1 M with $NaNO_3$. These solutions were spiked with the isotope concentrates such that the tracer concentrations, determined directly from the gamma activity of the solution, were approximately $10^{-8}$ M. Carrier europium in the $^{152}$Eu-traced aqueous solutions was present at ~$10^{-5}$ M.

Variable concentrations of dithiophosphinic acids and synergist ligand TBP were dissolved in toluene to give the concentration ranges needed for the extraction experiments. Extractions were performed by contacting equal volumes of aqueous and organic phase (1 to 2 mL) for 30 min at room temperature (22 to 24°C) using a mechanical shaker, followed by centrifugation for phase separation. (All equilibrations discussed in this paper were performed in this manner unless otherwise noted). Known aliquots (0.25 to 1.00 mL) of each phase were counted on a gamma counter for 10 min. Accountability for all extractions was 100 ± 10%. The distribution ratio (D) was calculated as the net cpm/mL in the organic phase divided by the net cpm/mL in the aqueous phase. The estimated errors in the D values are 8 to 10% for americium and 9 to 30% for europium. The pH of the aqueous phase was determined after extraction. The reversibility of the HDCHDTP extraction was measured by taking the organic phase after extraction and contacting it with an equal volume of fresh untraced aqueous solution. Back-extraction of americium and europium from HDCHDTP was done by contacting the organic phase after extraction with an equal volume of 0.01 M $HNO_3$ /1.0 M $NaNO_3$ solution.

Slope analysis was used to obtain information about the composition of the extracted species. The simplifying assumptions used in interpreting plots of log D versus pH at constant [ligand] and [synergist] and log D versus log [ligand] or log [synergist] at constant pH and [synergist] or [ligand] are discussed in detail by Cox and Flett.[8]

### Ligand Distribution and Stability Studies

**$^1$H NMR Studies of Ligand Distribution and Stability.** For the $^1$H NMR studies, $d_8$-toluene was used to prepare the organic phase and $D_2O$ containing 1% DSS (3-trimethylsilyl-1-propanesulfonic acid, sodium salt) as an internal standard was used for the aqueous phase. Organic solutions were prepared containing 0.50 M ligand. Equilibration with the aqueous phase was performed as described above. The $^1$H NMR of each phase was taken periodically.

Also, studies were done with variable aqueous phase pH values. Three solutions of 0.50 M HDPhDTP in $d_8$-toluene were prepared and were equilibrated with equal volumes of the aqueous phase as before. A 2.27 M solution of NaOH in $D_2O$ was slowly added to these solutions. After 15 min of shaking, the pH was determined, and the process was repeated until the desired pH was obtained. $^1$H NMR spectra were taken of each phase.

**Ligand Distribution and Stability Studies Using Gamma Counting.** For several weeks, a 10-mL solution of 0.40 M HDCHDTP and 0.20 M TBP in toluene was shaken in a Teflon centrifuge tube in air while in contact with aqueous acid containing 1 mM $HNO_3$, 0.1 M $NaNO_3$, and 0.01 M sulfanilic acid buffer. Organic phase samples (1.00 mL) were removed periodically and gamma-counted to determine the activity of americium and europium. The organic material was returned to the reaction mixture. Also, the pH of the aqueous phase was measured periodically.

## RESULTS AND DISCUSSION

### Dicyclohexyldithiophosphinic Acid (HDCHDTP) Studies

**Job's Plot.** A variety of pH values and concentrations of HDCHDTP and TBP were initially tested to find a parametric space where large separation factors (SF) between americium and europium might occur. It was found that a significant SF could be obtained near pH 3.50 with 0.50 M HDCHDTP and 0.10 M TBP. A Job's plot study for the Am (III) and Eu (III) ions (Figure 1) was used to optimize the SF at this pH and a total concentration of TBP and HDCHDTP equal to 0.60 M. The SF was generally high (>100) across the plot, dropping to low values (<10) only when the individual ligands (TBP or HDCHDTP) were present. The highest SF (>300) occurred in the mole fraction range of 0.17 to 0.33 TBP. The plot for americium is broad but peaks at approximately 0.33 mole fraction of TBP, indicating that the extracted complex contains a 1: 2 ratio of TBP : HDCHDTP ligands. However, the broadness of the peak may also indicate that two or more complexes with varying stoichiometries are involved in the extraction. The europium D values are low across the plot and therefore give a less reliable indication of the composition of the extractant complexes. Even though the solutions were at an initial pH of 3.60, the final aqueous solution pH values varied from 2.80 to 3.60. The lower pH values occurred for the higher HDCHDTP concentrations, indicating the possibility that the compound was partially distributing into the aqueous phase. The Job's plot data were corrected for the pH variation to a common pH of 3.00. The details of how this correction was applied are given below.

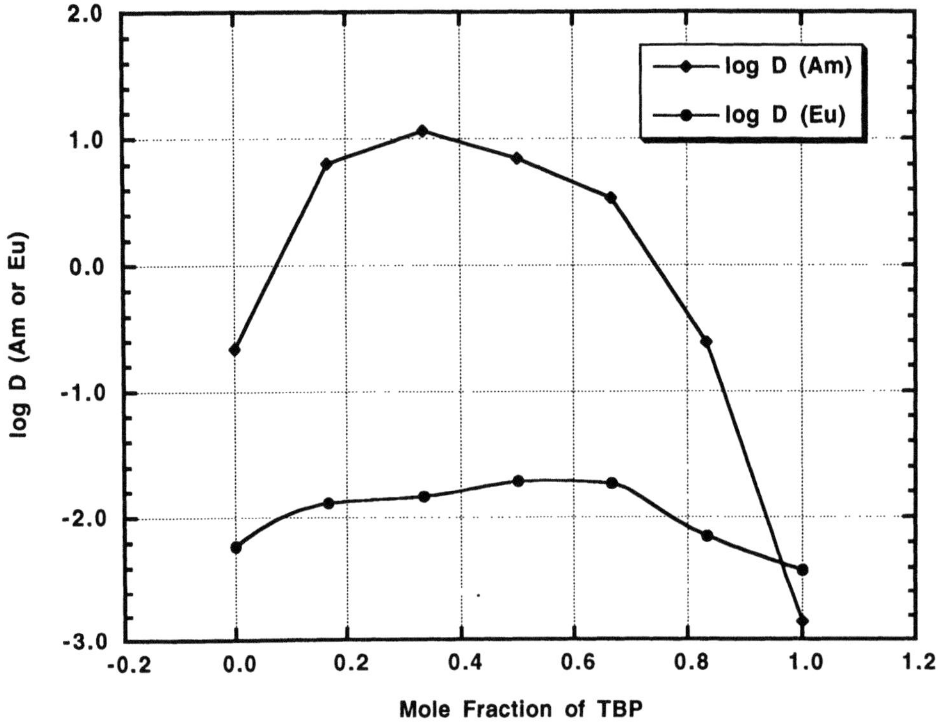


**Figure 1.** Job's Plot for $^{241}$Am and $^{152}$Eu Extraction for the System [HDCHDTP] + [TBP] = 0.60 M (data were corrected to pH = 3.00).

**pH Dependency.** The effect of the aqueous phase pH on the extraction of $^{241}$Am and $^{152}$Eu was studied at 0.28 M TBP and 0.25 M HDCHDTP. The initial aqueous phase pH values and the corresponding equilibrium pH values after the extraction contact are given in Table I. If it is assumed that the aqueous phase pH change is due mostly to partial solubility of the ligand in the aqueous phase, then a qualitative estimate of the solubility of the HDCHDTP can be obtained from the pH changes. For instance, at pH 4.00 the sulfanilic acid buffer (0.01 M) is 83% in the base form. A pH change to 3.18 reduces the amount of base form to 43%, requiring ~0.004 M protons. If the source of protons is the solubilized HDCHDTP, then the approximate distribution coefficient for HDCHDTP at this pH is 62 (or 1.5% distributes to the aqueous phase). The HDCHDTP is less soluble in the aqueous phase at lower pH. A pH-dependent solubility in the aqueous phase was also observed for

HDPhDTP, as discussed below. The plot of the log D versus the equilibrium pH is shown in Figure 2. Slopes were determined from a least squares fit of the data (R = 0.99 for americium and 0.80 for europium). The poorer fit of the europium data is due to the inherently greater inaccuracy in measuring the low count rates of $^{152}$Eu in the organic phase. Slope analysis gives a pH dependency of 2.9 for americium, indicating that 3 protons are released during the extraction process and, therefore, three HDCHDTP ligands may be involved in the americium-extracted complex. The europium pH dependency of ~1 is not as reliable due to the very low D values. The pH dependency plot shows that the Am/Eu separation factor increases with pH. The factor ranges from 40 at pH 2.53 to 630 at pH 3.20.

**Table I.** Aqueous Phase pH Before and After Extraction

| Initial pH | Equilibrium pH |
|---|---|
| 2.50 | 2.53 |
| 2.98 | 2.80 |
| 3.57 | 3.00 |
| 4.00 | 3.18 |
| 4.53 | 3.20 |

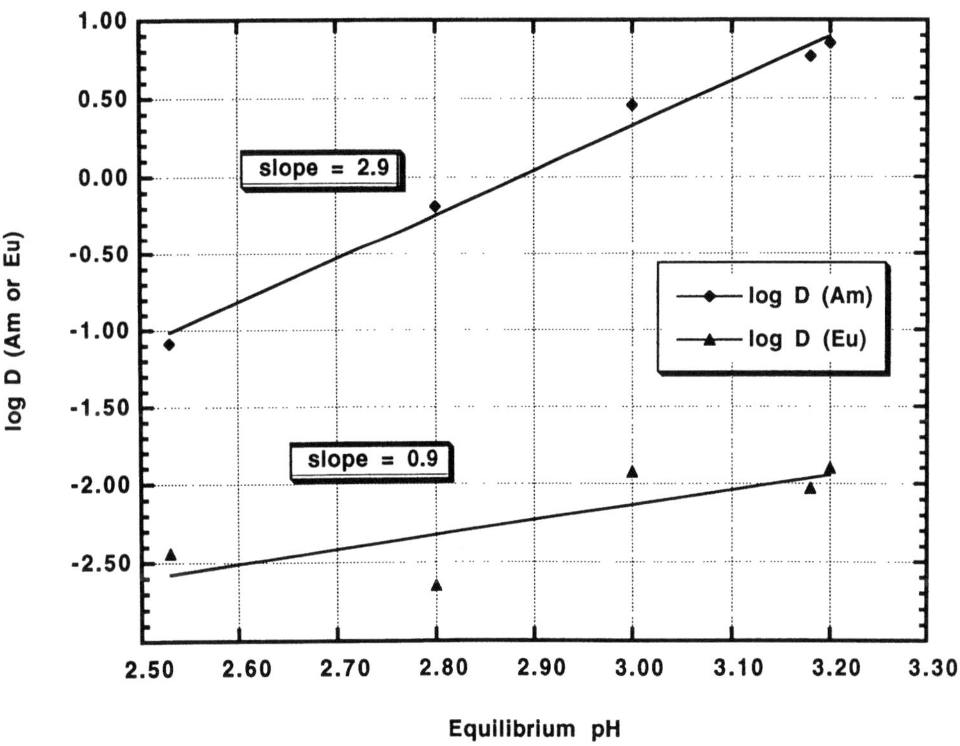


**Figure 2.** pH Dependency for $^{241}$Am and $^{152}$Eu Extraction at 0.25 M HDCHDTP and 0.28 M TBP.

**Dependency of the $^{241}$Am and $^{152}$Eu Extraction on the HDCHDTP Concentration.** Figure 3 shows a plot of log D versus log [HDCHDTP] at 0.50 M TBP and an initial pH of 3.66. The first four lines in Table II list the results for this experiment. The log of the distribution coefficients of both metals increase linearly with increasing log [HDCHDTP]. It should be noted that the HDCHDTP dependency for americium (1.5) is not consistent with the pH-dependency (2.9) because of the increasing solubility of HDCHDTP in the aqueous phase as its concentration in the organic phase increases. The HDCHDTP solubility is manifested in different equilibrium pH values which have substantial effects on the data. Thus the data in Figure 3 needs to be corrected for pH variation. Equations (1) and (2) were applied to the americium and europium data, respectively, and the data were corrected to a common pH of 3.00. The pH dependencies were obtained from the pH study described above (Figure 2) with one additional significant figure used in the calculations.

**Table II.** Summary of the Corrected $^{241}$Am and $^{152}$Eu Extraction Data Using HDCHDTP and TBP

| Equilibrium pH | [HDCHDTP] | [TBP] | D (Am) | D (Eu) | SF (Am/Eu) |
|---|---|---|---|---|---|
| 3.35 | 0.05 | 0.50 | 0.0424 | 0.0026 | 16 |
| 3.23 | 0.10 | 0.50 | 0.2858 | 0.0087 | 33 |
| 2.98 | 0.37 | 0.50 | 8.264 | 0.0483 | 171 |
| 2.93 | 0.50 | 0.50 | 16.98 | 0.0747 | 228 |
| 3 | 0.50 | 0.11 | 4.980 | 0.0061 | 819 |
| 3 | 0.50 | 0.20 | 9.010 | 0.0128 | 704 |
| 3 | 0.50 | 0.30 | 13.00 | 0.0200 | 650 |
| 3 | 0.50 | 0.40 | 14.20 | 0.0367 | 387 |
| 3 | 0.50 | 0.50 | 16.72 | 0.0694 | 241 |

$$\log D(\mathrm{Am})_{corr.} = \log D(\mathrm{Am}) + 2.85(\Delta \mathrm{pH}) \qquad \textbf{(1)}$$

$$\log D(\mathrm{Eu})_{corr.} = \log D(\mathrm{Eu}) + 0.94(\Delta \mathrm{pH}) \qquad \textbf{(2)}$$

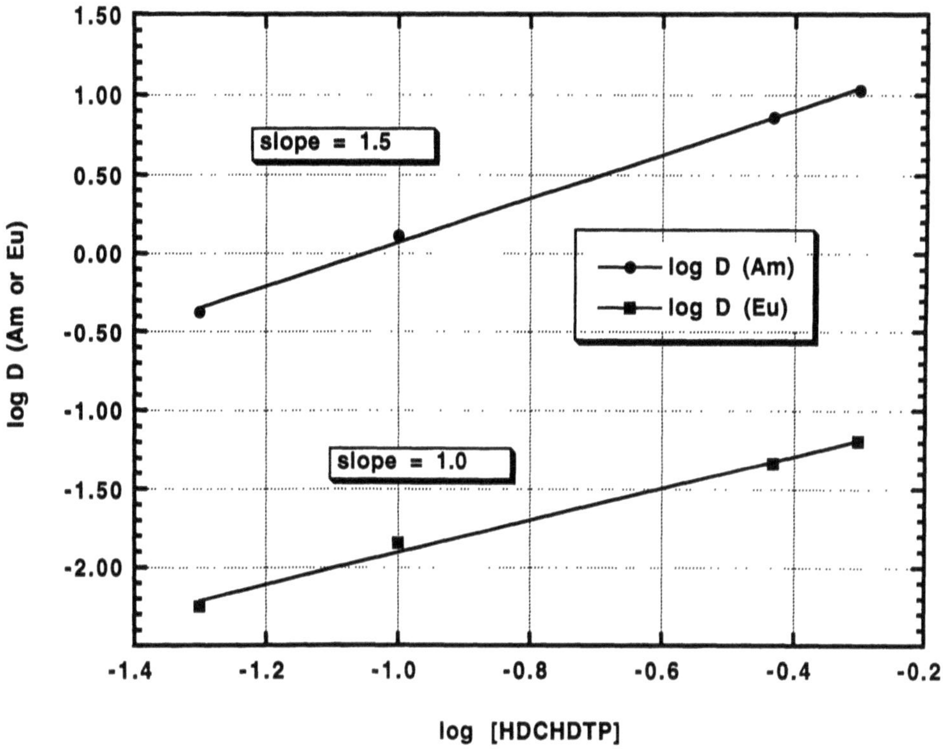


**Figure 3.** HDCHDTP Dependency for $^{241}Am$ and $^{152}Eu$ Extraction at 0.50 M TBP.

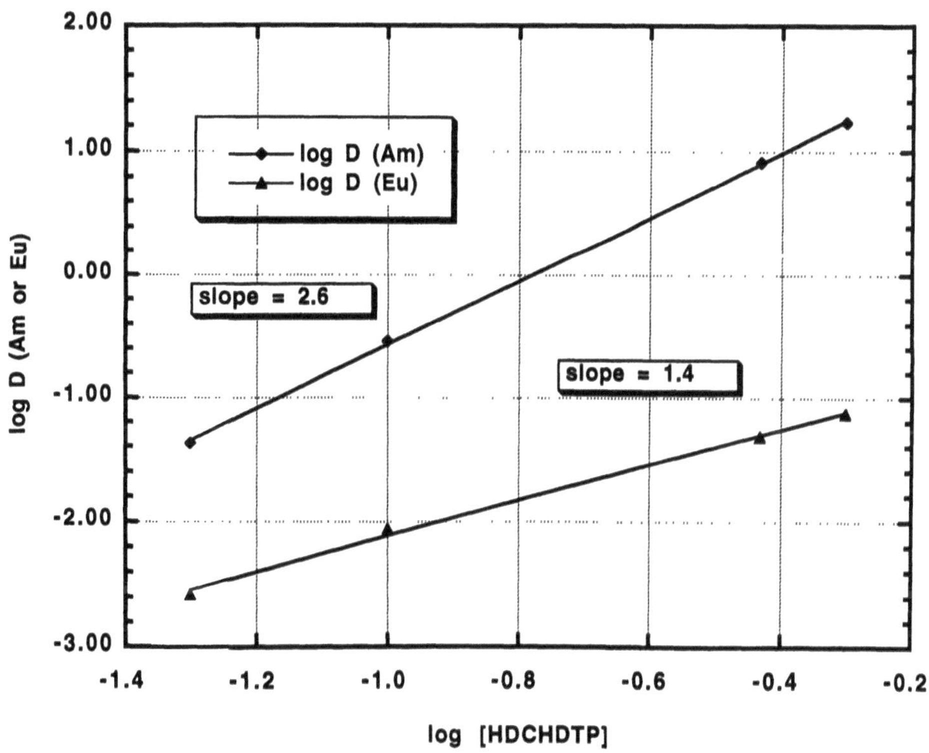


**Figure 4.** HDCHDTP Dependency for $^{241}Am$ and $^{152}Eu$ Extraction at 0.50 M TBP (corrected for aqueous distribution at pH = 3.00).

The ΔpH in this case would be $pH_{eqm}$ - 3.00. The corrected data are plotted in Figure 4. The slope of the HDCHDTP dependency for americium (2.6) is now more consistent with the pH dependency, but the dependencies are less consistent for europium (1.4 compared to 1.0). Although the least squares fit of the data are excellent for both elements (R = 0.99 for Am and R = 0.99 for Eu), the correction is probably less meaningful for the europium data because the 0.94 pH coefficient has a larger uncertainty associated with it. The ligand-dependency plot shows that the Am/Eu separation factor increases with the HDCHDTP concentration. The factor ranges from 16 to 228 across the ligand range studied.

**Reversibility of the Extraction.** Americium's extraction reversibility was studied between 0.05 and 0.37 M HDCHDTP at 0.50 M TBP by contacting the organic phase from the forward extraction with an equal volume of fresh aqueous solution containing no tracers. The corresponding equilibrium pH values measured for the forward and reverse reactions were within 0.05 pH units of each other. The D values for the forward and reverse extractions, after the pH correction was applied, agreed within 10%. The data for europium were unreliable because of the difficulty in measuring the low counts for the reverse extraction.

**Back-Extractions.** Americium and europium, after being extracted by 0.05 to 0.37 M HDCHDTP /0.50 M TBP solutions, were efficiently back-extracted by a 0.01 M $HNO_3$/ 1.0 M $NaNO_3$ solution. More than 99% of the americium and europium were removed from the organic phases.

**Dependency of the $^{241}$Am and $^{152}$Eu Extraction on the Synergist Ligand (TBP) Concentration.** The effect of the synergist ligand TBP on the extraction of $^{241}$Am and $^{152}$Eu was studied at 0.50 M HDCHDTP. A plot of the log D versus log [TBP] is shown in Figure 5. The results are also tabulated in the last five lines of Table II. The distribution coefficients are uncorrected because the equilibrium pH, although not measured, was expected to be ~3. The consistency in the americium D value at 0.50 M HDCHDTP and 0.50 M TBP in this experiment and at the corresponding concentrations in the log D versus HDCHDTP experiment (line 4 in Table II) supports this assumption. The slopes in Figure 5 give a TBP-dependency of 0.8 (R = 0.99) for the americium extraction and 1.5 (R = 0.98) for the europium extraction. The nonintegral slopes indicate that more than one extraction complex is formed. What is striking about this graph is the large separation factors across the TBP range studied, especially at low TBP concentrations. Table II lists the separation factors measured. An SF of about 800 was observed when 0.50 M HDCHDTP and 0.11 M TBP were used at a pH near 3.

**Ligand Distribution and Stability Studies.** It is believed that HDCHDTP distributes, to some degree, into the aqueous phase as observed by the decrease in pH with increasing HDCHDTP concentration in the organic phase. Although samples were not

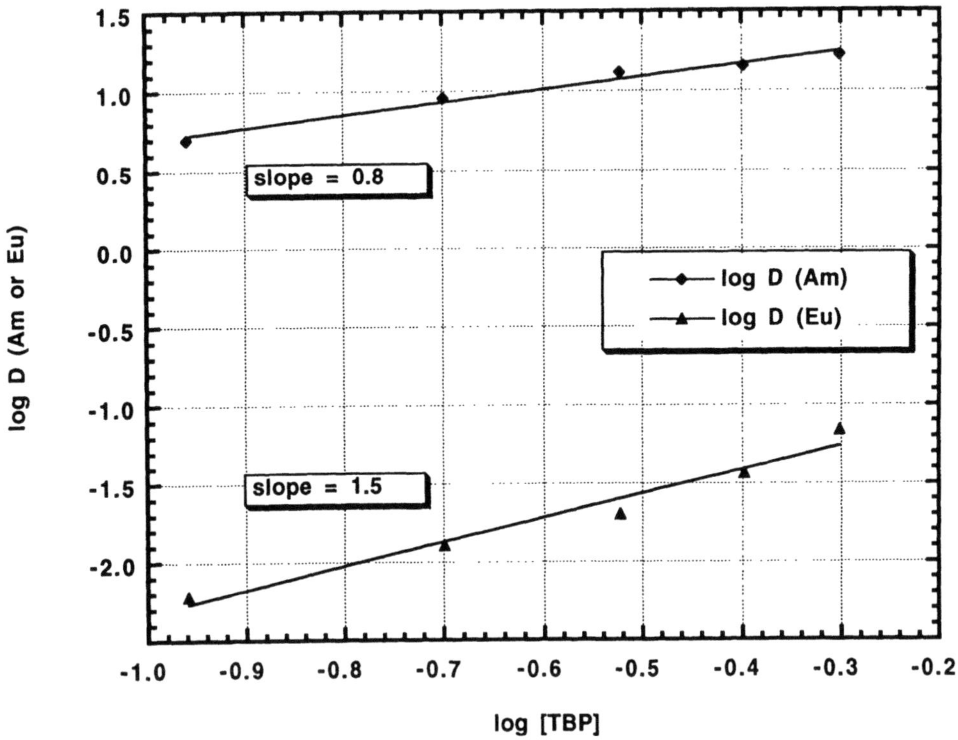


**Figure 5.** TBP Dependency for $^{241}Am$ and $^{152}Eu$ Extraction at 0.50 M HDCHDTP and pH = 3.00.

preequilibrated in the experiments reported here, previous scoping experiments where the samples were preequilibrated showed similar pH changes as a function of HDCHDTP concentration. Thus, to avoid ligand loss to the aqueous phase and uncertainty of the organic ligand concentration, all subsequent experiments were performed without preequilibration. It should be noted that it is not unreasonable that the HDCHDTP has some aqueous solubility, considering the high aqueous solubility shown with the diphenyldithiophosphinic acid (vide infra). An attempt was made to use $^1H$ NMR to determine the distribution coefficient of the ligand between the organic and aqueous phase. Also, $^1H$ NMR was used to observe any decomposition of the ligand over time under the conditions in which it was used. The limit of detection for $^1H$ NMR under the conditions employed (estimated at ~2 millimolar) was too high to provide convincing evidence of the presence of HDCHDTP in the aqueous phase, even at the higher organic concentrations. No decomposition products were observed over a 1-month period in the toluene-$d_8$ solution.

Another long-term stability study of the extractant solution at room temperature in contact with the aqueous solution (pH = 3.00) containing tracers was performed to determine if the ligand was degrading under these conditions. If hydrolysis of HDCHDTP was occurring, one would expect decreased separation factors and increased total extraction of both americium and europium. After 3 weeks, no decrease in the separation factor or increase in the total extraction was observed under the conditions of the experiment.

**Cyanex™ 301 Studies**

**Purification of Cyanex™ 301.** Cyanex™ 301, bis(2,4,4-trimethypentyl)dithiophosphinic acid is commercially available from American Cyanamid Corporation, but contains about 10% of an alkylphosphine oxide impurity. Because the impurity could act as a synergist and cause erroneous extraction results, the crude Cyanex™ 301 was purified. A zinc complex has been reported to be readily formed from the dithiophosphinic acid, as demonstrated by Rickelton,[9] but not by the alkylphosphine oxide impurity. Thus, the zinc complex of the dithiophosphinic acid can be readily separated from the impurity. The free ligand can then be obtained through release of the zinc ion by treatment with sulfuric acid. The purified Cyanex™ 301 showed evidence of some decomposition, which was indicated by development of a dark yellow color upon contact with 6 M HCl or when warmed to temperatures above 38°C. The colored species have not yet been identified. Under the conditions of the extraction experiments, decomposition of the Cyanex™ 301 does not appear to be a problem.

**Extraction Studies with Unpurified Cyanex™ 301.** Preliminary americium and europium extraction studies were performed at pH 1 and 2 at an ionic strength of 1.0 M $NaNO_3$ with unpurified Cyanex™ 301 alone (0.75 M) and with addition of TBP or trioctylphosphine oxide (TOPO) at 0.20 M. The D values ranged from 0.25 to 158. Separation of americium from europium was minimal in all cases and, for five of the six extractions studied, the europium extraction was slightly favored over americium. The TBP and TOPO enhanced the europium extraction at both pHs, but decreased the americium extraction, especially at a pH of 2.

**Extraction Studies with Purified Cyanex™ 301.** A Job's plot experiment, with purified Cyanex™ 301 and the synergist ligand TBP, was performed at pH 3.00 to determine the optimal ligand ratios. These conditions were chosen because they gave good separation factors for americium over europium in the HDCHDTP study. The Job's plot is shown in Figure 6. Separation factors were >10 for TBP mole fractions ranging between 0.17 and 0.70, with the highest factors (25) occurring between 0.30 and 0.50 mole fraction. Essentially no separation occurred when only the individual ligands (TBP or Cyanex™ 301) were present. The difference between the initial and equilibrium pH values was ≤ 0.08 pH units, thus, there did not appear to be a problem with the ligand distributing into the aqueous phase at this concentration range and initial pH.

**pH Dependency.** The effect of the aqueous phase pH on the extraction of $^{241}Am$ and $^{152}Eu$ was studied at 0.50 M TBP and 0.50 M Cyanex™ 301. The initial aqueous phase pH values and the corresponding equilibrium pH values were the same. The plot of the log D versus the equilibrium pH is shown in Figure 7. Slopes were determined from a least squares fit of the data (R = 0.99 for both americium and europium). Slope analysis gives a pH dependency of 2.8 for americium, indicating that 3 protons are released during the extraction process and, therefore, three Cyanex™ 301 ligands may be involved in the major extracted americium complex. For europium, the dependency is 1.4. It should be noted that the americium pH dependency is very similar for HDCHDTP and Cyanex™ 301. The pH-dependency plot shows that the Am/Eu separation factor increases with pH. The factor ranges from 20 at pH 2.72 to 250 at pH 3.50.

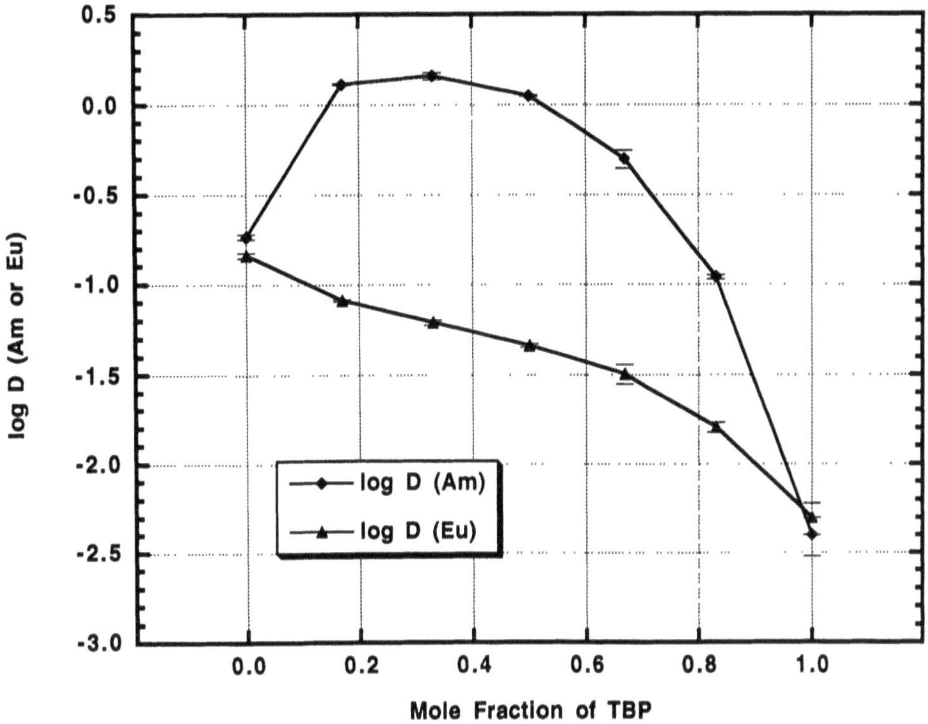


**Figure 6.** Job's Plot for $^{241}Am$ and $^{152}Eu$ for the System [Cyanex™ 301] + [TBP] = 0.60 M.

**Dependency of the $^{241}Am$ and $^{152}Eu$ Extraction on the Cyanex™ 301 Concentration.** Figure 8 shows a plot of log D versus log [Cyanex™ 301] with 0.10 M TBP at pH 3.10. The log of the distribution coefficients of both metals increases linearly with increasing Cyanex™ 301 concentration. The nonintegral slopes indicate that complexes with different stoichiometries are extracted. For europium, the slope is 2.4 (R = 0.99). The Am slope is 1.7 (R = 0.98), which is inconsistent with the pH dependency (2.8). This discrepancy may be caused by the narrow concentration range (0.4 - 0.6 M) studied for the Cyanex™ 301 dependency. The ligand dependency plot shows that the Am/Eu separation factor decreases slightly with the Cyanex™ 301 concentration; at 0.40 M the factor is 46, and the factor is 34 at a Cyanex™ 301 concentration of 0.60 M. The data are tabulated in the first five lines of Table III.

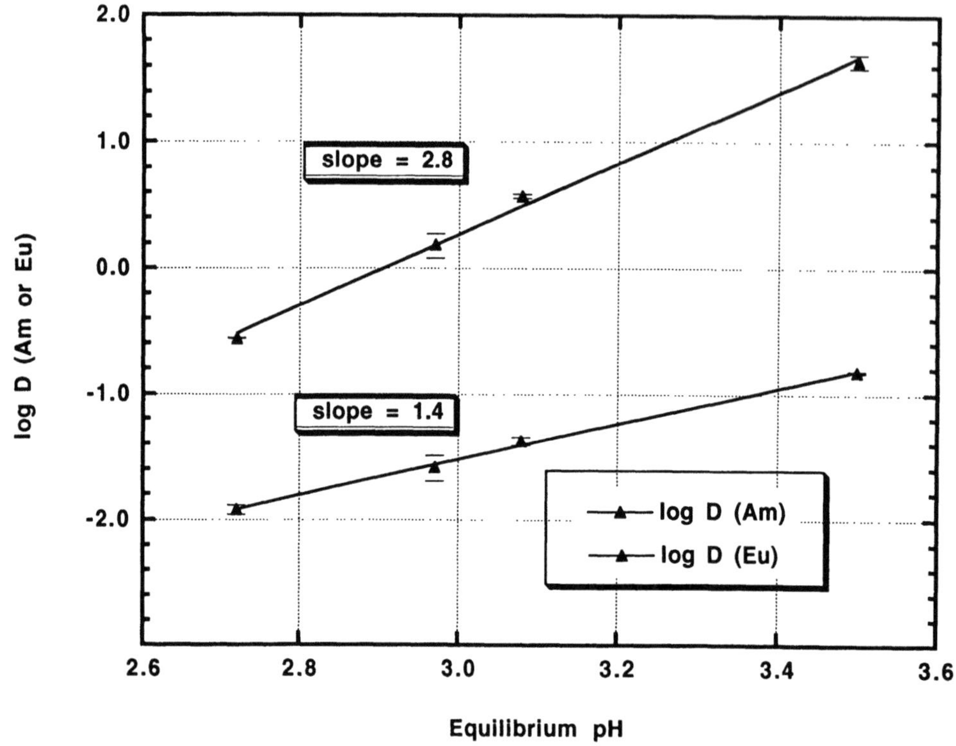


**Figure 7.** pH Dependency for $^{241}Am$ and $^{152}Eu$ Extraction at 0.50 M Cyanex™ 301 and 0.50 M TBP.

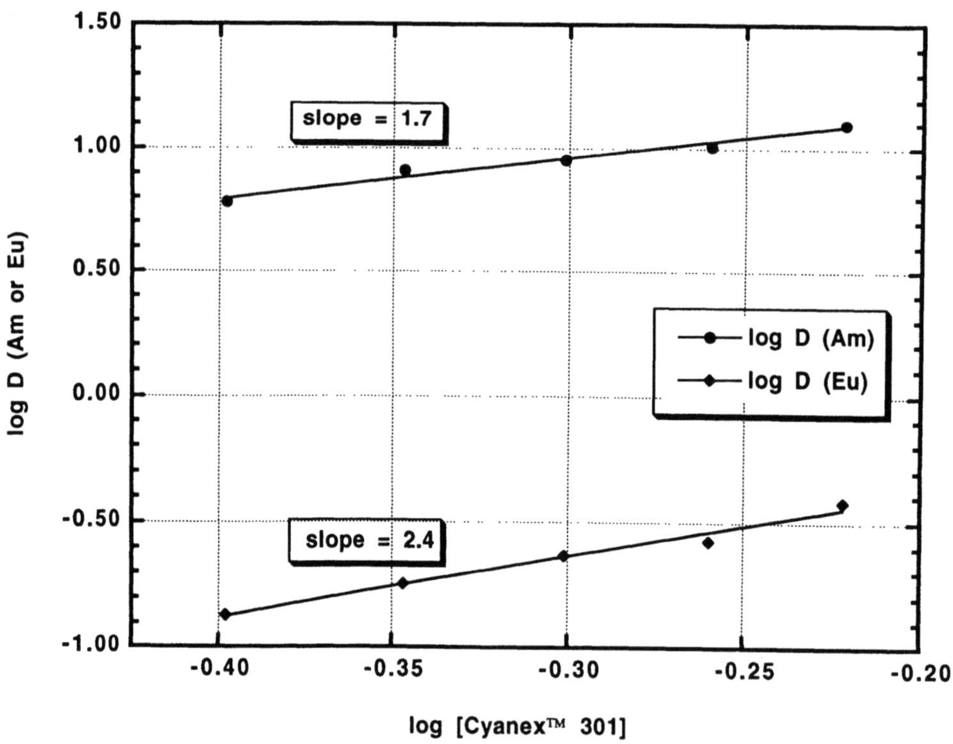


**Figure 8.** Cyanex™ 301 Dependency for $^{241}Am$ and $^{152}Eu$ Extraction at 0.10 M TBP and pH 3.10.

**Dependency of the $^{241}$Am and $^{152}$Eu Extraction on the Synergist Ligand (TBP) Concentration.** The effect of the synergist ligand TBP on the extraction of $^{241}$Am and $^{152}$Eu was studied at 0.50 M Cyanex™ 301 and a pH of 3.10. (The data are listed in the last five lines of Table III.) A plot of the log D versus log [TBP] is shown in Figure 9. The slopes show a TBP dependency of 1.0 (R = 0.99) for the americium extraction and 0.2 (R = 0.98) for the europium extraction. What is striking about this experiment is the almost constant D value for the Eu extraction. This is in contrast with the HDCHDTP, which had a dependency of 1.5. The separation factors in this system are not as large as the separation factors obtained with HDCHDTP. The factors range from 23 at 0.05 M TBP to 58 at 0.15 M TBP. A difference between these two systems is that the relative slopes are diverging with increasing TBP concentration in the Cyanex™ 301 system, whereas in the HDCHDTP system they are converging.

**Table III.** Summary of Separation Studies of $^{241}$Am and $^{152}$Eu Using Cyanex™ 301 and TBP

| **Equilibrium pH** | **[Cyanex™ 301]** | **[TBP]** | $D_{Am}$ | $D_{Eu}$ | $SF_{Am/Eu}$ |
|---|---|---|---|---|---|
| 3.08 | 0.40 | 0.10 | 6.1 | 0.134 | 46 |
| 3.09 | 0.45 | 0.10 | 8.2 | 0.177 | 46 |
| 3.09 | 0.50 | 0.10 | 9.1 | 0.232 | 39 |
| 3.07 | 0.55 | 0.10 | 10.3 | 0.268 | 38 |
| 3.10 | 0.60 | 0.10 | 12.8 | 0.380 | 34 |
| 3.09 | 0.50 | 0.05 | 4.8 | 0.208 | 23 |
| 3.09 | 0.50 | 0.075 | 6.8 | 0.214 | 32 |
| 3.09 | 0.50 | 0.10 | 9.3 | 0.226 | 41 |
| 3.10 | 0.50 | 0.125 | 13.3 | 0.238 | 56 |
| 3.09 | 0.50 | 0.15 | 14.1 | 0.242 | 58 |

### Diphenyldithiophosphinic Acid Studies

**Studies of the Effect of Ligand Concentration on $^{241}$Am(III) and $^{152}$Eu(III) Extraction.** The stability of HDPhDTP was somewhat lower than that of the other two dithiophosphinic acids. The odor of $H_2S$ was detectable and a greenish color was visible within a day or two after purification. After dissolution in toluene, the solution had a greenish tint. In addition, an interfacial scum formed in some of the extraction samples; even so, the accountability of the americium and europium tracers was >98%. Figure 10 shows a plot of log D versus log [HDPhDTP] at 0.50 M TBP and an initial pH of 3.57. Nonintegral slopes were obtained but could not be corrected using the pH dependencies. Control of the pH with this ligand was quite difficult because of its high distribution into the aqueous phase (discussed further below). The data from this study, listed in the first four lines of Table IV, indicates that a separation factor of up to 52 was possible using 0.50 M HDPhDTP and 0.50 M TBP.

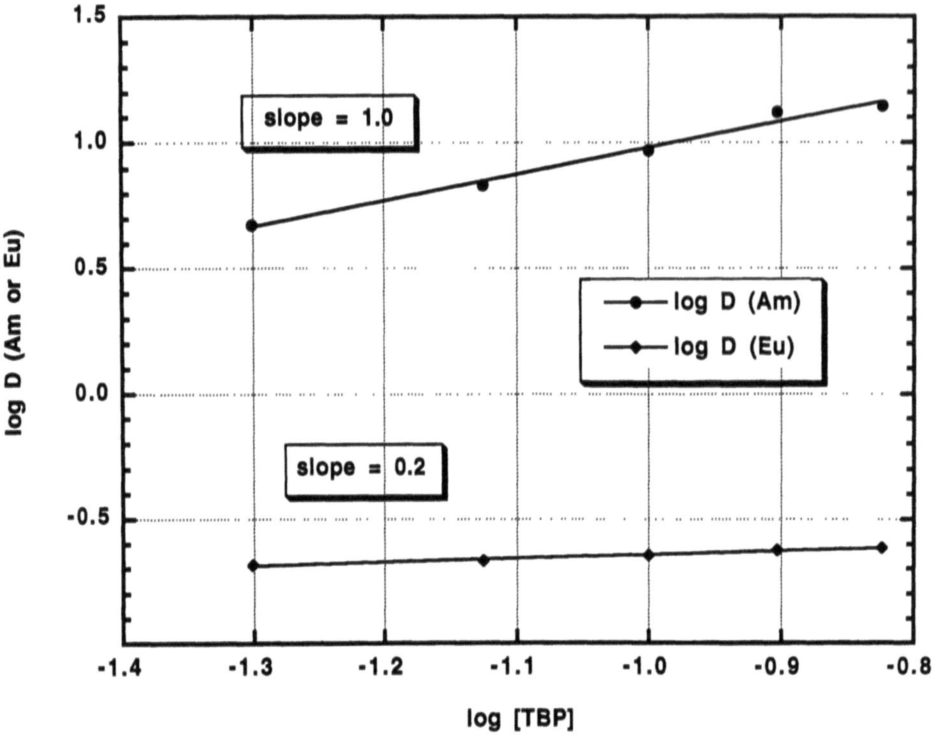


**Figure 9.** TBP Dependency for $^{241}Am$ and $^{152}Eu$ Extraction at 0.50 M Cyanex™ 301 and pH = 3.10.

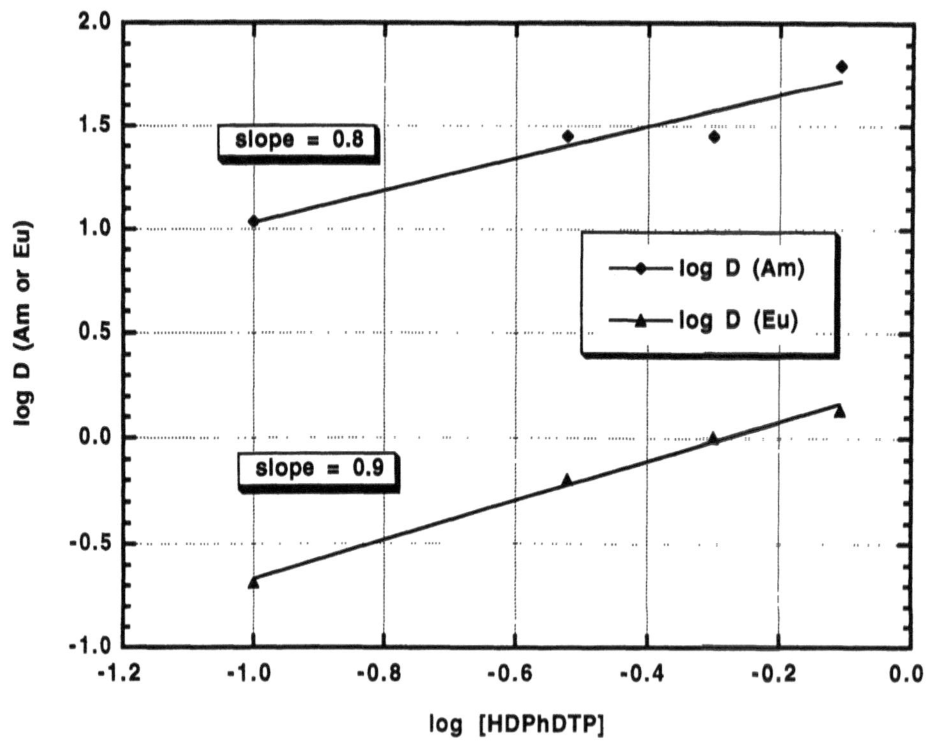


**Figure 10.** HDPhDTP Dependency for $^{241}Am$ and $^{152}Eu$ Extraction at 0.50 M TBP and pH ~1.6.

**Studies of the Effect of Synergist Concentration on $^{241}Am(III)$ and $^{152}Eu(III)$ Extraction.** Figure 11 shows a plot of log D versus log [TBP] at 0.50 M HDPhDTP and an initial pH of 3.57. Although the equilibrium pH values were not measured, the data in the third line of Table IV indicate that 0.50 M HDPhDTP systems have a final pH of ~1.6. The TBP dependencies for the extraction of americium and europium are nonintegral. This plot shows that the separation factor can be increased up to 332 by decreasing the [TBP] to 0.10 M and keeping the [HDPhDTP] at 0.50 M. Table IV lists the data from this study in the last four lines.

**Table IV.** Summary of Separation Studies of $^{241}Am$ and $^{152}Eu$ Using HDPhDTP and TBP

| Equilibrium pH | [HDPhDTP] | [TBP] | $D_{Am}$ | $D_{Eu}$ | $SF_{Am/Eu}$ |
|---|---|---|---|---|---|
| 1.94 | 0.10 | 0.50 | 10.89 | 0.21 | 52 |
| ----- | 0.30 | 0.50 | 28.18 | 0.63 | 45 |
| 1.58 | 0.50 | 0.50 | 28.26 | 1.01 | 28 |
| 1.47 | 0.78 | 0.50 | 63.14 | 1.36 | 46 |
| ----- | 0.50 | 0.10 | 3.32 | 0.01 | 332 |
| ----- | 0.50 | 0.30 | 20.59 | 0.16 | 129 |
| ----- | 0.50 | 0.50 | 28.26 | 1.01 | 28 |
| ----- | 0.50 | 0.77 | 61.53 | 5.16 | 12 |

**Studies of the Effect of Variable pH of the Aqueous Phase on $^{241}Am$ and $^{152}Eu$ Extraction.** It was not possible to obtain the pH dependencies of the extractions as had been done for HDCHDTP because most of the equilibrium pH values were at the same value: 1.56. Table V lists the initial pHs of the aqueous phases and the equilibrium pHs after extraction. A plot of log D versus equilibrium pH gives a cluster of points for which a linear fit is meaningless. For this reason, no attempt was made to correct for the aqueous solubility using the pH changes.

**Table V.** Initial and Equilibrium pH Values of the Aqueous Phase for the Extraction of $^{241}Am$ and $^{152}Eu$ from 0.50 M HDPhDTP and 0.50 M TBP

| Initial pH | Equilibrium pH |
|---|---|
| 2.50 | 1.48 |
| 2.98 | 1.55 |
| 3.57 | 1.57 |
| 4.00 | 1.56 |
| 4.53 | 1.56 |

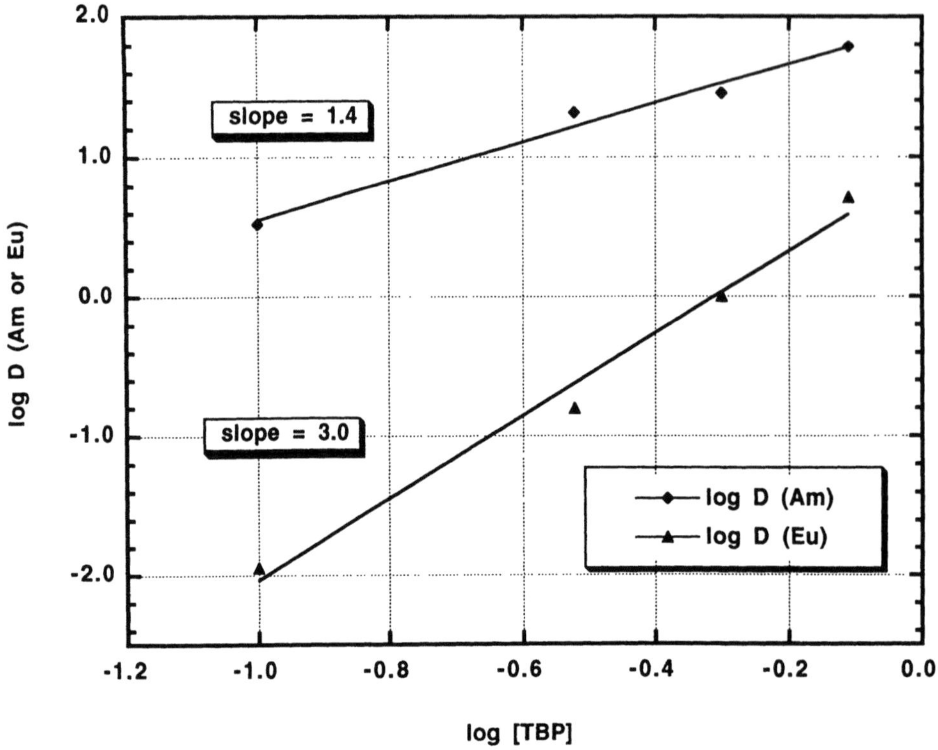


**Figure 11.** TBP Dependency for $^{241}Am$ and $^{152}Eu$ Extraction at 0.50 M HDPhDTP and pH ~1.6.

**Ligand Distribution Studies.** A $^{1}H$ NMR was taken of the organic (without TBP) and aqueous phases (initial pH 3.66) after a 30-min contact period to determine the distribution of HDPhDTP between the organic and aqueous phases (final pH ~1). The integrations of the peaks were normalized to the peaks of the internal standard, DSS, and the concentration of HDPhDTP in the aqueous phase was calculated relative to the standard. This amount was subtracted from the original concentration to determine the equilibrium concentration in the organic phase. Then D was calculated roughly to be $[HDPhDTP]_{org}/[HDPhDTP]_{aq} = 30$. When TBP was added at ~1 M concentration, D decreased to 26. No decomposition products were observed in either case.

The equilibrium pH of the aqueous phase was adjusted using NaOH in $D_2O$ to a pH of 2.40, 3.90, and 12.3, and the distribution tests were repeated. Note that the ionic strength is not 0.1 M. The aqueous phase solubility of the ligand increases greatly as the pH is increased; by a pH of 4, the ligand is essentially completely soluble ($D \leq 0.01$). At a pH of 12.3, a thick scum formed that was not isolated or analyzed. However, no soluble decomposition products were observed in the $^{1}H$ NMR spectra of the solutions.

Changes in the distribution of a 0.34 M HDPhDTP solution were observed over time through $^{1}H$ NMR. The aqueous solubility increases with contact time from a D of 30 after 30 min to D = 23 after 2 weeks to D = 22 after 1 month. An interfacial scum was noted to have grown over time; however, no impurities were observable in the $^{1}H$ NMR spectra of the $d_8$-toluene solutions.

Because of its instability and relatively high solubility in the aqueous phase, HDPhDTP will not be used in further studies unless it can be derivatized to increase its stability and decrease its aqueous solubility.

**Comparison of the Ligands.** From the studies described in this paper, it appears that the HDCHDTP - TBP and Cyanex™ 301 - TBP extraction systems have promise for separating trivalent actinides from lanthanides. The greater instability and substantial aqueous solubility of HDPhDTP made this ligand appear less useful for this potential application. Both Cyanex™ 301 and HDCHDTP appear to be stable under the testing conditions. Although Cyanex™ 301 showed less solubility in the aqueous phase, HDCHDTP yielded the greatest separation factors for americium over europium. Table VI lists the relative Am/Eu extraction separation factors for the three compounds studied under similar concentrations of dithiophosphinic acid and TBP; however, the pH value of the HDPhDTP system was different than that of the other two systems. The complex $AmL_3(TBP)$ appears to be a major complex in the organic phase, as indicated by the slope analysis of the extraction data for the HDCHDTP - TBP and Cyanex™ 301 - TBP extraction systems. This agrees with the data from the French work on the di-2-ethylhexylphosphoric acid - TBP extraction system.[3, 10] However, the slope values for the conditions studied in this paper are generally further from integral values, suggesting that complexes with other compositions are contributing substantially to the observed distribution ratios. The composition of the europium extraction complexes is much less certain since the D values under the conditions studied were quite low.

**Table VI.** Separation Factors (SF) for 0.50 M Dithiophosphinic Acids and 0.125 M TBP

| Compound | Equilibrium pH | SF |
|---|---|---|
| HDCHDTP/TBP | 3.0 | 1000 |
| HDPhDTP/TBP | 1.6 | 250 |
| Cyanex™ 301/TBP | 3.1 | 58 |

## CONCLUSIONS

The americium and europium distribution ratios of three dithiophosphinic acid/tributylphosphate extraction systems were measured. Very large separation factors for americium over europium were observed for these compounds, again demonstrating the usefulness of the soft donor concept for obtaining systems that selectively extract trivalent actinides. The stability studies indicated that these compounds, except perhaps for diphenyldithiophosphinic acid, have considerably greater stability than the dialkyldithiophosphoric acid compounds studied by Musikas and coworkers. Further testing will be needed to determine if these compounds have potential for analytical or process applications.

Future studies will include extraction measurements with a wider range of lanthanides, actinides, and other metal ions, use of other synergists and solvents, and more extensive evaluation of the chemical and radiation stability of these compounds.

## ACKNOWLEDGMENTS

Work performed under the auspices of the Efficient Separations and Processing Integrated Program of the Office of Technology Development (EM-50), U. S. Department of Energy under contract number W-7405-ENG-36.

## REFERENCES

1. Diamond, R.M., K. Street, and G.T. Seaborg, *An Ion-exchange Study of Possible Hybridized 5f Bonding in the Actinides.* Journal of the American Chemical Society, 1954. **76**(6): p. 1461-1469.
2. Nash, K.L., *A Review of the Basic Chemistry and Recent Developments in Trivalent f-Elements Separations.* Solvent Extraction and Ion Exchange, 1993. **11**(4): p. 729-768.
3. Fitoussi, R., C. Musikas, and H. Ramariveic, *Method for Separating Actinides from Lanthanides in an Acidic Aqueous Solution*1984, U. S. Patent No. 4,461,747.
4. Tjioe, T.T., P.F.M. Durville, and G.M. van Rosmalen, *Extraction of Cadmium from Phosphoric Acid with Diorganyldithiophosphinic Acids.* Solvent Extraction and Ion Exchange, 1989. **7**(3): p. 435-459.
5. Rickelton, W.A., *Novel Uses for Thiophosphinic Acids in Solvent Extraction.* Journal of Organometallic Chemistry, 1992. (May): p. 52-54.
6. Rauhut, M.M., H.A. Currier, and V.P. Wystrach, *The Preparation of Dithiophosphinates from Secondary Phosphines and Sulfur under Alkaline Conditions.* Journal of Organic Chemistry, 1961. **26**: p. 5133-5135.
7. Higgins, W.A., P.W. Vogel, and W.G. Craig, *Aromatic Phosphinic Acids and Derivatives. I. Diphenylphosphinodithoic Acid and Its Derivatives.* Journal of the American Chemical Society, 1955. **77**: p. 1864-1867.
8. Cox, M. and D.S. Flett, *Metal Extraction*, in *The Handbook of Solvent Extraction*, T.C. Lo and M.H.I. Baird, Editors. 1983, Wiley Interscience: New York. p. 53-90.
9. Rickelton, W.A. and R.J. Boyle, *The Selective Recovery of Zinc with New Thiophosphinic Acids.* Solvent Extraction and Ion Exchange, 1990. **8**(6): p. 783-797.
10. Pattee, D., *et al.*, *Extraction des Lanthanides et Actinides Trivalents par l'Acide Di-2-ethylhexyldithiophosphorique: Structure des Complexes Organiques.* Journal of Less-Common Metals, 1986. **122**: p. 295-302.

# DESIGN AND SYNTHESIS OF F-ELEMENT SELECTIVE LIGANDS

R.T. Paine

Department of Chemistry
University of New Mexico
Albuquerque, NM 87131

## ABSTRACT

Classical coordination chemistry principles provide a crucial foundation for the systematic construction of improved extractants for f-element ion separations. Such principles have been used recently to design ligands that contain both pyridine N-oxide and organophosphine oxide donor groups that act as bidentate or tridentate chelators on Ln(III) ions. The diverse structural features of these complexes are reviewed.

## INTRODUCTION

At the present time, there is a good deal of renewed interest in the coordination chemistry of f-element ions and in the design of new ligands that will selectively bind, with a wide range of stabilities, to these interesting species. This attention stems from a diverse collection of potential applications for f-element complexes in materials science, separations chemistry, environmental chemistry, and medicinal chemistry. Some of these are discussed in greater detail elsewhere in this volume.

The general coordination behavior of f-element ions was established during intense studies in the period 1940-70.[1] From that work, it is known that the ions typically are:

- relatively large with large charge to ionic radius ratios,
- "hard" and prefer to bind with oxygen and nitrogen donors in neutral or anionic ligands that can compete against water in the inner-coordination sphere,
- stabilized in complexes with high coordination numbers without dominant stereo-regulation induced by metal orbital orientations, and
- stabilized by ligand chelation effects.

Consequently, it appears that the most thermodynamically stable ligand/metal coordination conditions should contain multidentate O-donors efficiently packed around the f-element ion in such a way to provide maximum numbers of ligand donor-metal interactions with a

*Separations of f Elements*, Edited by K.L. Nash
and G.R. Choppin, Plenum Press, New York, 1995

minimum of steric strain. Based on discussions provided by Hancock and Martell,[2] it is anticipated that the overall stability of complexes of the highly charged f-elements will be enhanced in a "first-order" variation by the presence of negatively-charged oxygen atom donor centers relative to neutral oxygen donors. In "second-order" variations, it is expected that ligand selectivity may be improved by introduction of neutral O-donor or N-donor centers and perhaps softer donors such as P and S. It is also likely that geometric "preorganization" of the donor groups in a multifunctional ligand will be energetically favorable to both complex stability and selectivity. Indeed, the influences of these types of electronic and architectural features are intertwined and both play important roles in the design of new ligands for selective metal binding.

Significant efforts are in progress in many groups[2-15] today to develop fundamental information on the interplay of geometric and electronic effects through more sophisticated ligand design and molecular modeling. For example, with most simple d-block metal ions, regular octahedral coordination fields are preferred, and complexes with bidentate ligands having five-membered chelate rings are often energetically favored over chelate structures having six-membered or larger rings, all else being equal. However, with f-element ions that can adopt large (CN=7-12), irregular coordination fields, multidentate ligands easily utilize six, seven, and even eight-membered chelate rings. This allows more varied geometrical flexibility in developing preferred ligand environments on f-element ions. It is also expected that f-element specificity may be influenced by a greater range of variations of steric and electronic effects on a multifunctional ligand. For example, the introduction of one "softer" donor site, by charge or element modifications, on a multifunctional/multidentate ligand or the combination of different chelate ring sizes in a single ligand may lead to enhanced specificity for f-elements.

Obviously, there is a diverse collection of ligand architectures that might be pursued in the hunt for selective f-element chelators. One early design scheme was proposed by Siddall,[16-21] and this involved the development of so-called CMP (**1**) and CMPO (**2**) ligands. He reasoned that, since monofunctional organo carbonyls and phosphoryls are good ligands

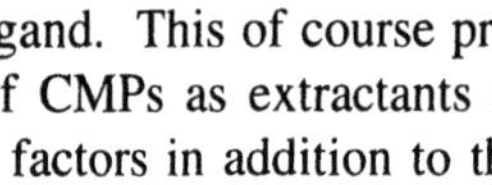

**1**

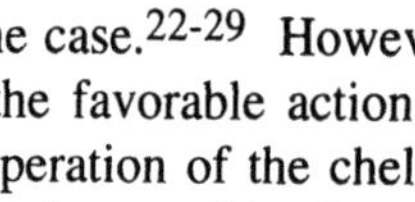

**2**

toward Ln(III) and An(III) ions, then a ligand containing both functional groups in a geometric arrangement that permits favorable chelate ring formation should be an even more strongly binding ligand. This of course proved to be the case.[22-29] However, it is also true that the efficacy of CMPs as extractants stems from the favorable action of several other solution chemistry factors in addition to the possible operation of the chelate effect. In an effort to expand on this design concept, we examined other combinations of donor groups that might be assembled on a hydrocarbon backbone. We initially attempted to prepare ligands with P=O and S=O groups as well as P=O and $SO_2$ groups, but these compounds were not especially attractive since they are generally unstable in aqueous acid solution.

Subsequent examination of the solvent extraction literature revealed that pyridine N-oxide derivatives have found some use as extractants since the N-oxide group is stable in acid solutions and an excellent oxide donor toward hard Ln(III) and An(III) ions. We therefore chose to explore methods for the synthesis of potentially chelating ligands containing both phosphoryl and N-oxide donor centers. Prior to our work, there appears to be few reports of the formation of chelating ligands containing both N–O and P=O donors. Some examples of pyridine compounds substituted with phosphide $-PR_2$ or phosphonate –

$P(O)(OR)_2$ groups were known and that chemistry has recently been reviewed.[30] Initially, we selected the bifunctional 2-(dialkoxyphosphono)pyridine N,P-dioxides **3** (R=OR) and **4** (R=Ph) as targets. A molecular structure determination for (2–diphenylphosphino)quinoline N,P–dioxide **4** has been completed (Fig. 1: H atoms here and in other figures are removed for clarity.),[31] and it is interesting that the N–O and P=O donor groups have a more or less *anti* orientation in the free ligand state. This configuration obviously would not permit bidentate metal coordination, and the desired *syn* conformation requires an apparently hindered rotation about the P-C bond involving the quinoline fragment.

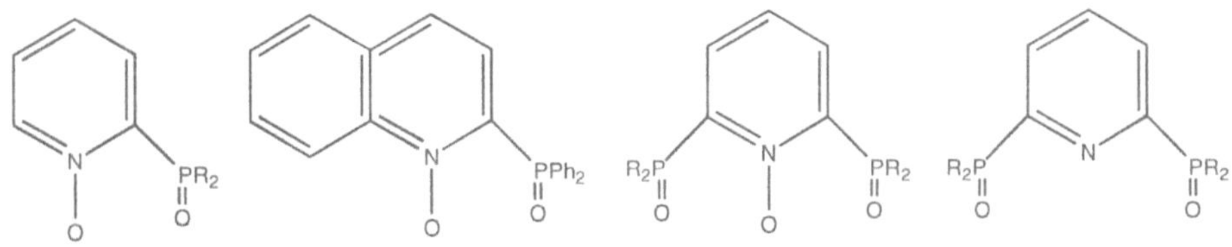


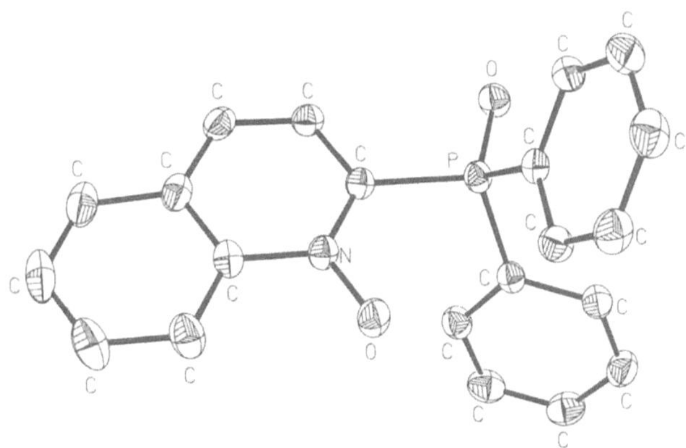


Figure1. Molecular structure of **4**.

The coordination chemistry of **3** and **4** has been partially explored.[32,33] It was found, for example, that stable complexes of the general types $UO_2(NO_3)_2$•L and $Ln(NO_3)_3$•2L are produced, and spectroscopic and X-ray crystallographic data indicated that the isolated complexes contain bidentate bonded ligands. Clearly, therefore, the ligands are able to undergo reorganization of the *anti* conformation. Liquid-liquid extraction data for compound **3** were also collected, and they show some favorable behavior partitioning Ln(III) ions from weakly acidic aqueous solutions into the organic phase.[34]

The initial success with ligands of types **3** (R=EtO, i-PrO, BuO, HexO, Ph) and **4** (R=Ph) encouraged us to attempt to prepare examples of compound **5**. It was reasoned that such compounds might function as tridentate ligands and display even stronger ligation behavior. Unfortunately, it has so far proven impossible to obtain these compounds. The intermediates **6** are obtained in good yield, but all attempts to N-oxidize **6** gave rise to pyridine ring decomposition. A limited amount of coordination chemistry has been accomplished for **6**, and this in fact indicates that **5** might not serve as a tripodal ligand even if it were prepared. The molecular structure of $UO_2(NO_3)_2$·**6** (R=OEt) (Fig. 2),[35] for example, forms an infinite chain in which $UO_2(NO_3)_2$ groups are bridged by the ligand. In this structure, the two P=O bond vectors are turned out and away from the pyridine N atom. Therefore, without the N–O donor group, this ligand does not function in a chelating fashion, and it is sterically improbable that the N-oxidized version would serve as a chelating ligand.

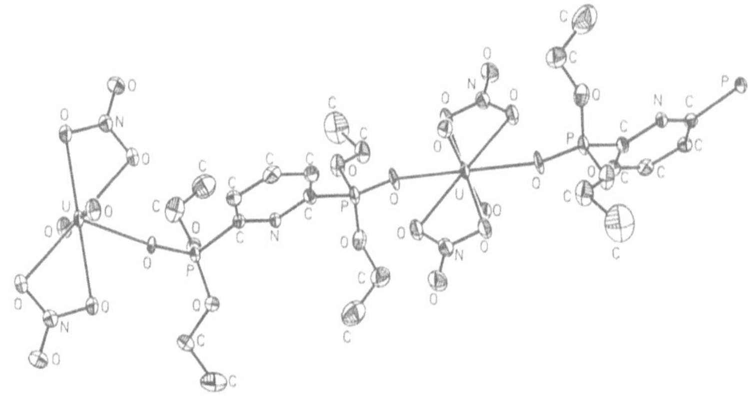

Figure 2. Molecular structure of $UO_2(NO_3)_2$ • **6**.

Attention was then directed toward modifying these ligand systems. Although ligand **3** formed complexes containing a normally favored six-membered chelate ring, the N-O and P=O bond vectors are in fact not coplanar. The chelate ring adopts a "twisted" chair conformation. Subsequent molecular modeling suggested that this chelate ring geometry is somewhat strained. It was reasoned that the strain might be relieved by introduction of a methylene group between the pyridine ring and the phosphoryl group; therefore, the next synthesis targets became ligands **7** and **8**. We succeeded in preparing these new ligands, and the molecular structures of **7** and **8** (R=Ph) have been determined (Figs. 3 and 4).[36-38]

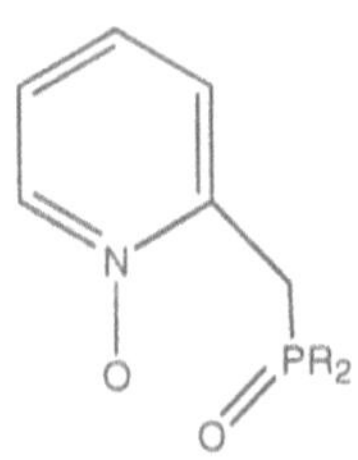


7

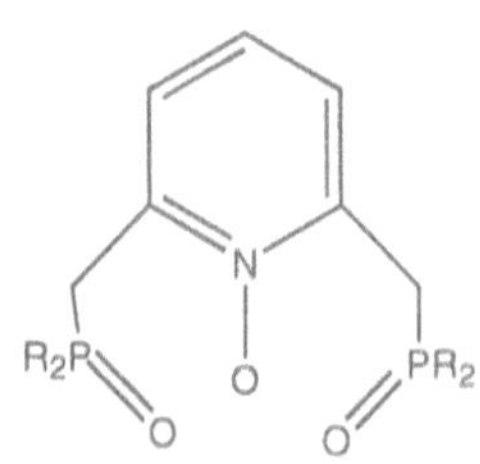


8

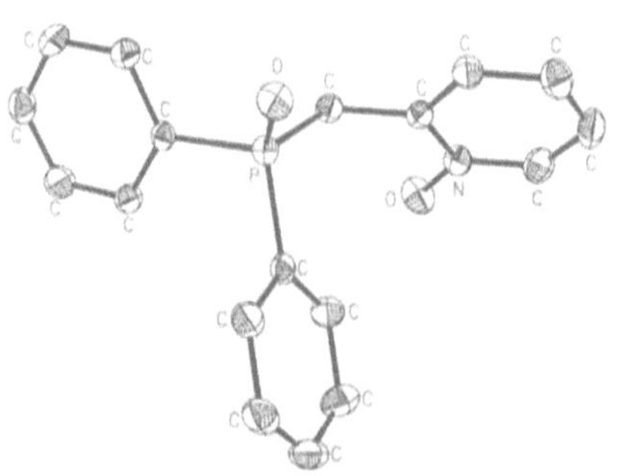

Figure 3. Molecular structure of **7**.

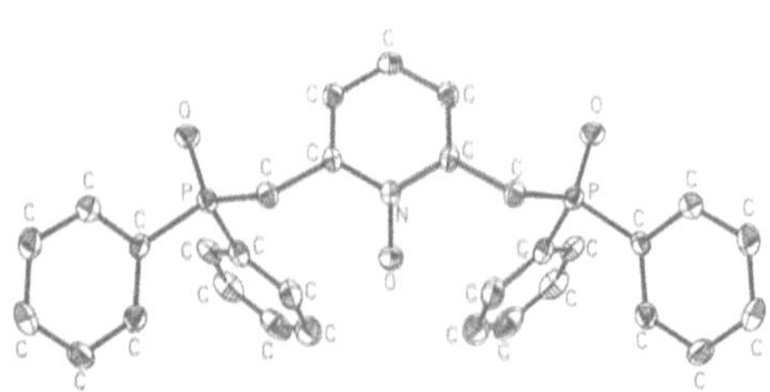

Figure 4. Molecular Structure of **8**.

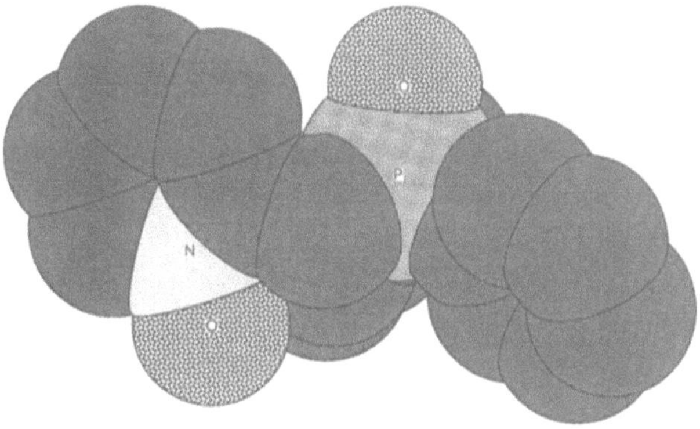

Figure 5. Space-filling representation of **7**.

Several structural features are worth summarizing. The P=O and N–O bond vectors in the free ligands once more are oriented in an *anti* conformation, and molecular modeling suggests that the $Ph_2P(O)$ groups are in a low strain condition. Despite a rather congested appearance in a space-filling representation of **7** (Fig. 5), molecular modeling suggests that the $P–CH_2$ and $CH_2–C$(ring) bonds are relatively free to rotate. It can be surmised that the *anti* conformation in the free ligands is preferred as a result of strong P=O and N-O dipole–dipole repulsion. Coordination of these dipoles should allow them to align in a *syn*–like conformation. Indeed, ligand **7** forms a series of lanthanide complexes with L/M ratios 1:1, 2:1, and 4:1.[38] In the 1:1 and 2:1 complexes, the counter anions (typically $NO_3^-$) remain in the inner coordination sphere. The molecular structure of one complex $Pr(NO_3)_3{\cdot}(\mathbf{7})_2$(R=Ph) (Fig. 6) reveals that **7** forms a bidentate chelate ring on Pr(III). In 4:1 complexes, on the

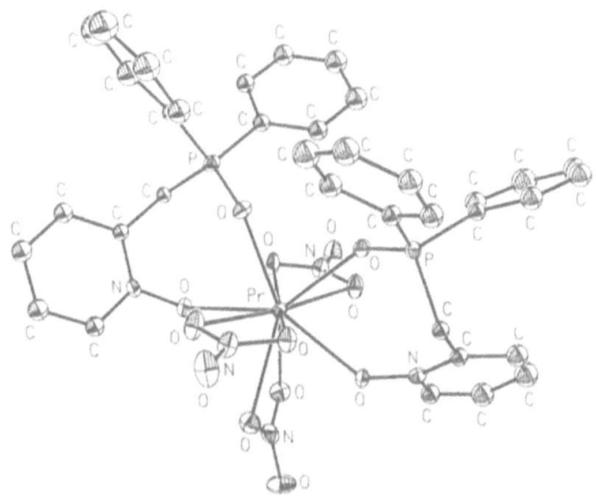

Figure 6. Molecular structure of $Pr(NO_3)_3 \bullet (\mathbf{7})_2$.

other hand, four molecules of **7** completely displace the anions to the outer coordination sphere, and the ligands adopt a bidentate chelate coordination geometry that, by molecular modeling, appears unstrained. The eight-membered coordination polyhedron forms a square antiprism. It can be concluded that the $CH_2$ group provides geometric flexibility without apparent introduction of a large strain energy. A full discussion of the structural parameters for the complex $[Tb(\mathbf{7})_4](NO_3)_3$ has been published,[38] and a view of the closely related $[Pr(\mathbf{7})_4](NO_3)_3$ complex is shown in Figure 7. The extraction properties of this ligand are

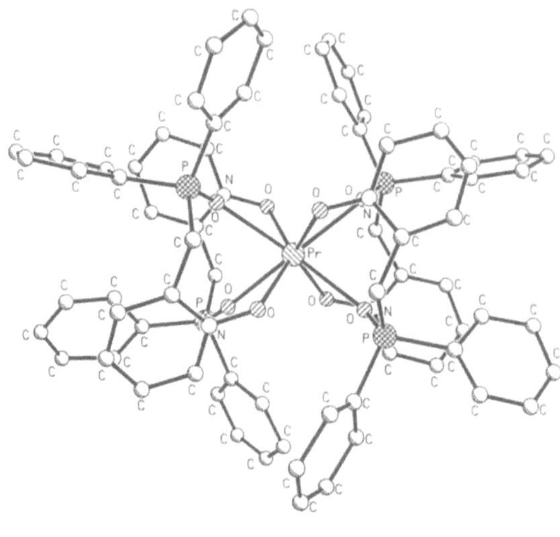

Figure 7. Molecular structure of $[Pr(\mathbf{7})_4](NO_3)_3$.

under study, and the initial results indicate that **7** is a moderately good extractant for Ln(III) and An(III) ions in highly concentrated $HNO_3$ solutions.

The trifunctional ligand **8** (R=Ph) forms a series of 1:1 and 2:1 lanthanide complexes in which the ligands are bonded in a tridentate fashion. The 1:1 complexes retain the three counter anions (e.g., $NO_3^-$) in the inner coordination sphere; therefore, the central metal ion has a coordination number of 9. However, in the 2:1 complexes, two of the $NO_3^-$ ions are displaced to the outer sphere, and the metal ion coordination number is 8. The molecular structures of the 1:1 and 2:1 complexes of Yb have been published previously,[38] and the closely related structures containing Tb (1:1) (Fig. 8), Y (2:1) (Fig. 9), and Pr (2:1) (Fig. 10) ions are presented here. With thorium(IV), a 2:1 stoichiometry is achieved, and two nitrate ions are displaced to the outer sphere.[38] Initial distribution studies confirm that the

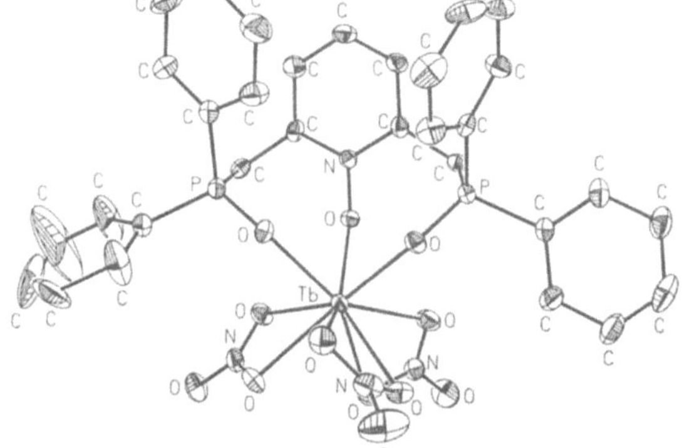

Figure 8. Molecular structure of $Tb(\mathbf{8})(NO_3)_3$.

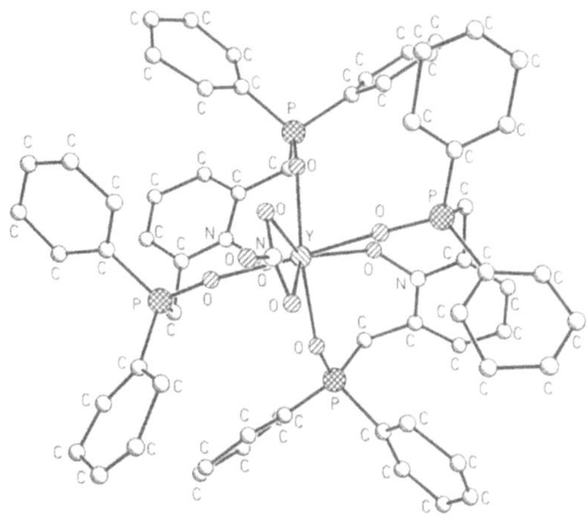

Figure 9. Molecular structure of [Y $(\mathbf{8})_2(NO_3)](NO_3)_2$.

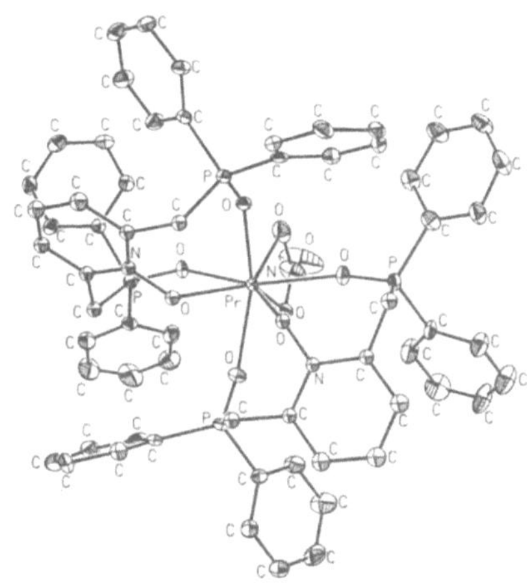

Figure 10. Molecular structure of [Pr$(\mathbf{8})_2(NO_3)](NO_3)_2$.

complexes formed under extraction conditions, namely in the presence of a large excess of ligand, are very stable, and Ln(III) and An(III) ions are extracted from 6 M $HNO_3$ solutions with Ds >500.

Finally, we have recently observed that combination of $Bi(NO_3)_3$ with **8**, even in excess, results only in formation of a 1:1 complex (Fig. 11). This is interesting since the Bi(III) ion is larger for example than the Yb(III) ion which forms the *bis*–tripod coordination complex. Comparison of the triangular coordination "footprint" for **8** in these complexes (Figs. 12 and 13) shows, as expected, that the triangle is significantly larger in the Bi(III)

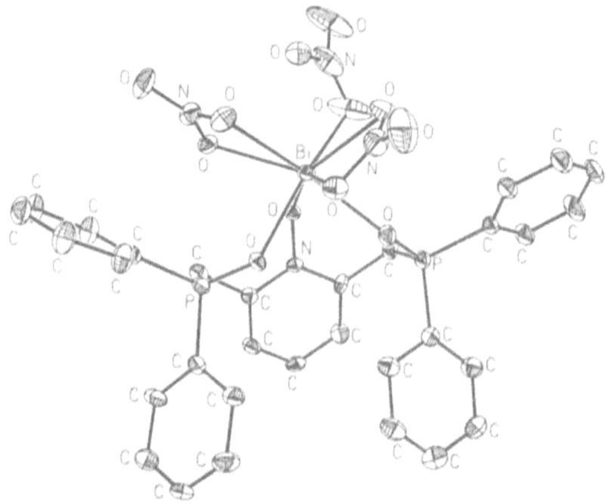

Figure 11. Molecular structure fo Bi(**8**)($NO_3$)$_3$.

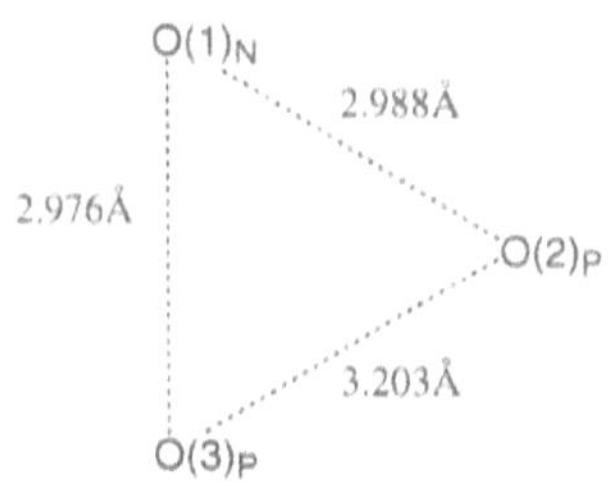


Figure 12. Coordination footprint for ligand **8** on Bi(III).

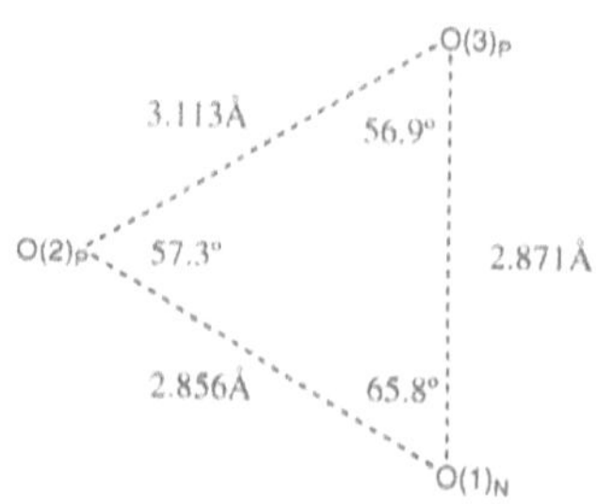


Figure 13. Coordination footprint for ligand **8** on Yb(III).

complex. We are attempting to use molecular modeling to determine if this larger coordination footprint induces unfavorable chelate ring strain that might "block" addition of the second ligand molecule on Bi(III).

The derivatives of **7** and **8** with R=Ph are stable toward acid decomposition in 1M $HNO_3$. However, we have found that, if **7** (R=OEt) and **8** (R=OEt) are N-oxidized under too vigorous conditions or are allowed to react with $Me_3SiBr$ and then hydrolyzed in water, the respective phosphonic acids **9** and **10** are obtained.[39] These ligands have been full charact-

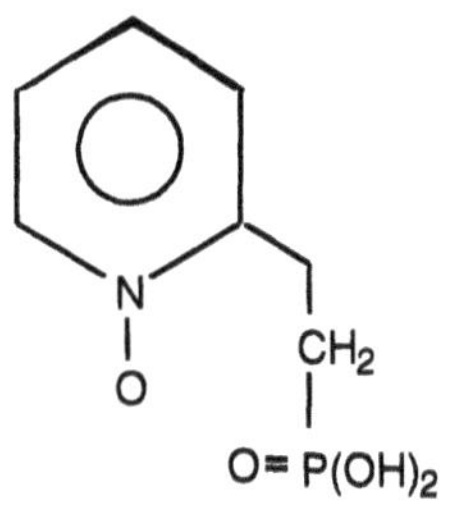

**9**

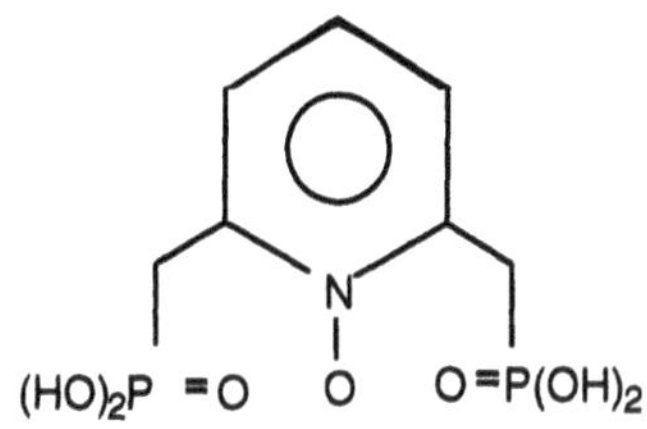

**10**

erized, and their coordination chemistry is under investigation. Since these ligands easily ionize (**9**: $pK_a$ = 1.8 and 6.9; **10**: $pK_a$ = 2.4, 6.8, 7.8), they form a complicated array of coordination complexes, depending on the pH of the solution. A speciation plot for the system Nd(III)/**10** in a 1:2 ratio as a function of pH is shown in Figure 14. Efforts are in progress to isolate one or more of the complexes for single crystal structure analysis. The relatively large formation constant for the $ML_2^{5-}$ species indicates that the anionic ligands are very strongly bound, and they may have some practical solution scrubbing utility if they can be attached to a chromatographic support.

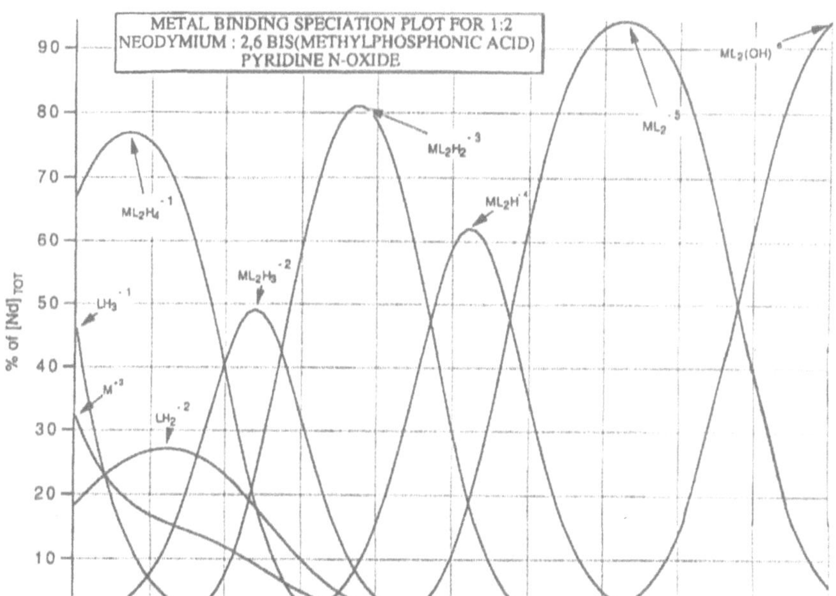


Figure 14. Speciation plot for the combination of **10** and Nd(III) as a function of pH.

We have also considered the possibility of making derivatives like **9** and **10** based on polypyridine fragments. To this end, we have recently prepared a novel ligand **11** based on a bipyridyl fragment.[40] A limited amount of coordination chemistry has been

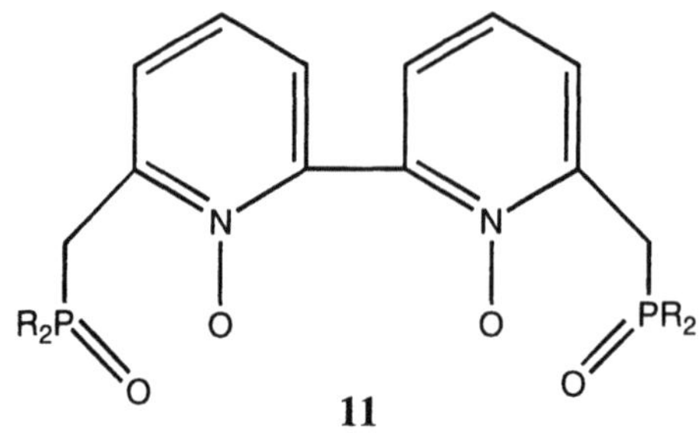


**11**

completed, and only the 2:1 ligand/metal stoichiometry has been achieved on Ln(III) ions. The molecular structure of the Nd complex has been solved, and it shows that **11** acts as a tetradentate ligand (Figs.15 and 16) that displaces all of the nitrate ions from the inner

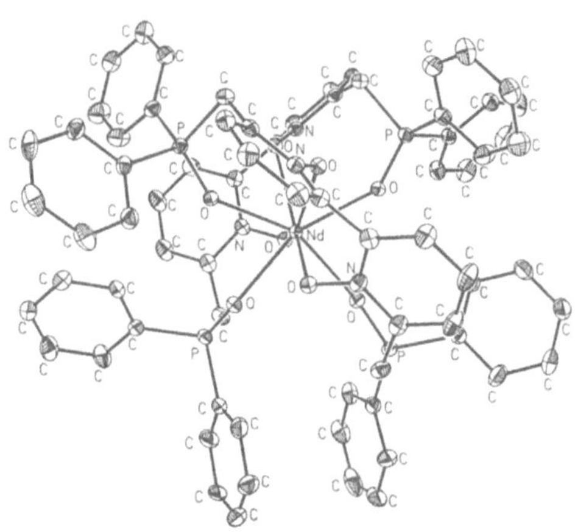

Figure 15. Molecular structure of $[Nd(\mathbf{11})_2](NO_3)_3$.

coordination sphere. Survey solvent extraction distribution measurements with this ligand have also been completed. The data show that **11** is a good extractant in $CHCl_3$ solution although it does not appear to be as powerful as ligand 8. It is also clear that in order to realize the full extraction potential of **7**, **8** and **11** it will be necessary to prepare new derivatives containing substituent groups on the phosphorus atom, the phenyl rings and/or the pyridine rings that make these compounds more soluble in aromatic or preferably in aliphatic diluents. Such synthetic efforts are in progress in our group at this time. In addition, we are also exploring methods for chemisorbing these ligands on macroreticular resins as well as covalently immobilizing the ligands on various solid supports. These materials of course should be useful in the development of chromatographic separations for actinide ions.

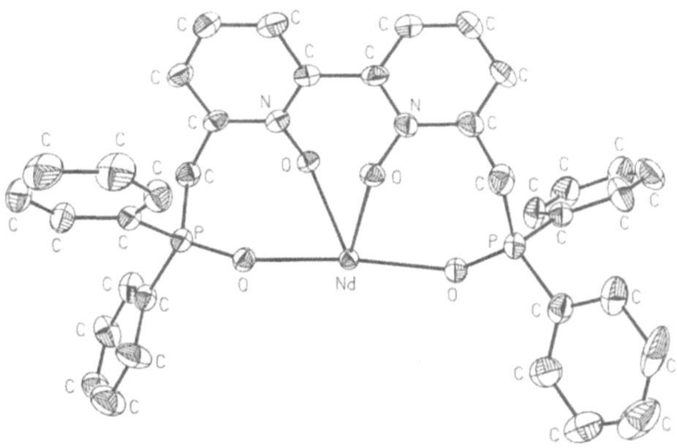


Figure 16. Simplified view of [Nd(**11**)$_2$] showing only the coordination condition of one ligand.

## SUMMARY


This overview of recent studies of pyridinylphosphonate N,P-oxides in our group reveals some unique ligand design features about these ligands and their complexes. It is clear that the compounds **7-11** are potentially useful extractants, and work is progressing to characterize this practical utility. In addition, the fundamental features revealed in the coordination chemistry are stimulating further ligand design advances that should produce new ligands with even greater f-element selectivity.


## REFERENCES


1. N.N. Greenwood and A. Earnshaw, "Chemistry of the Elements," Pergamon Press, Oxford (1984).
2. R.D. Hancock and A.E. Martell, Ligand Design for Selective Complexation of Metal Ions in Aqueous Solution, *Chem. Rev.* 89:1875 (1989).
3. J.-M. Lehn, Supramolecular Chemistry: Receptors, Catalysts, and Carriers, *Science* 227:849 (1985).
4. J.-M. Lehn, Supramolecular Chemistry -- Scope and Perspectives Molecules, Supermolecules, and Molecular Devices (Nobel Lecture), *Angew. Chem. Intl. Ed. Engl.* 27:90 (1988).
5. D.J. Cram, The Design of Molecular Hosts, Guests, and Their Complexes (Nobel Lecture), *Angew. Chem. Intl. Ed. Engl.* 27:1009 (1988).
6. D.J. Cram, The Design of Molecular Hosts, Guests, and Their Complexes, *Science* 240:760 (1988).
7. K. Wolfgang and F. Vögtle, Functionalized, Oligocyclic Large Cavities -- A Novel Siderophore, *Angew. Chem. Intl. Ed.* 23:714 (1984).
8. J.S. Bradshaw, K.E. Krakowial, and R.M. Izatt, "Azacrown Macrocycles," Taylor, E.C. Ed., J. Wiley, New York (1993).
9. G.W. Gokel, S.H. Korzeniowske, "Macrocyclic Polyether Synthesis," Springer-Verlag, Berlin (1982).

10. T.M. Garrett, P.W. Miller, and K.N. Raymond, 2,3-Dihydroxytetrephthalamides: Highly Efficient Iron(III)-Chelating Agents, *Inorg. Chem.* 28:128 (1989).
11. K.N. Raymond, T.J. McMurry, and T.M. Garrett, Macrocyclic catechol-containing ligands, *Pure Appl. Chem.* 60:545 (1988).
12. K.N. Raymond and T.J. Garrett, Sequestering agents specific for high oxidation state cations, *Pure Appl. Chem.* 60:1807 (1988).
13. T.S. Franczyk, K.R. Czerwinski, and K.N. Raymond, Stereognostic Coordination Chemistry. 1. The Design and Synthesis of Chelators for the Uranyl Ion, *J. Am. Chem. Soc.* 114:8138 (1992).
14. C. Musikas, Solvent Extraction for the Chemical Separations of the 5f Elements, *Inorg. Chim. Acta* 140:197 (1987).
15. R.D. Hancock, Molecular Mechanics Calculations as a Tool in Coordination Chemistry, *Prog. Inorg. Chem.* 37:187 (1989).
16. T.H. Sidall, Bidentate Organophosphorus Compounds as Extractants -- I: Extraction of Cerium, Promethium, and Americium Nitrates, *J. Inorg. Nucl. Chem.* 25:883 (1963).
17. T.H. Siddall, Bidentate Organophosphorus Compounds as Extractants -- II: Extraction Mechanisms for Cerium(III) Nitrate, *J. Inorg. Nucl. Chem.* 26:1991 (1964).
18. W.E. Stewart and T.H. Siddall, Bis-(Diisopropoxy-phosphinyl)-Methane Complexes with Rare-Earth Nitrates and Chlorides, *J. Inorg. Nucl. Chem.* 30:1513 (1968).
19. W.E. Stewart and T.H. Siddall, Diisopropyl-N,N-Diethylcarbamylphosphonate Complexes with Lanthanide Nitrates and Chlorides, *J. Inorg. Nucl. Chem.* 30:3281 (1968).
20. M.L. Good and T.H. Siddall, Some New Organophosphorous Chelates, *J. Inorg. Nucl. Chem. Lett.* 2:337 (1966).
21. W.E. Stewart and T.H. Siddall, Diisopropyl-N,N-Diethylcarbamylmethylene-Phosphonate Complexes of Lanthanide Nitrates, *J. Inorg. Nucl. Chem.* 32:3599 (1970).
22. W.W. Schulz and L.D. McIsaac "Transplutonium Elements," W. Muller and R. Lindner Eds., North Holland Publ. Co., Amsterdam (1976).
23. "Transplutonium Elements-Production and Recovery," W.W. Schulz and J.D. Navratil Eds., Amer. Chem. Soc. Symp. Ser., Washington, D.C. (1981).
24. "Actinide Separations," W.W. Schulz and J.D. Navratil Eds., Amer. Chem. Soc. Symp. Ser., Washington, D.C. (1980).
25. D.G. Kalina, E.P. Horwitz, L. Kapland, and A.C. Muscatello, The Extraction of Am(III) and Fe(III) by Selected Dihexyl N,N-Dialkylcarbamoylmethyl-Phosphonates, -Phosphinates and -Phosphine Oxides from Nitrate Media, *Sep. Sci. Tech.* 16:1127 (1981).
26. E.P. Horwitz, D.G. Kalina, L. Kaplan, G.W. Mason, and H. Diamond, Selected Alkyl(phenyl)-N,N-dialkylcarbamoylmethylphosphine Oxides as Extractants for Am(III) from Nitric Acid Media, *Sep. Sci. Tech.* 17:1261 (1982).
27. E.P. Horwitz, A.C. Muscatello, D.G. Kalina, and L. Kaplan, The Extraction of Selected Transplutonium(III) and Lanthanide(III) Ions by Dihexyl-N,N-diethylcarbamoylmethylphosphonate from Aqueous Nitrate Media, *Sep. Sci. Tech.* 16:417 (1981).
28. E.P. Horwitz, D.G. Kalina, and A.C. Muscatello, The Extraction of Th(IV) and U(VI) by Dihexyl-N,N-diethylcarbamoylmethylphosphonate from Aqueous Nitrate Media, *Sep. Sci. Tech.* 16:403 (1981).
29. A.C. Muscatello, E.P. Horwitz, D.G. Kalina, and L. Kaplan, The Extraction of Am(III) and Eu(III) from Aqueous Ammonium Thiocyanate by Dihexyl-N,N-diethylcarbamoylmethylphosphonate and Related Compounds, *Sep. Sci. Tech.* 17:859 (1982).

30. G.R. Newkome, Pyridylphosphines, *Chem. Rev.* 93:2067 (1993).
31. A.A. Russell, E.N. Duesler, and R.T. Paine, 2-(Diphenylphosphino) quinoline N,P dioxide, *Acta Cryst. C.*, submitted.
32. D.J. McCabe, A.A. Russell, S. Karthikeyan, R.T. Paine, and R.R. Ryan, Synthesis and Coordination Chemistry of 2-(Diethoxyphosphino)- and 2-(Diphenylphosphino)pyridine N,P-Dioxides. Crystal and Molecular Structures of Bis(nitrate)[2-(diethoxyphosphino)pyridine N,P-dioxide]dioxouranium(VI) and Bis(nitrato)[2-(diphenyl-phosphino)pyridine N,P-dioxide]dioxouranium(VI), *Inorg. Chem.* 26:1230 (1987).
33. G.S. Conary, A.A. Russell, R.T. Paine, J.H. Hall, and R.R. Ryan, Synthesis and Coordination Chemistry of 2-(Diisopropoxyphosphino)pyridine N,P-Dioxide. Crystal and Molecular Structure of Bis[2-diisopropoxy-phosphino)pyridine N,P-dioxide]lanthanum Nitrate, *Inorg. Chem.* 27:3242 (1988).
34. S.L. Blaha, D.J. McCabe, R.T. Paine, and K.W. Thomas, Extraction of Lanthanides with 2-(Dihexylphosphino)-Pyridine N,P-Dioxide, *Radiochim. Acta* 46:123 (1989).
35. A.A. Russell, R.L. Meline, E.N. Duesler, and R.T. Paine, Synthesis and Coordination Chemistry of 2,6-Bis(Diethoxyphosphino) Pyridine P,P-Dioxide. Crystal and Molecular Structure of $UO_2(NO_3)_2\{[(EtO)_2P(O)]_2C_5H_3N\}$, *Inorg. Chim. Acta*, accepted.
36. A.A. Russell, E.N. Duesler, and R.T. Paine, 2-(Diphenylphosphinomethyl) pyridine N,P-dioxide, *Acta Cryst. C.*, submitted.
37. B.M. Rapko, D. Frutos, E.N. Duesler, and R.T. Paine, Synthesis and Structures of Bismuth Nitrate Complexes with 2,6-Bis(phosphinomethyl) pyridine N,P,P-Oxides, *Polyhedron*, accepted.
38. B.M. Rapko, E.N. Duesler, P.H. Smith, R.T. Paine, and R.R. Ryan, Chelating Properties of 2-((Diphenylphosphino)methyl)pyridine N,P-Dioxide and 2,6-Bis((diphenylphosphino)methyl)pyridine N,P,P'-Trioxide toward f-Element Ions, *Inorg. Chem.* 32:2164 (1993).
39. B.M. Rapko, J. Fox, P.H. Smith, and R.T. Paine, unpublished results.
40. X. Gan, E.N. Duesler, P.H. Smith, and R.T. Paine, Synthesis and Coordination Properties of 6,6'-bis(diphenylphosphinomethyl)-2,2'-bipyridine N,N,P,P' Tetraoxide, *Inorg. Chim. Acta*, submitted.

# SYNTHESIS AND EVALUATION OF POLYHYDROXAMATE CHELATORS FOR SELECTIVE ACTINIDE ION SEQUESTRATION

Aravamudan Gopalan[1]*, Vincent Huber[1], Nirmal Koshti[1], Hollie Jacobs[1], Orhan Zincircioglu[1], Paul Smith[2]* and Gordon Jarvinen[2]

[1]Dept. of Chemistry and Biochemistry, New Mexico State University Las Cruces, NM 88003-8001
[2]Los Alamos National Laboratory, Los Alamos, NM 87545

## ABSTRACT

The overall goal of our research program is to design, synthesize and evaluate organic chelators for the specific binding/removal of actinides in a variety of environmentally relevant situations. Such chelators would be useful to selectively remove actinide ions such as plutonium from a variety of waste forms including soils and waste streams. We have identified a new class of polyhydroxamates as potential chelating agents for actinides based on computer modeling, solubility properties and other important features including ease of synthesis. Several members of this class of tetrahydroxamate chelators have been synthesized in our laboratory and evaluated for the binding of actinides and other metal ions in solution. Some of the hydroxamate chelators that we have developed have also been evaluated for their plutonium(IV) binding and the results are very encouraging. Detailed studies of the complexation behavior of this class of chelators are currently in progress and the goal of these experiments is to develop an understanding of the efficiency and nature of the metal complexation chemistry. This in turn should allow the further modification of this class of chelators to obtain agents with higher specificity for the actinide ions.

Another major goal of this program is to develop polymer supported, ion specific extraction systems for removing actinides and other hazardous metal ions from wastewaters. Selected ligands from our ongoing efforts are being incorporated into polymeric backbones to be evaluated for their abilities to selectively remove the target metal ions from process waste streams. The synthesis of some chelating polymers and results of their preliminary evaluation are described.

*Separations off Elements*, Edited by K.L. Nash
and G.R. Choppin, Plenum Press, New York, 1995

## INTRODUCTION

Many facilities within the United States Department of Energy (DOE) complex are burdened with high and low level radioactive, transuranic, and mixed wastes that need to be characterized, separated and remediated. Previous waste disposal practices have proven inadequate and aggressive action is required to avoid the possibility of contamination of soil and water at some of these sites.[1,2] In addition, at several sites, containers with radioactive waste are decomposing, resulting in contamination of the adjacent soils. The DOE has made a commitment to cleaning and restoring its nuclear research and production sites within the next 30 years and also ensuring that risks to the public are essentially eliminated. This in turn has necessitated the development of a variety of new technologies for use in radioactive waste remediation.

One of the key constituents in radioactive waste is the class of radionuclides known as actinides. The actinides pose a long term hazard to the biosphere making their isolation from the environment an important public concern. The separation of actinides from various waste forms is a complex and challenging task that is strategically important in radioactive waste remediation. A significant reduction in the volume of transuranic waste destined for long term storage in geological repositories should result in substantial cost savings during the 'clean-up' effort. Cost effective and efficient methods are needed to remove actinides from process waste streams to ultra-low levels in order to meet increasingly stringent regulatory limits. Also, the remediation of actinides in soils is an issue that has to be addressed in a timely manner. Some actinides form soluble and therefore mobile species which can migrate rapidly into the ground water thus further enlarging the scope of the contamination! In contrast, other actinides form insoluble oxides and hydroxides which are difficult to solubilize making their recovery from the environment difficult.

The overall goal of our research program is to design, synthesize and evaluate organic chelators for the specific binding of actinides for use in environmental remediation. A number of actinide chelators have been developed previously for the biological decorporation of plutonium. Representative examples of some chelators that have been synthesized are shown in Figure 1. Their structures and binding properties provide valuable precedents in the task of design and synthesis of more efficient actinide chelators.[3]

Researchers have prepared chelators that are capable of taking advantage of the higher number of coordination sites possessed by the actinide ions due to their larger size relative to most main group and transition metals. Preorganization of the ligand groups around a spacer in order to provide the appropriate coordination geometry is expected to enhance both actinide ion selectivity and binding. It is important to consider not only cyclic but also acyclic spacers for anchoring the ligand moieties. It has been proposed that in some cases the greater stereochemical freedom of linear backbones, may lead to more effective chelation.[4,5]

The choice of ligands to be incorporated into the chelator is also an important factor. Based on hard acid-hard base interactions, it is clear that multidentate oxoligands such as catecholates, hydroxamates, polyaminocarboxylates and hydroxypyridinonates show promise for the binding of the relatively hard actinide ions present in aqueous process solutions.[6,7,8] $Fe^{3+}$ and $Pu^{4+}$ have similar charge to ionic radius ratios (46 and 42 e $nm^{-1}$ respectively) and show a preference to similar types of ligands.[9,10] Raymond and others have used this observation to develop plutonium chelators that contain structural features similar to known iron sequestering agents such as siderophores.[11,12] Further, $Th^{4+}$ and $U^{4+}$ complexes have been studied by X-ray crystallography and the structural information has been utilized in the design of new actinide chelators.[13]

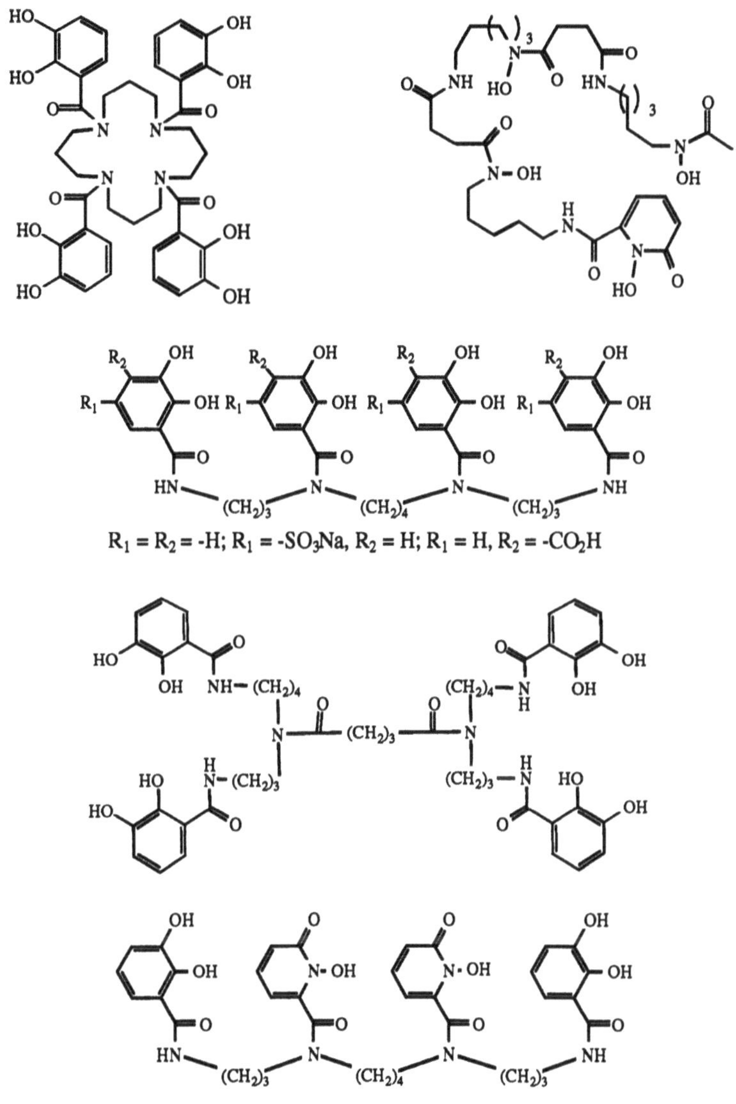


**Figure 1.** Representative actinide chelators.

A number of chelators, having both cyclic or acyclic backbones have been synthesized and examined for their actinide binding properties.[14,15,16] Raymond and coworkers have synthesized and evaluated a number of tetracatecholates (including sulfonate and carboxylate analogs, see Figure 1) for their actinide sequestration properties. These tetracatecholate chelators are, in general, more efficient in the coordination of the plutonium ion under alkaline pH conditions. Only three of the four catecholate ligands present in this class of chelators appear to bind the metal ion at neutral or lower pH values.[17] In order to improve the efficiency of this class of chelators one or more of the catecholate ligands have been replaced with hydroxypyridononate ligands.[18] However, in these chelators also, only three of the ligand groups appear to readily coordinate the metal ion under biological conditions. Raymond and coworkers have also studied a macrotricyclic chelator with four endocyclic catechol groups[19] and derivatives of the siderophore desferrioxamine having an additional ligand group.[20]

Most of the chelators developed so far are yet to be examined for use in actinide remediation where the challenges are quite formidable. A successful chelator must be able to remove small amounts of actinides from soils and other waste forms in the presence of more abundant and competing metal ions. In addition to high actinide selectivity, the chelator must also possess the requisite solubility, stability and complexation kinetics critical for 'clean-up' applications.

## RESULTS AND DISCUSSION

Research in our laboratory over the last few years has led to the identification of a relatively simple class of tetrahydroxamates, such as **1-6**, as potential agents for the specific complexing of actinides.[21] These chelators contain the hydroxamate ligand which is well known for its high affinity for highly charged metal ions such as plutonium. Both simple space filling models and computer modeling suggest that these chelators could be potentially useful for actinide chelation. Figure 2 shows a computer generated model of the complex formed between chelator **2** and plutonium.

In these tetrahydroxamate chelators, the ligands are arranged so that they can complex the actinide ion in an octadentate manner and thus bind it selectively over a six-coordinate transition metal. The presence of the tertiary amine groups which are generally protonated at neutral pH and below is expected to enhance both the solubility of the chelator and its resultant actinide complexes, particularly under slightly acidic conditions. In these ligands, the diamine backbone as well as the chelating chain length are easily varied providing a method for systematically evaluating the effects of structural variation on their metal binding properties.

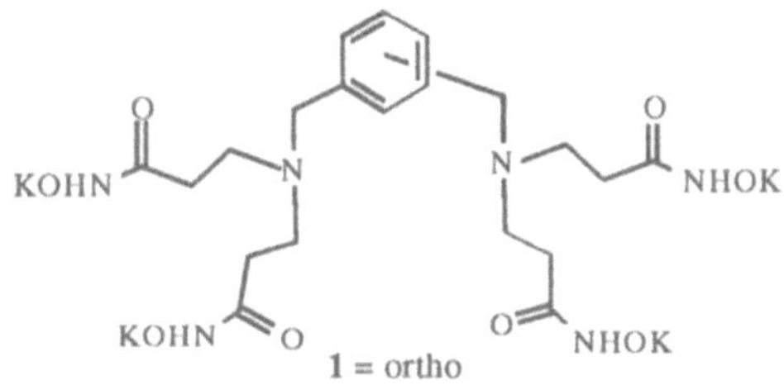


1 = ortho
2 = meta
3 = para

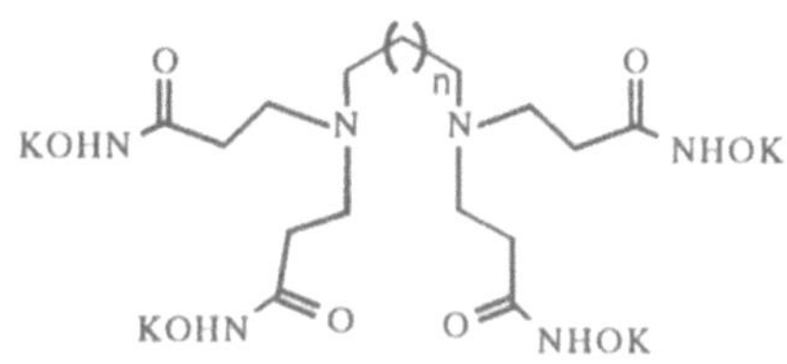


4 (n = 1)
5 (n = 4)

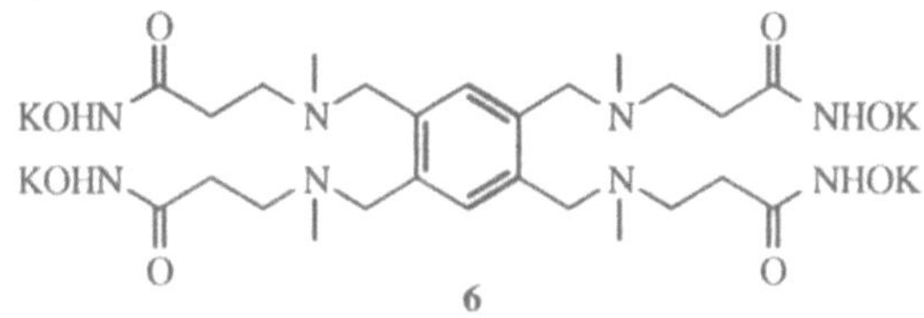


6

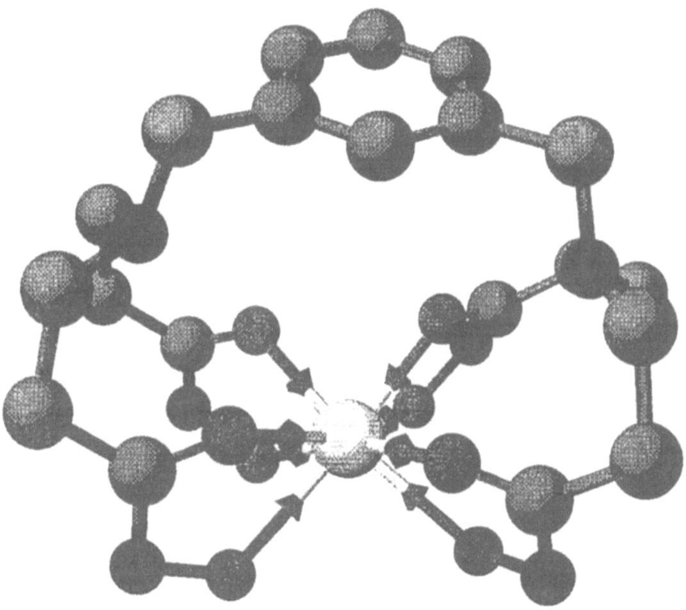

**Figure 2.** Computer model of the chelator **2**-Plutonium(IV) complex.

Multigram quantities of these tetrahydroxamates can be easily prepared from readily available starting materials using standard organic procedures. The general synthetic pathway used for the preparation of tetrahydroxamates **1-5** is shown in Scheme 1. Reaction of the appropriate diamine with excess ethyl acrylate gives the corresponding tetraester in good yields. The tetraesters are then reacted with excess hydroxylamine in basic solution to give the desired hydroxamic acids which are isolated as potassium salts. Chelator **6** was prepared by a multistep sequence from the corresponding 1,2,4,5-tetrakis-(bromomethyl)benzene.

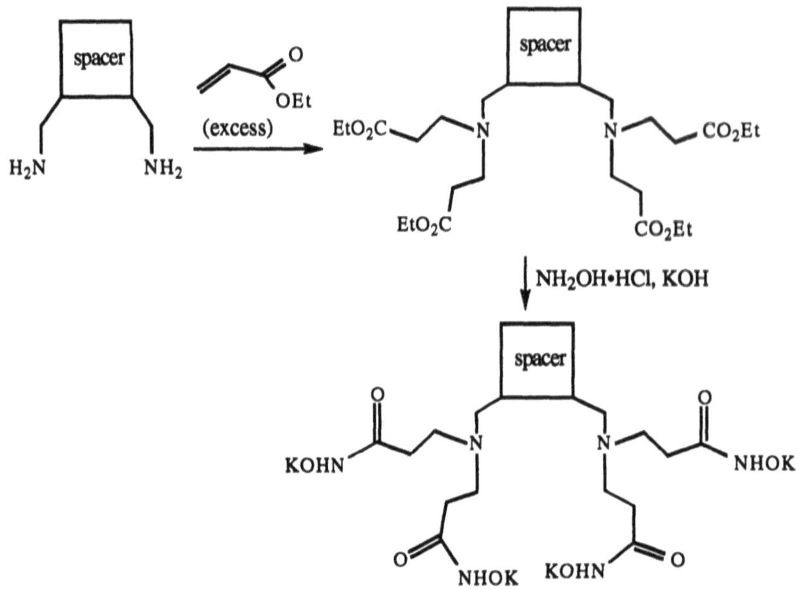


Scheme 1

The binding properties of these tetrahydroxamates have been evaluated using potentiometric methods. The same basic experimental approach was used to evaluate each ligand. First, the ligand protonation constants were measured potentiometrically. The titration curves that were obtained were then analyzed using the non-linear least-squares program BETA.[22] The ligands were then evaluated potentiometrically in the presence of the metal ions of interest, and the preliminary metal binding constants and protonation constants of the complex determined.

Our preliminary studies indicated that this class of chelators bind $Th^{4+}$, a representative actinide ion, strongly. The protonation constants for **2**, **3**, and **6**, are shown in Table 1 and their metal ion binding constants are shown in Table 2. The overall binding constant of $10^{31.88}$ determined for chelator **2** with thorium is comparable to some of the highest binding constants known. For example, the $Th^{4+}$ binding constants with DTPA (Diethylenetriamine pentaacetic Acid) and TTHA (Triethylenetetraamine hexaacetic Acid) are $10^{27.5}$ and $10^{31.9}$, respectively.[23] Our studies indicate that chelator **2** is moderately selective for thorium over iron and binds both ions strongly. Neodymium(III) which is used as a model for americium(III) is not bound as well.[24] It should be pointed out that the actinide ion binding constants with these chelators are quite high. Hence a method other than potentiometry may be necessary for their precise determination. We plan to do competition studies with these chelators to measure their actinide binding constants more accurately.

**Table 1.** Protonation constants for chelators **2**, **3** and **6** at 25.0 °C, 0.10 mol dm$^{-3}$ ($KNO_3$).

| Quotient, $K^b$ | $\log K^a$ | | |
|---|---|---|---|
| | **2** | **3** | **6** |
| $[LH]/[L][H]$ | 10.18 | 10.22 | 10.28 |
| $[LH_2]/[LH][H]$ | 9.64 | 9.60 | 9.95 |
| $[LH_3]/[LH_2][H]$ | 8.99 | 9.05 | 9.42 |
| $[LH_4]/[LH_3][H]$ | 8.17 | 8.31 | 8.75 |
| $[LH_5]/[LH_4][H]$ | 6.38 | 6.40 | 8.28 |
| $[LH_6]/[LH_5][H]$ | 5.36 | 5.35 | 7.31 |
| $[LH_7]/[LH_6][H]$ | | | 4.09 |
| $[LH_8]/[LH_7][H]$ | | | 3.01 |

[a]All constants determined by the program BETA. [b]Charges have been omitted for simplicity. Standard deviations are ± 0.01 log units as indicated by the program BETA.

The positive results of the binding studies with thorium clearly encouraged the examination of these chelators with plutonium(IV). A number of problems had to be addressed prior to initiating potentiometric titrations involving plutonium. The propensity of plutonium to exist as a colloid even at pH 1 and to readily undergo hydrolysis is known. The fact that the stock solution of plutonium is prepared in strong acid further complicates the potentiometric analysis. The excess acid in the plutonium stock solution was determined by titrating the $Pu^{4+}$-DTPA complex to its endpoint. Also, plutonium is known to readily disproportionate under a variety of conditions. The precise oxidation state of the plutonium sample was confirmed by comparison of the UV-Vis spectrum of our plutonium stock solution to a spectrum of an authentic sample.[25,26] The preliminary results obtained for some of these chelators with plutonium and other metals are shown in Table 2.

The analysis of the data obtained so far clearly confirms the fact that the meta- and para-tetrahydroxamate chelators **2** and **3** developed in our laboratories bind plutonium strongly and show significant selectivity for $Pu^{4+}$ over $Fe^{3+}$. We have yet to evaluate the plutonium binding properties of chelator **6** or any of our other tetrahydroxamate chelators. However, thorium binding studies on chelator **6** suggest it may also be selective for plutonium over iron.[27] Figure 3 shows the amount of uncomplexed metal, pM, (iron, thorium and plutonium) in the presence of a 10-fold excess of synthetic chelators **2**, **3** and **6** as a function of pH. The pM value is defined as -log[M], where [M] is the amount of free metal ion present in solution at any given pH. This figure shows the potential of these

**Table 2.** Formation constants for chelators **2**, **3**, and **6** at 25.0 °C, 0.10 mol dm$^{-3}$ ($KNO_3$)

| | logQ[a] | | | | | | | |
|---|---|---|---|---|---|---|---|---|
| | $Pu^{4+}$ | | $Th^{4+}$ | | | $Fe^{3+}$ | | |
| Quotient, Q[b] | **2** | **3** | **2** | **3** | **6** | **2** | **3** | **6** |
| $[ML]/[M][L]$ | 35.74(3) | 35.53(3) | 31.88(6) | | 27.44(3) | | | 21.1(2) |
| $[MLH]/[M][L][H]$ | 40.88(3) | 40.29(3) | 39.65(5) | 36.85(6) | 37.57(4) | | | 31.9(3) |
| $[MLH_2]/[M][L][H]^2$ | 44.10(2) | 43.35(3) | 43.16(4) | 42.13(6) | 46.86(3) | | 40.50(7) | 40.9(2) |
| $[MLH_3]/[M][L][H]^3$ | 46.92(2) | 46.27(3) | 45.92(6) | 45.18(5) | 54.63(2) | 43.41(2) | 45.21(7) | 49.9(2) |
| $[MLH_4]/[M][L][H]^4$ | | | | 48.28(7) | 59.98(2) | 47.43(2) | 48.70(7) | 56.8(2) |
| $[MLH_5]/[M][L][H]^5$ | | | | | | | | 60.0(1) |
| $[MLH_6]/[M][L][H]^6$ | | | | | | | | 64.5(2) |
| $[M_2L_2H_3]/[M]^2[L]^2[H]^3$ | | | | | 88.01(8) | 74.16(2) | | |
| $[M_2L_2OH]/[M]^2[L]^2[OH]$ | | 68.74(6) | | | | | | |
| p[M][c] | 29.2 | 29.4 | 25.9 | 22.6 | 23.6 | 17.6 | 19.0 | 18.3 |

[a]All constants determined by the program BETA. [b]Charges have been omitted for simplicity. Numbers in parenthesis are the standard deviations of the formation constants as determined by the program BETA. [c]$p[M] = -\log [M^{n+}]$ and were calculated at $p[H] = 7.4$ from the previously determined stability constants where $[L]_T = 0.001$ M and $[M]_T = 0.0001$ M.

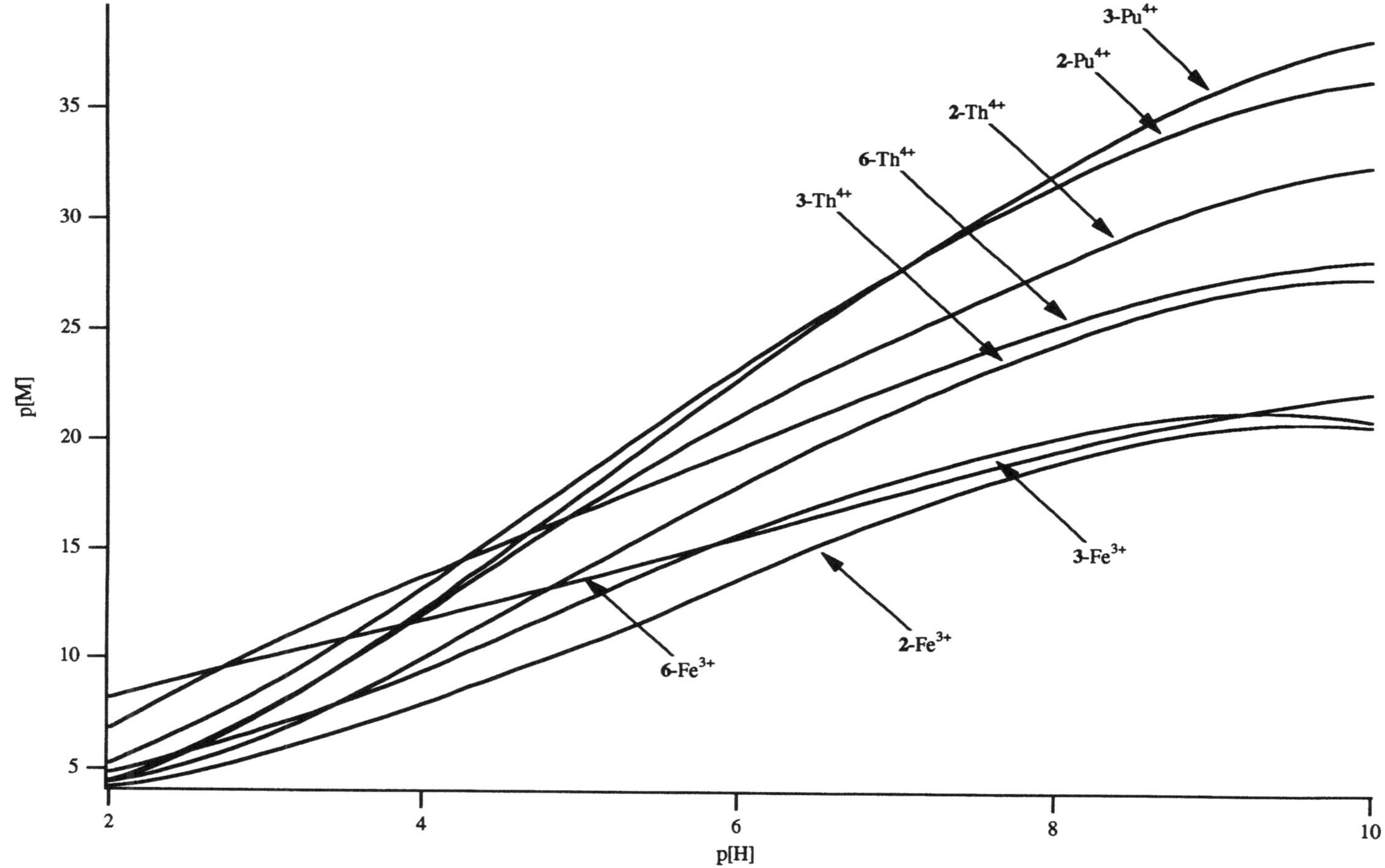


**Figure 3**. Amount of free $Pu^{4+}$, $Th^{4+}$, and $Fe^{3+}$ in the presence of chelators **2**, **3**, and **6**.

chelators at pH values greater than 6 to reduce the amount of uncomplexed plutonium to ultra-low levels. It also demonstrates these ligands selectivity for plutonium over iron.

The species plots for ligand **2** with iron and plutonium are shown in Figure 4. The protonated complexes of the iron and plutonium chelates of **2** are readily observed in the species plots. The dominant species around pH 4.8 appears to be $MLH_3$ for iron. Chelator **2** has two amine groups that are expected to be protonated at lower pH values. Hence, we believe that $MLH_3$ corresponds to a ferric complex in which both nitrogens are protonated and one hydroxamate group is undissociated. In the case of plutonium, the dominant species around pH 4.2, MLH, appears to be a tetrahydroxamate complex with one of the amines protonated.

In order to improve the selectivity of our synthetic chelators, it is important to understand their mode of complexation with actinides and competing metal ions like iron. For high actinide affinity, it is important that the four arms of a chelator such as **2**, are optimized so that the ligand groups are able to fully coordinate the metal and form a 1:1 complex. Hence, systematic studies are underway to ascertain the extent to which the four ligand groups of the representative tetrahydroxamate **2** are coordinating the thorium or plutonium ion. For this purpose, we have synthesized compounds **7**, **8** and **9**, the dihydroxamate and trihydroxamate analogs of **2** and are evaluating their binding properties. It is our plan to correlate the various metal/complex equilibria with the potentiometric data using spectroscopic techniques (e.g., NMR, UV-vis). The goal of these experiments is to develop an understanding of the efficiency and nature of the metal chelation process which in turn permits further structural refinements.

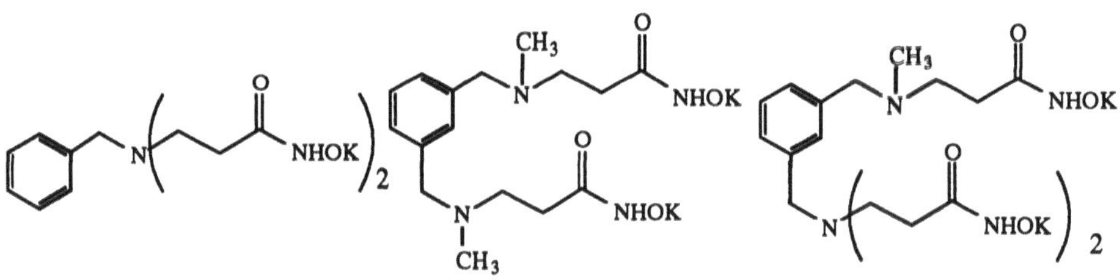


**7** **8** **9**

A particularly convenient method for probing the nature of the selectivity of our chelators for actinides over iron involves the spectrophotometric analysis of the iron complexes. For example, the absorption maxima of the iron complexes with our chelators can be used to determine the number of hydroxamate ligands that are coordinated to the metal ion.[28, 29, 30] Figure 5 shows the results of a spectrophotometric titration of chelators **2** and **3** with ferric ion. Typically, tris(hydroxamato)iron(III) complexes are known to have an absorption maximum at 430 nm. The ferric ion complexes of both **2** and **3** have an absorption maximum at about 430 nm around pH 5. This strongly suggests that both these chelators form trihydroxamate complexes at this pH thus supporting the potentiometric data.

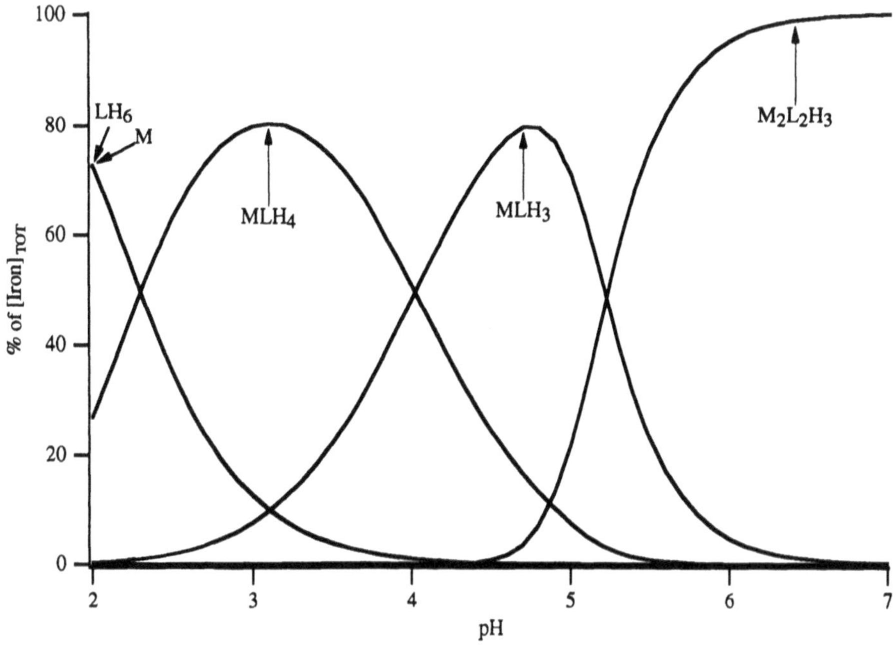


**Figure 4a**. Speciation diagram for **2**-Iron(III). [L] = [M] = 0.001 M

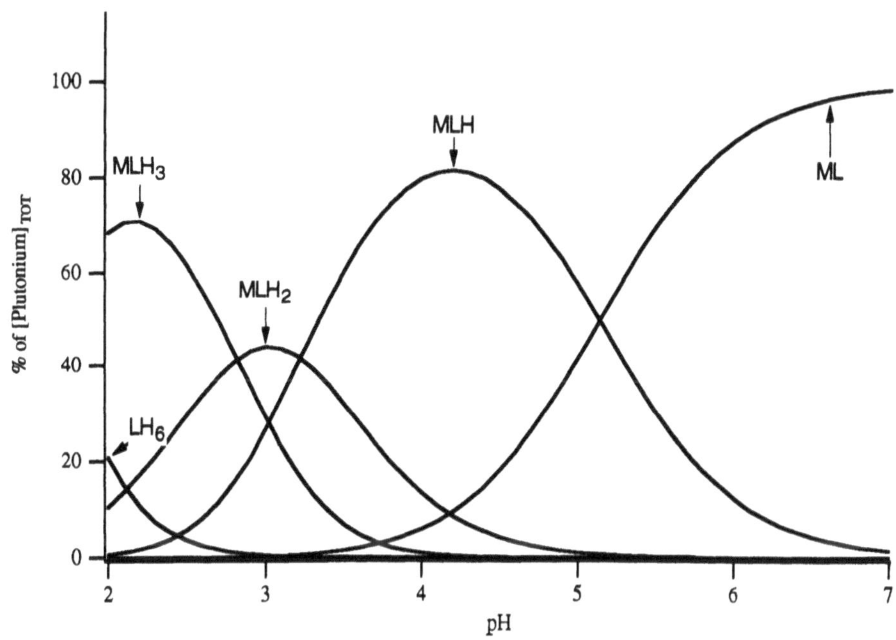


**Figure 4b**. Speciation diagram for **2**-Plutonium(IV). [L] = [M] = 0.001 M

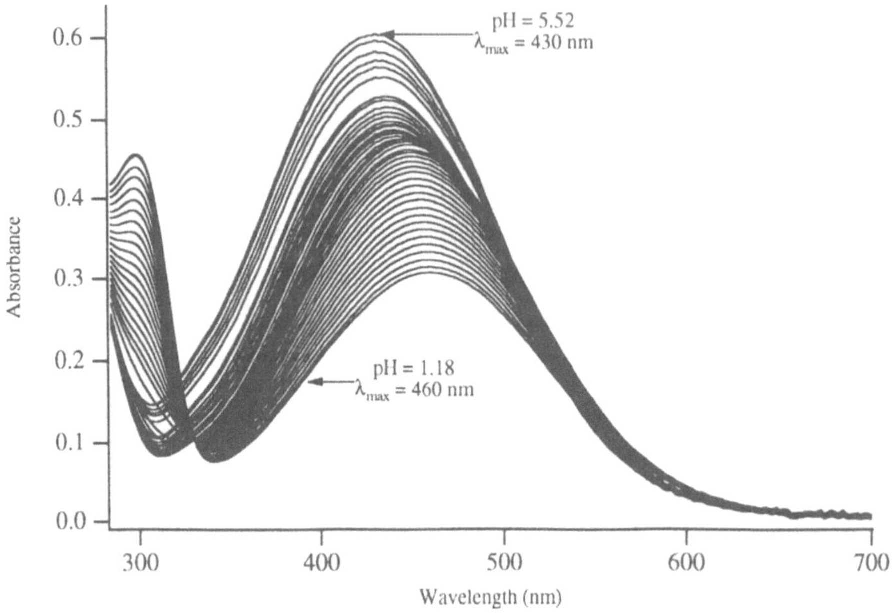


**Figure 5a**. Spectrophotometric titration of **2**-Iron(III). [L] = [M] = 0.23 mM.

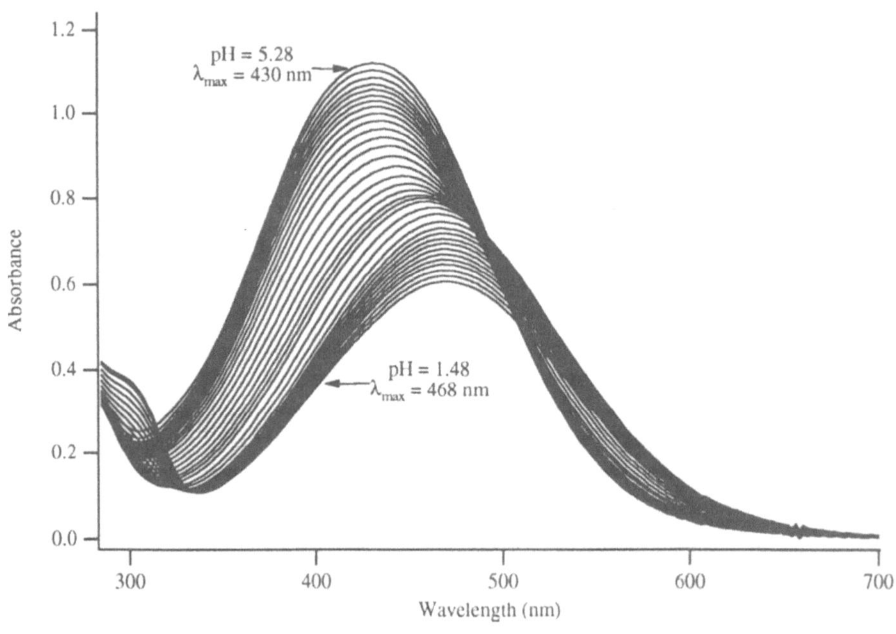


**Figure 5b**. Spectrophotometric titration of **3**-Iron(III). [L] = [M] = 0.40 mM.

In general, all of the chelators that have been studied so far exhibit high binding constants for thorium, show favorable solution kinetics, and form soluble complexes under slightly acidic conditions. The evaluation of our synthetic polyhydroxamate chelators for their ability to selectively bind thorium and/or plutonium is being continued. We believe that this class of chelators shows much promise and needs further development. These studies will allow us to identify the most promising chelators for incorporation into polymeric matrices for use in actinide remediation.

**Synthesis of other actinide chelators**

As mentioned before, it is our intention to identify new, and promising structures that show high selectivity for the binding of actinide ions and to subsequently incorporate these features into polymeric chelators which are specific for actinide binding. In fact, we have recently developed synthetic methodology that has allowed the preparation of N-methyl hydroxamic acid analogs of chelators **2** and **3**.[31] It is of interest to see whether these chelators show increased and more selective plutonium binding relative to the corresponding primary hydroxamic acids. The methodology developed by us for their preparation is described in Scheme 2 and should also be useful for the functionalization of polymeric amines. This in turn should facilitate the preparation of some of our polymeric hydroxamate targets.

The reagents **10**, **11**, and **12** can be made in one step by the coupling of acryloyl chloride with the appropriate protected hydroxylamine derivative. They have been reacted successfully with a number of amines to give the corresponding adducts in good yields. The Michael addition reactions were generally carried out by refluxing the amine with a slight excess of the reagent in THF. Deprotection of the Michael adducts could be accomplished using either catalytic hydrogenation or acid as shown in Scheme 2. A number of polyhydroxamates have been prepared using this methodology.

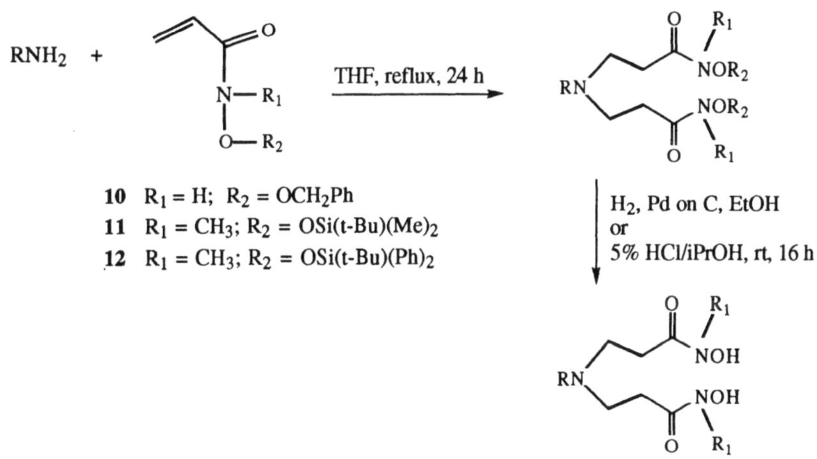


Scheme 2

Molecular modeling shows that two new chelators, the cyclam tetrahydroxamate, CYTROX, and the cyclam tetraacetonylacetone derivative, CYTAC, have potential for the binding of plutonium(IV). The structures of these chelators are shown in Figure 6. The synthesis of these two chelators has been achieved using short sequences from commercially available cyclam. Both the details of the molecular modeling and the synthetic route to these molecules have been recently published.[32] CYTAC is different from most of our other chelators as it has acetonylacetone as the ligand moiety rather than hydroxamate.

Some of the other tetrahydroxamates that have been synthesized and are undergoing evaluation are also shown in Figure 6. The ability of these hydroxamates to bind metal ions other than actinides is also being assessed in order to gauge their selectivity in the competitive situations present in environmental applications.

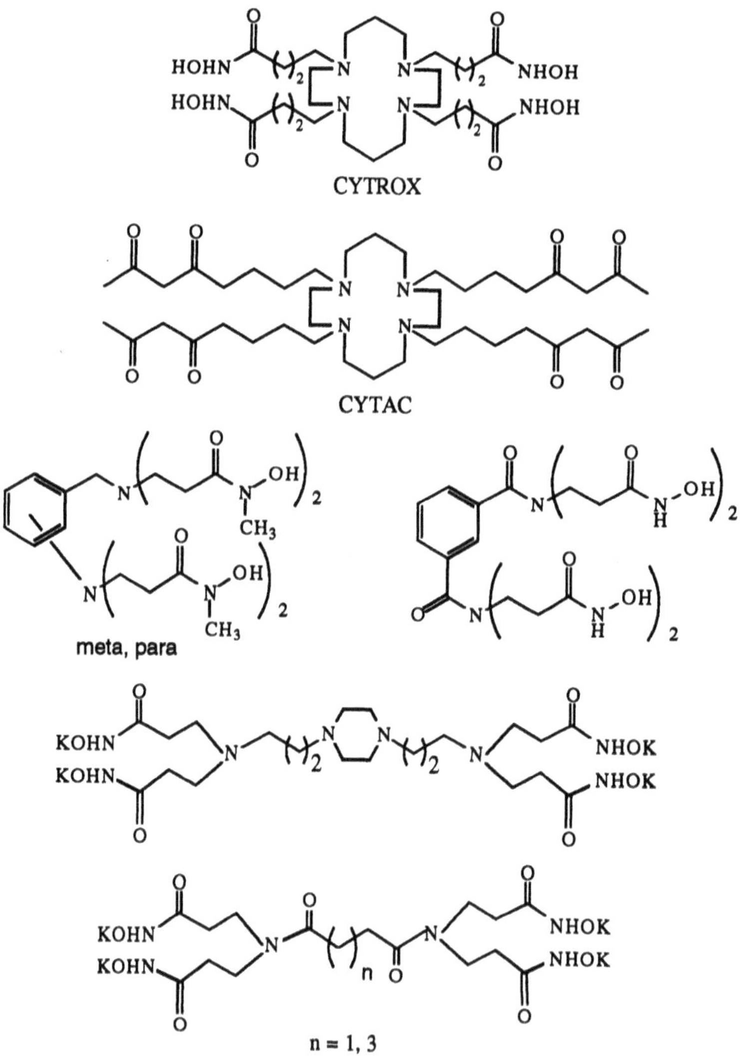


**Figure 6.** Other synthetic actinide chelators.

**Development of Chelating Polymers for Actinides**

Over the past 20 years, the development of chelating polymers that are selective for various metal ions, has received much attention.[33,34] Chelating polymers are the basis of a number of successful industrial separations including removing calcium to very low levels from brine and removing radioactive cesium from alkaline waste waters.[35,36,37] Much of the work so far has involved transition and main group metals and very little work has focused on the actinides, with the exception of uranium.[3]

Incorporation of chelating moieties that are specific for actinides into insoluble polymeric matrices would allow the development of new materials that can be utilized to remediate contaminated groundwater, or waste water from nuclear processes. Even if the initial cost of synthesizing such polymers is somewhat high, the possibility of regenerating and recycling them could make them cost effective. An advantage of the chelating polymers, relative to liquid-liquid extraction methods, is that the immobilization of the chelator in the polymeric matrix prevents the introduction of a potential organic pollutant into the system.

Various ligand systems may be incorporated into polymeric matrices (beads, fibers, membranes, etc.) to provide an elegant and efficient method for the separation of actinide ions from a liquid phase. One simple method of preparing a chelating polymer is the polymerization of monomers containing pendant ligand groups. However, in such cases, the efficiency of binding may depend on obtaining intrapolymeric cooperation of the monomeric binding sites, which is often unpredictable. Additionally, spacer groups may have to be varied to ensure that the chelating ligands have the conformational mobility necessary for binding the metal ion.

An alternate strategy is to synthesize a multidentate chelator which possesses the correct orientation to bind an actinide ion, and then attach it to the polymer backbone.[38,39,40] In this case, all the components necessary to form a complete and stable complex are present in one ligand. These polymers, which may be more complex to prepare, do not rely on intrapolymeric cooperation of binding sites.

Many commercially available polymers have the requisite functional handles which can be used directly for the subsequent introduction of ligands. The ideal procedure for polymer modification is to preassemble the whole ligand and then attach it to the polymer in one step. It is our intention to attach some of our synthetic chelators onto such polymeric systems, using either substitution or condensation reactions as appropriate. The chelating polymers can then be evaluated for their ability to bind actinides and other metal ions.

A number of factors are critical to the success of a chelating polymer. Because the actinide ions are usually present in relatively low concentration in waste solutions with respect to other metal ions, it is essential that the polymer possess both high binding and selectivity for actinides. Also, formation of the complex must be fast in order to develop viable separation processes. The final factor that is essential is the wettability of the polymer. Suitable hydrophilic groups must exist in the polymer to achieve this.

Our laboratory has been involved in both the synthesis and evaluation of new polyhydroxamate chelating polymers[41,42,43] for actinides. In connection with this effort, the hydroxamate polymer **13** was prepared by functionalization of commercially available chloromethylpolystyrene (Scheme 3). This polymer has served as the model substrate for establishing the procedures for metal ion binding studies. Reversibility of the protonation and deprotonation behavior of this polymer has been shown by acid/base titrations. We have determined from the shape of the titration curve that there are 2.7 milliequivalents of hydroxamic acid per gram of this polymer. This capacity is comparable to commercial resins.

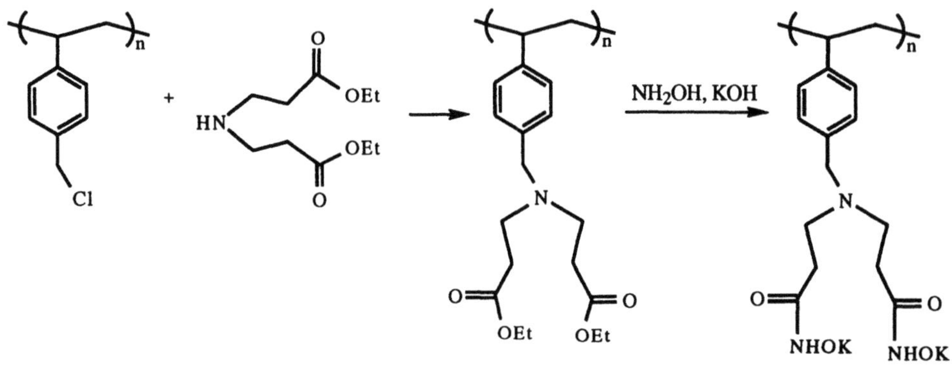


**13**

Scheme 3

Some studies regarding the metal ion binding ability of this polymer have been performed.[44] We have titrated the polymer in the presence of $Nd^{3+}$, $Th^{4+}$, and $Fe^{3+}$ and the results are shown in Figure 7. The titration curves in the presence of the metal ions are depressed relative to those in the absence of metal ions. This depression in the titration curves is similar to those observed for the corresponding water soluble chelators described earlier and is an indication of the degree of metal ion binding. It should be noted that the depression of the curve is much more pronounced in the presence of $Th^{4+}$ and $Fe^{3+}$ than it is in the presence of $Nd^{3+}$.[45] Based solely on this observation, it appears that this polymer has the following binding selectivity: $Nd^{3+} < Th^{4+} < Fe^{3+}$. It is encouraging that our model chelating polymer does show strong binding for the metal ions of interest. Also it is important to note that the behavior of this polymer is quite similar to that of the corresponding water soluble dihydroxamate analog **7** that has been synthesized and analyzed. Our binding studies have shown that dihydroxamate **7** has a binding constant of approximately of $10^{18.0}$ for $Fe^{3+}$ (ML) and $10^{18.4}$ for $Th^{4+}$. It appears that this chelator has very little selectivity between these ions. However, this data supports our contention that both the hydroxamate groups of a polymeric chelator, such as **13**, can simultaneously coordinate the actinide ion.

The binding properties of polymer **13** with plutonium have also been examined. When a sample of this insoluble polymer in 0.1 M nitric acid was contacted with a dilute sample of $Pu^{4+}$ solution, about 98 % of the $Pu^{4+}$ was removed from the solution after 30 minutes of contact time. This is a very promising result which strongly encourages further development of this class of polymeric chelators.

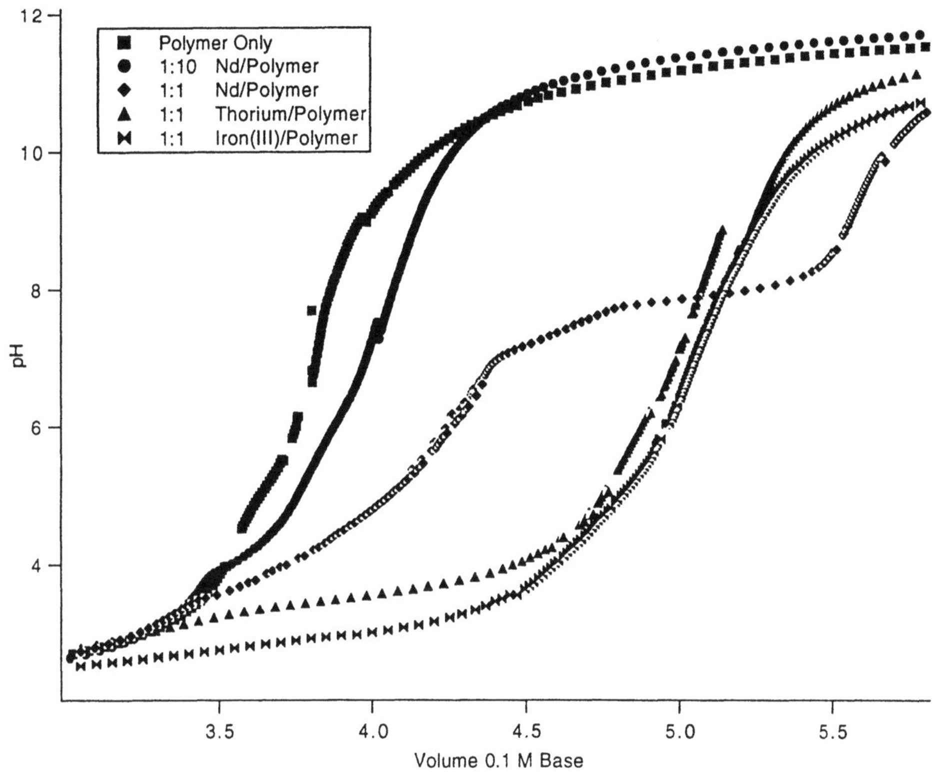


**Figure 7.** Titration of chelating polymer **13** with various metal ions.

Another chelating polymer that has been prepared is the polyacrylamide **14**. This polymer has been prepared in two steps from commercially available polyacryloyl chloride as shown in Scheme 4. It is anticipated that binding studies on this novel polymer will be initiated soon.

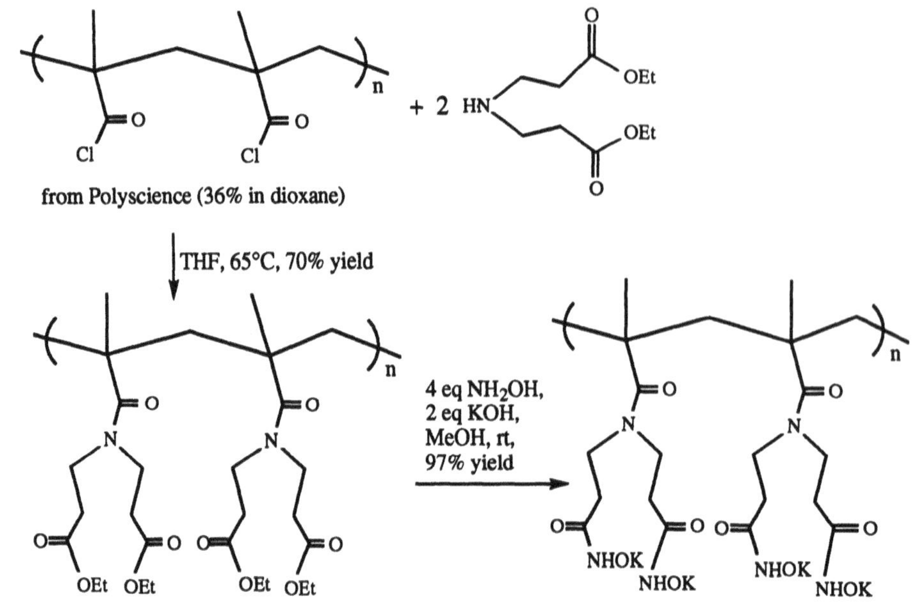


**14**

Scheme 4

The development of water soluble chelating polymers for applications using ultrafiltration as the mode of actinide ion separation is also of interest. Polyethyleneimine has a hydrophilic backbone and is available in a wide range of molecular weights from commercial sources. A simple route for the preparation of water soluble hydroxamate polymer **15** from polyethyleneimine has been developed (Scheme 5). This polymer is currently being evaluated for use with ultrafiltration membranes developed for metal ion separation.

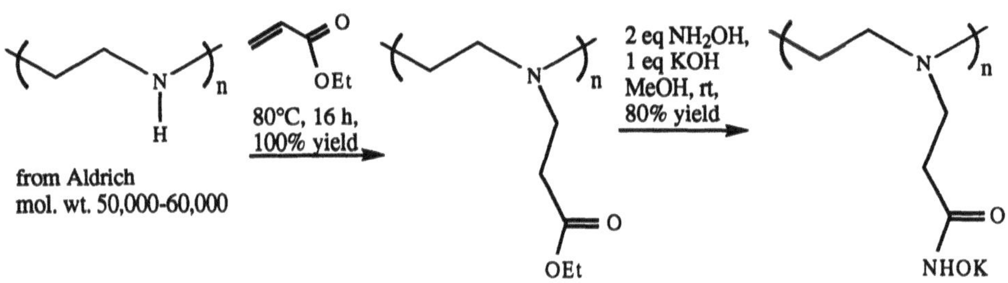


**15**

Scheme 5

In a preliminary study, known quantities of radioactive americium-241 and europium-152 were added to a solution of polymer **15** (0.05% by weight) in 0.1 N $NaNO_3$ or NaCl (100 mL). The pH of this solution was adjusted to different values between 0.5 and 10. At each pH value, the sample was allowed to equilibrate and the initial radioactivity measured (liquid scintillation and gamma) by analyzing an aliquot of the solution. Then a small volume was withdrawn and filtered under pressure through a cellulose filter with polypropylene backing (molecular weight cutoff of about 10,000). The gamma counts in both the permeate and the retentate were measured to estimate the amount of americium or europium that is impermeable to the filter due to binding to the polymer. The results indicate that at pH = 6.5, in the presence of the polymer, retention of the americium and europium was 95 % and 93 %, respectively. At pH values below 6, the efficiency of the metal ion separation with this polymer decreases rapidly. Further studies are in progress to more thoroughly assess the chelating properties of the polyethyleneimine hydroxamate polymer.

The synthesis of water soluble hydroxamate polymer **16** from commercially available polyallylamine has also been accomplished. The synthetic route for this new chelating polymer is described in Scheme 6. The binding properties of this polymer are expected to be studied in the near future.

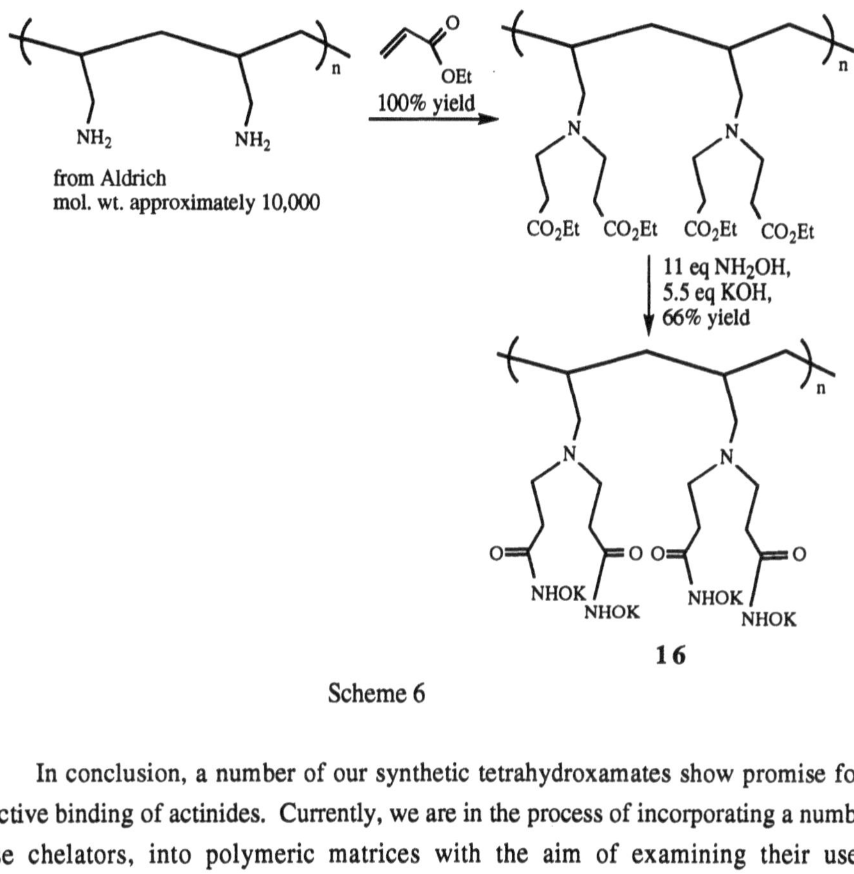


**16**

Scheme 6

In conclusion, a number of our synthetic tetrahydroxamates show promise for the selective binding of actinides. Currently, we are in the process of incorporating a number of these chelators, into polymeric matrices with the aim of examining their use for environmental applications. We believe that by incorporation of more complex and preorganized chelators into polymers one can synthesize new polymeric materials having the desired selectivity and high binding required for actinide remediation. Methods for the synthesis of a number of analogs of our parent water soluble chelator **2** have been successfully accomplished. The hydroxamate chelators that have been developed continue to be evaluated for their ability to bind thorium(IV) and plutonium(IV). The results so far have been very encouraging. Currently, detailed studies of the complexation behavior of this class of chelators are in progress in order to identify key structural features that may allow further improvement of their actinide complexation abilities.

## Acknowledgments

We would like to thank the Waste-management Education and Research Consortium and the Department of Energy and the Efficient Separations and Processing Integrated Program (ESPIP) for their financial support of this program. We thank Doris Ford, Wendy Bisset, Patrick Martin, Dr. Barbara Smith, and Dr. Ken Bower for their help and technical contributions. Dr. David Clark (LANL) is thanked for helpful discussions.

## REFERENCES


1. Berlin, R. E. and C. C. Stanton. "Radioactive Waste Management", John Wiley:New York, 1989.

2. U. S. Department of Energy Office of Environmental Restoration and Waste Management, DOE/EM-0109P, February 1994.

3. A. S. Gopalan, V. J. Huber, and H. K. Jacobs, in *Waste Management: From Risk to Remediation*. R. Bhada Ed., ECM, in press and references cited therein.

4. R.D. Hancock and A.E. Martell, *Chem. Rev.*, **89**, 1875, (1989).

5. K. N. Raymond, G. E. Freeman, and M. J. Kappel, *Inorg. Chim. Acta*, **94**, 193, (1984).

6. M. Streater, P.D. Taylor, R.C. Hider, and J. Porter, *J. Med. Chem.*, **33**, 1749, (1990).

7. P.S. Dobbin, and R.C. Hider, *Chemistry in Britain*, 565, (1990).

8. G. Wilkinson, R. D. Gillard and J. A. McLeverty, Eds. *Comprehensive Coordination Chemistry*, Pergamon Press:NY, 1987, Vol 1-6.

9. K. N. Raymond and P. W. Durbin, Proceedings of the First Hanford Separation Science Workshop, July 23-25, 1991, Richland Washington, II.15, (1993).

10. P. Yakirevitch; N. Rochel, A. M. Albrecht-Gary; J. Libman, and A. Shanzer, *Inorg. Chem.* **32**, 1779, (1993).

11. K. N. Raymond and T. M. Garrett, *Pure and Appl. Chem.* **60**, 1807, (1988).

12. M.J. Miller, *Chem. Rev.* , **89**, 1563, (1989).

13. W. L. Smith and K. N. Raymond, *J. Am. Chem. Soc.*, **103**, 3341, (1981).

14. R.J. Bergeron, S.J. Kline, J.D. Navratil, and C.M. Smith, *Radiochimica Acta*, **35**, 47 (1984).

15. F.L. Weitl, K.N. Raymond, W.L. Smith, and J.R. Howard, *J. Am. Chem. Soc.*, **100**, 1170, (1978).

16. F.L. Weitl and K.N. Raymond, *J. Am. Chem. Soc.*, **102**, 2289, (1980).

17. M. J. Kappel, H. Nitsche, and K. N. Raymond, *Inorg. Chem.* **24**, 605, (1985).

18. L. C. Uhlir, P. W. Durbin, N. Jeung, and K. N. Raymond, *J. Med. Chem.*, **36**, 504, (1993).

19. J. Xu, T. D. P. Stack, and K. N. Raymond, *Inorg. Chem.*, **31**, 4903, (1992).

20. A. Hou, D. W. Whisenhunt, Jr., J. Xu, and K. N. Raymond, *J. Am. Chem. Soc.* **116**, 840, (1994).

21. A. Gopalan, O. Zincircioglu and P. Smith, *Radioactive Waste Management and the Nuclear Fuel Cycle Journal*, **17/3-4**, 161, (1993)

22. W. R. Harris and K. N. Raymond, *J. Am. Chem. Soc.*, **101**, 6534, (1979).

23. A. E. Martell, R. M. Smith, and R. J. Motekaitis. *NIST Critical Stabilitiy Constants of Metal Complexes Database*. 1993.

24. A. Gopalan, V. Huber, O. Zincircioglu, and P. Smith, *J. Chem. Soc., Chem. Commun.*, 1266, (1992).

25. J. M. Cleveland, *The Chemistry of Plutonium*, Gordon and Beach:New York, 1970.

26. J. J. Katz, G. T. Seaborg and L. R. Morss, Eds. *The Chemistry of the Actinide Elements. 2nd Edition*, Chapman and Hall:London, 1986.

27. The iron and plutonium hydrolysis constants (MOH) were not used in the determination of the metal-ligand binding constants or subsequent calculations. Spectrophotometric data did not indicate the presence of these species in detectable concentrations.

28. Y. Sun and A. E. Martell, *Tetrahedron*, **46**, 2725, (1990).

29. S. Konetschny-Rapp, G. Jung, K. N. Raymond, J. Meiwes and H. Zähner, *J. Am. Chem. Soc.*, **114**, 2224, (1992).

30. C. Y. Ng, S. J. Rodgers and K. N. Raymond, *Inorg. Chem.* **28**, 2062, (1989).

31. N. M. Koshti, H. K. Jacobs, P. A. Martin, P. H. Smith, and A. S. Gopalan, *Tetrahedron Lett.*, **35**, 5157, (1994).

32. N. Koshti, V. Huber, P. Smith, and A. S. Gopalan, *Tetrahedron*, **50**, 2657, (1994).

33. D.C. Sherrington and D. Hodge, Eds. *Synthesis and Separation Using Functional Polymers*, John Wiley:NY, 1988.

34. A. Warshawsky, *Ion Exchange and Sorption Processes in Hydrometallurgy*, M. Streat and D. Naden, Eds., John Wiley:NY, 1987, pp 166-225.

35. C. Kantipuly, S. Katragadda, A. Chow, and H.D. Gesser, *Talanta,* **37**, 491, (1990).

36. K. Geckler, G. Lange, H. Eberhardt, and E. Bayer, *Pure and Appl. Chem.*, **52**, 1883, (1980).

37. C. Calmon, *J. Am. Water. Work Assc.,* **73**, 652, (1981).

38. P. Hodge and D. C. Sherrington, Eds. *Polymer-supported Reactions in Organic Synthesis*, John Wiley:NY, 1980.

39. E. Marechal, *Comprehensive Polymer Science*, G. C. Eastmond, A. Ledwith, S. Russo, and P. Sigwalt, Eds., Pergamon Press:NY, 1989, Vol. 6, pp 1-47.

40. F. G. Thorpe, *New Methods of Polymer Synthesis*, J. R. Ebdon, Ed., Blackie & Son Ltd.:London, 1991, pp 139-161.

41. T. Hirotsu,S. Katoh, K. Sugasaka, M. Sakuragi, K. Ichimura, Y. Suda, M. Fujishima, Y. Abe and T. Misonoo, *J. Polym. Sci. Part A: Polym. Chem.* **24**, 1953, (1986).

42. F. Vernon, *Pure and Appl. Chem.*, **54**, 2151, (1982).

43. C. Y. Liu, M. J. Chen, N. M. Lee, H. C. Hwang, S. T. Jou and J. C. Hsu, *Polyhedron*, **11**, 551, (1992).

44. A. S. Gopalan, P. Smith, G. Jarvinen and D. Ford, unpublished results.

45. The effect of metal ion hydrolysis on the shapes of these titration curves has not yet been determined.

# SYNTHESIS, CHARACTERIZATION AND ACTINIDE EXTRACTION BEHAVIOR OF BRIDGE-MODIFIED CARBAMOYLMETHYL PHOSPHONATES AND PHOSPHINE OXIDES

Brian M. Rapko

Pacific Northwest Laboratory
Richland, Washington 99352

## ABSTRACT

Modification of carbamoylmethyl-phosphonates (CMPs), $(RO)_2P(O)CH_2C(O)NR'_2$, and of octyl(phenyl)-N,N diisobutylcarbamoylmethyl-phosphine oxide (CMPO), $(n\text{-}C_8H_{17})(Ph)P(O)CH_2C(O)N(i\text{-}Bu)_2$, was accomplished by replacing a hydrogen on the central methylene carbon atom with either a primary alkyl group (n-octyl) or a secondary alkyl group (isobutyl). The ability of these modified compounds to extract nitric acid, uranyl ions, plutonium(IV), and americium(III) was evaluated as a function of aqueous nitric acid concentration in a liquid-liquid extraction system. Modifying the central methylene carbon atom appears to have little influence on the amount of nitric acid extracted under comparable extraction conditions. The ability to extract uranyl ions is also unaffected by the methylene modifications. However, in all cases, the ability to extract americium(III) is reduced. Plutonium(IV) is an intermediate case, with the carbamoylmethyl-phosphonate systems showing an effect, but with the carbamoylmethyl-phosphine oxide system being unaffected. These results are interpreted as being a reflection of the steric demands on the metal for a given extraction system.

## INTRODUCTION

Bifunctional organophosphorus compounds have been examined as extractants for actinide elements in a variety of oxidation states from acidic radioactive waste solutions. The earliest studies of these compounds were performed by Siddall and coworkers,[1-3] who studied the f-block element extraction properties and mechanism of carbamoylmethyl-phosphonates (CMPs, Figure 1a), $(RO)_2P(O)CH_2C(O)NR'_2$. Such studies on bifunctional actinide extractants have been continued by many other research groups during the intervening years. In particular, the effects of systematic modifications on the ancillary ligands of the phosphoryl group, on the alkyl groups of the amide functionality, and on the effects of modifying the number of carbon atoms that bridge the phosphoryl and amide

*Separations of f Elements*, Edited by K.L. Nash
and G.R. Choppin, Plenum Press, New York, 1995

functionalities have been extensively studied. In this context, of special note is a series of studies by the group of Horwitz and coworkers at Argonne National Laboratory.[4-8] As a result of that research, the compound octyl(phenyl)-N,N-diisobutylcarbamoylmethyl-phosphine oxide (CMPO, Figure 1b) was developed and is now the extractant used in the transuranic extraction (TRUEX) process, a liquid-liquid extraction process for actinide removal from acidic radioactive waste solutions.[9,10]

Although such systematic studies have been performed on the effects of modifying the phosphoryl and amide groups in these bifunctional organophosphorus compounds on the extraction of f-block elements, relatively less study has been devoted to the effects following modification of the methylene group. Paine and coworkers have replaced a hydrogen atom on the central methylene of dihexyl-N,N-diethylcarbamoylmethyl phosphonate (DHDECMP) with a methyl and a benzyl group, respectively, and have studied how these modifications affect the extraction of early and late lanthanides from acidic nitrate solutions.[11] Horwitz and coworkers have replaced a hydrogen atom on the central methylene of CMPO with a pentyl group and have studied how this modification affects the extraction of Am(III) from acidic nitrate solutions using a modified-CMPO/tributyl phosphate extractant system.[12] In related work, Russian workers have examined the effects of bridge modifications on f-block element extraction with a variety of bifunctional organophosphorus compounds,[13-15] and French workers have examined the effects of bridge-modifications on actinide and lanthanide extraction with bifunctional organic extractants derived from malonamide,[16-19] which possesses structural features similar to CMPs and CMPOs.

The effect of these bridge modifications depends on the specific system studied. Paine found that, while bridge substitution in general decreased the ability of chloroform solutions containing DHDECMP to extract lanthanides at a given nitric acid concentration, the magnitude of the changes differed.[11] Replacing the hydrogen with a methyl group resulted in only a modest relative decrease in the distribution values as a function of nitric acid concentration while replacing the hydrogen with a benzyl group resulted in an extractant that not only showed a substantial decrease in the distribution values, but, at low acid, a change in the slope's sign as well. Horwitz[12] found that with tetrachloroethylene or tridecane solutions containing CMPO, replacement of the methylene hydrogen with a pentyl group resulted in a decrease in the compound's distribution values for Am(III). In this instance, the shape of a plot of the distribution values as a function of nitric acid concentration remained similar to that of the unmodified extractant. Myasoedov found that with dichloroethane solutions containing diphenyl,-N,N-diethylcarbamoylmethyl-phosphine oxide, replacing a methylene hydrogen with a heptyl group results in a small decrease in the compound's ability to extract Am(III), but replacing a methylene hydrogen with a methyl group results in a substantial decrease.[13,14] In both cases, the shapes of a plot of distribution values for Am(III) as a function of nitric acid concentration appeared similar. Little change was observed in the extractant's ability to select for Am(III) over Eu(III), although the ability to select for Pu(IV) and $UO_2^{2+}$ over Am(III) increased upon substitution: the methyl substituted compound was more selective than the heptyl substituted compound. The malonamides respond somewhat differently to bridge substitution. Musikas and coworkers have shown that bridge substitution with alkyl or oxy-alkyl groups generally seems to increase the distribution values for actinides with malonamides in aliphatic and aromatic solvents, although the shape of a plot of distribution values as a function of nitric acid concentration changes somewhat, with the bridge-modified malonamides being relatively enhanced for Am(III) extraction at high nitric acid concentrations.[16,17]

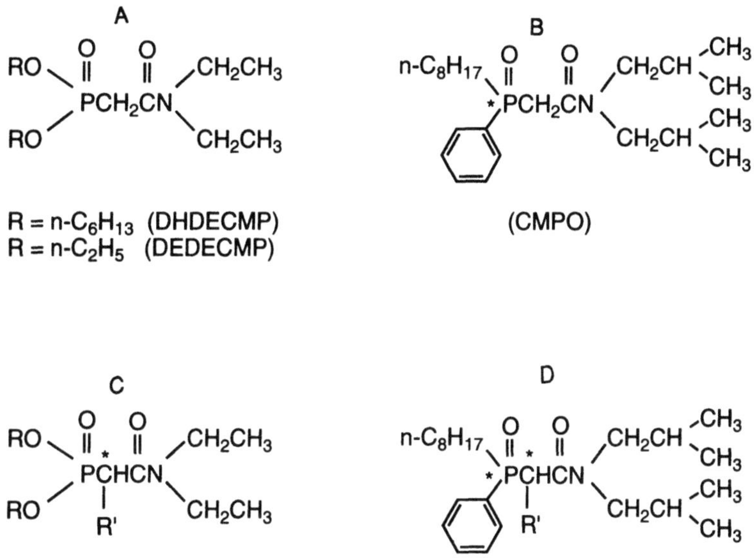


**Figure 1.** Structures of carbamoylmethyl-phosphonates (**A**), carbamoylmethyl-phosphine oxides (**B**), and their bridge-modified analogues (**C&D**, respectively). The asterisks refer to the presence of chiral centers in the molecule.

Our group has a standing interest in evaluating potential extractants for selectively removing transuranic elements (TRUs) from acidic nuclear waste streams.[20-22] The interpretation of the changes in these actinide extractants that result from bridge modification is complicated by differing solvents and differing types of bridge substitutions. Therefore, a study was initiated to try to evaluate the impact of bridge substitution on carbamoylmethyl-phosphonates and a carbamoylmethyl-phosphine oxide system where changes in distribution values for a variety of actinides could be evaluated for the same type of bridge modification in the same solvent. It was hoped that such a study might lead to a clearer understanding of how bridge modifications impact these bifunctional organophosphorus compounds' ability to extract actinides.

## EXPERIMENTAL

### General

The DHDECMP used in this study was provided courtesy of Dr. Barbara Smith of Los Alamos National Laboratory. The DEDECMP used in this study was provided courtesy of Professor R. T. Paine of the University of New Mexico. Tert-butyl benzene (Aldrich Chemical Company), CMPO (Strem Chemicals), and all other bulk chemicals were of reagent grade and were used as received. Tetrahydrofuran was dried by distillation under nitrogen over potassium/benzophenone.

### Generic Synthetic Procedure for CMP and CMPO Bridge Substitutions

In a dry box, a 100% mole excess of potassium hydride (KH) was loaded into a Schlenk flask and an addition funnel was attached. The system was capped with a rubber septum, and the apparatus was removed from the dry box and attached to a Schlenk line. Approximately 50 mL of dry THF was transferred via stainless steel cannula to the addition funnel and then was added to the THF. Meanwhile, the respective CMPO, DHDECMP or DEDECMP (25 to 75 mmol) was weighed out in the dry box and placed in a 250-mL Schlenk flask. The flask was capped with a rubber septum, removed from the dry box, and attached to the Schlenk line. Approximately 100 to 150 mL of dry THF was transferred via cannula into the flask containing the respective CMPO or CMP. The resulting solution was then transferred to the addition funnel. The respective CMPO or CMP flask was rinsed with 2- by 10-mL portions of dry THF, and these washings were also transferred to the addition funnel.

At room temperature under a nitrogen atmosphere, the respective CMPO or CMP solution was added dropwise to the well-stirred THF/KH suspension. This suspension gradually darkened and turned yellow to orange as the addition proceeded. Gas evolution occurred as the CMP/CMPO solution was added to the KH suspension. This addition required about 2 hours to complete. After the addition was completed, the addition funnel was removed, the Schlenk flask was capped with a glass stopper, and the system was allowed to stir at room temperature overnight.

Next, a 200-mL Schlenk filter flask with a medium sintered-glass frit was attached to a 500-mL Schlenk flask equipped with a Teflon® magnetic stirring bar. The system was attached to the Schlenk line and evacuated and backfilled with nitrogen three times. Meanwhile, the stirring of the respective CMPO or CMP/KH suspension was stopped, and the system was allowed to settle for several hours. The supernatant was then transferred via cannula into the filter flask and filtered into the receiving container. The solids on the KH/CMP or KH/CMPO flask were washed twice with 25 mL of dry THF and these washings were also transferred to the filter flask and filtered into the receiving flask.

The filter flask was removed, and an addition funnel was attached to the receiving flask. The addition funnel was charged with a 10% mole excess of the indicated alkyl halide and 100 mL of dry THF. Dropwise addition of the alkyl halide solution under nitrogen occurred over about 2 hours. After the addition was completed, the addition funnel was replaced with a glass stopper, and the system was stirred at room temperature overnight. A yellow to orange suspension resulted. Volatile materials were removed at room temperature under vacuum. The residue was suspended in 100 to 200 mL of chloroform and washed with four equal portions of water. The organic phase was dried over 4Å molecular sieves and filtered, and the volatile materials were removed under vacuum, leaving the product as a light yellow to bright yellow oil.

**Characterization**

All nuclear magnetic resonance (NMR) spectra were recorded on a Varian Instruments NMR using 5 mm sample tubes and operating at a nominal field of 300 MHz ($^1H$). For $^1H$ and $^{13}C$ NMR spectra, chemical shifts were referenced to tetramethylsilane (TMS) either directly or by using known solvent resonances. For the $^{31}P$ NMR spectra, chemical shifts were referenced using the substitution method to dilute $CHCl_3$ solutions of CMPO or DHDECMP using the known literature values for these secondary standards.[23] Infrared (IR) spectra were recorded on a Nicolet 510P Fourier Transform-Infrared spectrometer at 2 $cm^{-1}$ resolution using NaCl plates for neat samples or as $CHCl_3$ solutions using NaCl solutions cells of nominal 0.1 mm thickness. Thin layer chromatography (TLC) data were collected using 5 cm by 10 cm silica gel on aluminum plates (Sigma-Aldrich). These plates were 0.2 mm thick, with a particle diameter 2 to 25 μm and a mean pore diameter of 60Å. Samples were developed using $I_2$ vapor. Gas chromatography/mass spectroscopy (GC/MS) was obtained on a HP5890A GC instrument with a HP 5971 selective mass detector. Samples for GC/MS study were injected as $CH_2Cl_2$ solutions onto a J&W Scientific DB5 column (0.25 μ thick, 30 m length, 0.25 mm id). A 1-μL injection was employed using a split ratio of 35:1. Other parameters include detector temp = 275°C, injector temp = 300°C, and temperature profile: 5 minutes at 100°C, followed by an increase to 275°C at 10°C/minute, with the temperature then held for 30 minutes at 275°C.

**CMPO-octane.** Alkyl halide used in synthesis - 1 bromooctane. Yield - 97%: $R_F$ 0.64 and 0.76 (Ethyl Acetate); $^1H$ NMR (300 MHz, $CDCl_3$) δ 7.9-7.7 (m, 2H), 7.6-7.4 (m, 3H), 3.6-2.8 (m, 5H), 2.7-1.5 (m, 8H), 1.5-1.0 (m, 22H), 1.0-0.8 (m, 18H); $^{13}C$ NMR (75.4 MHz, $CDCl_3$) δ 170.22, 169.7, 131.43, 131.40, 131.33, 131.22, 131.11, 130.28, 130.21, 128.17, 128.09, 128.03, 127.94, 56.67, 56.09, 55.40, 47.23, 46.81, 46.41, 46.02, 33.49, 32.47, 31.39, 31.36, 30.80, 30.60, 30.44, 29.35, 29.15, 28.93, 28.81, 29.77, 28.71, 28.64, 28.55, 28.37, 27.81, 26.67, 26.55, 25.74, 25.56, 24.84, 24.65, 22.23, 20.54, 20.49, 20.21, 20.07, 19.96, 19.87, 19.70, 19.58, 19.34, 13.71; $^{31}P$ NMR (121.4 MHz, $CDCl_3$) δ 40.21 (s), 40.00 (s); IR (neat) 3057, 2955, 2921, 2870, 2854, 1634, 1591, 1468, 1438, 1422, 1387, 1367, 1342, 1292, 1239, 1199, 1176, 1139, 1110, 1071, 1029, 998, 943, 923 $cm^{-1}$; MS {GC/MS EI (70EV)} *m/e* Parent ion = 519.42. Parent ion not found., 420.45 (3.7%), 391.4 (14.6%), 237.25 (36.9%), 282.4 (100%), 128.2 (21.4%); GC retention times = 37.31 and 39.26 minutes.

**CMPO-isobutyl.** Alkyl halide used in synthesis - 1-bromo-2-methylpropane. Yield - 89%: $R_F$ 0.61 and 0.73 (Ethyl Acetate); $^1$H NMR (300 MHz, $CDCl_3$) $\delta$ 8.9-8.6 (m, 2H), 7.6-7.4 (m, 3H), 3.7-2.7 (m, 4H), 2.7-1.4 (m, 8H), 1.4-1.1 (m, 12H), 1.0-0.8 (m, 21H); $^{13}$C NMR (75.4 MHz, $CDCl_3$) $\delta$ 170.7, 170.2, 131.63, 131.56, 131.52, 131.45, 130.54, 128.41, 128.37, 128.27, 128.22, 56.8, 56.0, 55.4, 45.5, 45.1, 44.7, 44.4, 37.6, 37.5, 31.67, 31.09, 30.95, 30.89, 30.76, 28.93, 28.75, 27.15, 27, 26.87, 26.82, 26.75, 26.68, 26, 25.1, 23.48, 23.03, 22.53, 22.34, 21.98, 20.82, 20.77, 20.52, 20.44, 20.36, 20.19, 20.04, 19.67, 14; $^{31}$P NMR (121.4 MHz, $CDCl_3$) $\delta$ 41.04 (s), 40.46 (s); IR (neat) 3058, 2959, 2930, 2870, 2860, 1634, 1591, 1468, 1438, 1421, 1387, 1368, 1341, 1296, 1248, 1198, 1172, 1140, 1109, 1072, 1029, 998, 943, 924 $cm^{-1}$; MS {GC/MS EI (70EV)} *m/e* Parent ion = 463.36. Found: 463.55 (0.9%), 364.35 (5.6%), 335.3 (23.9%), 237.25 (49.2%), 226.3 (100%), 128.2 (40.3%), 407.45 (6.7%); GC retention times = 25.26 and 26.06 minutes.

**CMPO.** Used as received. $R_F$ 0.34 (Ethyl Acetate); $^1$H NMR (300 MHz, $CDCl_3$) $\delta$ 7.9-7.7 (m, 2H), 7.6-7.4 (m, 3H), 3.3-3.0 (m, 6H), 2.4-2.1 (m, 2H), 2.0-1.6 (m, 2H), 1.5-1.1 (m, 12H), 0.9-0.7 (m, 15H); $^{13}$C NMR (75.4 MHz, $CDCl_3$) $\delta$ 161.2 (d, 3.5), 132 (d, 94.8), 131.8 (d, 2.7), 130.8 (d, 9), 128.5 (d, 11.7), 56.8, 54.3, 38.2, 37.4, 31.7, 30.9, 30.7, 30.1, 29.2, 29.0, 28.2, 26.6, 22.6, 21.1, 21.0, 20.2, 19.9, 14.1; $^{31}$P NMR (121.4 MHz, $CDCl_3$) $\delta$ 37.7 (s); IR ($CHCl_3$ solution) 3062, 2964, 2931, 2873, 2858, 1627, 1468, 1451, 1438, 1389, 1370, 1342, 1290, 1265, 1172, 1105, 861, 808 $cm^{-1}$; MS {GC/MS EI (70EV)} *m/e* Calculated = 407.3. Found 407.45 (5.5%), 308.25 92.9%, 351.35 6.9%, 292.2 100%, 237.25 84.6%, 128.2 94.5%; GC retention time = 24.96 minutes.

**DHDECMP-octane.** Alkyl halide used in synthesis - 1 bromooctane. Yield - 95%: $R_F$ 0.64 (Ethyl Acetate); $^1$H NMR (300 MHz, $CDCl_3$) $\delta$ 4.2-4.0 (m, 4H), 3.7-3.0 (m, 5H), 2.2-1.7 (m, 2H), 1.7-1.5 (m, 4H), 1.5-1.0 (m, 26H), 0.9-0.8 (m, 9H); $^{13}$C NMR (75.4 MHz, $CDCl_3$) $\delta$ 167.32 (d, 4.4), 66.5 (d, 6.6), 66.26 (d, 7.0), 42.98, 42.42, 41.22, 40.97, 33.87, 32.70, 31.67, 31.24, 30.44, 30.37, 29.36, 29.20, 29.11, 28.98, 28.60, 28.51, 28.30, 28.05, 27.85, 27.79, 25.05, 22.50, 22.42, 14.54, 13.95, 13.86, 12.80; $^{31}$P NMR (121.4 MHz, $CDCl_3$) $\delta$ 26.36 (s); IR (neat) 2958, 2929, 2870, 2857, 1645, 1461, 1430, 1380, 1362, 1252, 1221, 1136, 1062, 1040, 997, 962 $cm^{-1}$; MS {GC/MS EI (70EV)} *m/e* Parent ion = 475.38. Found 475.55 (1.2%), 403.45 (5%), 226.3 (35.9%), 72.1 (100%); GC retention time = 24.04 minutes.

**DHDECMP-isobutyl.** Alkyl halide used in synthesis - 1-bromo-2-methylpropane. Yield - 80%: $R_F$ 0.64 (Ethyl Acetate); $^1$H NMR (300 MHz, $CDCl_3$) $\delta$ 4.1-3.9 (m, 4H), 3.7-3.1 (m, 5H), 2.1-1.9 (m, 1H), 1.7-1.4 (m, 6H), 1.3-1.0 (m, 18H),

0.9-0.7 (m, 12H); $^{13}C$ NMR (75.4 MHz, $CDCl_3$) δ 167.0 (d, 4.5), 66.3 (d, 6.5), 65.9 (d, 7.0), 42.14, 40.76, 40.56, 39.01, 36.37, 36.30, 30.94, 30.19, 30.16, 30.11, 30.08, 26.52, 26.33, 22.67, 22.12, 21.52, 14.04, 13.52, 12.35; $^{31}P$ NMR (121.4 MHz, $CDCl_3$) δ 26.69 (s); IR (neat) 2959, 2932, 2872, 2860, 1644, 1462, 1432, 1381, 1364, 1319, 1250, 1220, 1139, 1118, 1099, 1062, 1040, 998, 962, 924 $cm^{-1}$; MS {GC/MS EI (70EV)} *m/e* Calculated 420.3. Parent ion not found, 391.4 (0.74%), 348.35 (2.1%), 170.2 (34.4%), 72.1 (100%); GC retention time = 19.16 minutes.

**DHDECMP.** Used as received. $R_F$ 0.34 (Ethyl Acetate); $^1H$ NMR (300 MHz, $CDCl_3$) δ 4.2-4.0 (m, 4H, O-$CH_2$), 3.5-3.3 (m, 4H, N-$CH_2$), 3.02 (d, J=22.1 Hz, 2H, central methylene), 1.8-1.6 (m, 4H), 1.4-1.1 (m, 18H), 0.89 (t, J=6.4 Hz, 6H); $^{13}C$ NMR (75.4 MHz, $CDCl_3$) δ 163.59 (d, 5.9), 66.23 (d, 6.4), 42.7 (40.18), 32.98 (d, 134.2), 31.05, 30.15 (d, 5.9), 24.85, 22.24, 13.90, 13.68, 12.62; $^{31}P$ NMR (121.4 MHz, $CDCl_3$) δ 23.6 (s); IR (neat) 2957, 2931, 2872, 2860, 1646, 1464, 1433, 1381, 1363, 1313, 1257, 1223, 1100, 1063, 1044, 1000, 965, 919, 882, 817, 792, 727 $cm^{-1}$; MS {GC/MS, EI (70EV)} *m/e* Calc. 363.25. Found. 363.35 (3.3%), 291.25 (1.2%), 114.1 (0.7%), 72.1 (100%), 351.35 (6.9%); GC retention time = 18.76 minutes.

**Octyl-DEDECMP.** Alkyl halide - 1-bromooctane. Yield 74%: $R_F$ 0.30 (Ethyl Acetate); $^1H$ NMR (300 MHz, benzene-$d_6$) δ 4.3-3.9 (m, 4H), 3.6-3.3 (m, 2H), 3.25-3.0 (m, 2H), 2.95-2.75 (m, 1H), 2.55-2.35 (m, 1H), 2.05-1.85 (m, 1H), 1.5-0.7 (m, 27H); $^{13}C$ NMR (75.4 MHz, benzene-$d_6$) δ 167.83 (d, 3.7), 63.5 (d, 5.9), 61.94 (d, 6.4), 44.24, 43.05, 42.51, 41.60, 32.63, 30.40, 30.22, 30.09, 29.52, 29.32, 29.23, 29.16, 23.47, 17.08, 17.01, 16.94, 15.21, 14.76, 13.63; $^{31}P$ NMR (121.4 MHz, benzene-$d_6$) δ 27.67 (s); IR (neat) 2962, 2929, 2975, 2856, 1644, 1483, 1461, 1452, 1431, 1380, 1363, 1251, 1220, 1164, 1137, 1099, 1056, 1027, 964, 903, 855, 787, 764, 723, 678 $cm^{-1}$; MS {GC/MS EI(70EV)} *m/e* Parent ion = 363.25. Found:, 363.45 (1.2%), 335.4 (1.4%), 291.25 (9.3%), 226.3 (36.7%), 137.1 (3.9%), 72.05 (100%); GC retention time = 17.73 minutes.

**DEDECMP.** Used as received. $R_F$ 0.21 (Ethyl Acetate); $^1H$ NMR (300 MHz, benzene-$d_6$) δ 4.02 (m, 4H), 3.2-3.0 (m, 4H), 2.89 (d, J=22 Hz, 2H), 1.08 (m, 6H), 0.93 (m, 3H), 0.76 (m, 3H); $^{13}C$ NMR (75.4 MHz, benzene-$d_6$) δ 164.23 (d, 5.4), 62.66 (d, 6.3), 43.38, 40.83, 34.38 (d, 132), 16.89 (d, 6.2), 14.57, 13.56; $^{31}P$ NMR (121.4 MHz, benzene-$d_6$) δ 23.7 (s); IR (neat) 2980, 2936, 2911, 2876, 1642, 1460, 1452, 1434, 1383, 1366, 1286, 1256, 1223, 1164, 1100, 1055, 1028, 968, 919, 865, 832, 783, 722, 667 $cm^{-1}$; MS{(GC/MS EI(70EV)} m/e Parent ion = 251.13. Found:, 251.25 (6.1%), 223.3 (1.3%), 222.35 (0.45%), 179.05 (21.2%), 137.1 (2.6%), 114.1 (1.7%), 72.15 (100%); GC retention time = 12.78 minutes.

**Distribution Measurements**

Plutonium(IV) was available in house as a stock solution in 7M $HNO_3$. The +4 oxidation state was verified by adding 1 μL of the Pu solution to 1 mL of 0.3M $HNO_3$, then contacting this aqueous phase with 1 mL of 0.5M thenoyltrifluoroacetone in xylene. A $D_{Pu}$[24] of greater than 200 was found, indicating >99% of the Pu was in the +4 oxidation state. The aqueous solutions for all of the Pu(IV) distribution measurements contained 0.01M sodium nitrite and 0.005M sodium metavanadate to maintain Pu in the +4 oxidation state.[25] Americium(III) was present as a stock solution in 0.3M $HNO_3$. Uranium(VI) was introduced from a 0.05M stock solution in 2M $HNO_3$, which was prepared by dissolving the appropriate amount of solid uranyl nitrate hexahydrate into a nitric acid solution.

Metal distribution coefficients were determined by mixing 1 mL of the CMP or CMPO organic solution with 1 mL of the appropriate aqueous phase for 60 seconds. After 60 seconds of centrifugation at 3000 rpm to ensure complete phase separation, aliquots of the aqueous phase were taken for analysis. In the case of the Am and Pu measurements, aliquots of the organic phase were also taken.

The relative Am and Pu concentrations of each phase at equilibrium were determined by liquid scintillation counting. The counting was performed by mixing a 0.1-mL aliquot of the appropriate phase with 5 mL of either OPTI-FLUOR® or ULTIMA GOLD® liquid scintillation cocktail (Packard Instrument Co., Downers Grove, Illinois). To compensate for possible quenching effects, 0.1 mL of the tracer-free other phase was added. The relative alpha activity of each phase was determined using a Packard TRI-CARB 2500 Liquid Scintillation Analyzer.

The distribution of $UO_2^{2+}$ was determined in a manner similar to that used for Am and Pu, with a 1 μL aliquot of the uranyl nitrate solution added to the aqueous phase before contact with the CMP or CMPO extractant. Analysis of the uranium concentration in the aqueous phase was done using laser fluorimetry before and after contact with the extractant. The organic phase $UO_2^{2+}$ concentration was determined by the difference between the initial and equilibrium aqueous $UO_2^{2+}$ concentrations.

Nitric acid concentrations were determined by potentiometric titration with a standardized NaOH solution. Aliquots (ranging from 20 μL to 1 mL) of the solution to be analyzed were added to 10 mL of deionized water. The resulting solutions were titrated with 0.05M NaOH using a Mettler DL21 automatic titrator. To obtain the $HNO_3$ distribution measurements, the aqueous $HNO_3$ solutions were analyzed before and after contact with an equal volume of the organic phase. The organic phase $HNO_3$ concentration was determined by the difference between the initial and equilibrium aqueous $HNO_3$ concentrations. As a control, for some of the extractant contacts, small aliquots (ranging from 10 μL to 1 mL) of the organic phase also were added to 10 mL of deionized water and titrated. Material balance between the initial aqueous $HNO_3$ concentration and the sum of the organic and aqueous phase $HNO_3$ concentration in these cases was excellent, with typically >98% and often >99% of the initial $HNO_3$ accounted for. Agreement among the distribution coefficients obtained by each method was excellent, with no significant differences observed, provided small organic-phase aliquots were used.

## RESULTS AND DISCUSSION

### Synthesis and Characterization of Bridge-Modified CMP's and CMPO

#### A) Synthesis

The synthetic route used to replace a hydrogen atom of the central methylene in CMP's and CMPO with an alkyl group is shown in equation 1. It is well known that one of the hydrogen atoms on the central methylene is relatively acidic, and that deprotonation by base occurs preferentially at this site.[23a] Nucleophilic attack of the resulting anion with an alkyl halide results in salt elimination and formation of the desired bridge-modified product. This approach has been used previously to prepare bridge modified CMPs[26] and CMPO.[12] The reaction appears general and quite successful for this series of molecules, with typical isolated yields of greater than 70% and, in selected cases, quantitative yields were obtained. Purity appears excellent, considering the simplicity of the workup, with purities exceeding 95% in all instances and with >99% in many cases as determined by GC/MS and $^{31}P(^{1}H)$ NMR spectroscopy (see below). The only significant impurity detected was unreacted starting material, and, given its low abundance, further purification was deemed unnecessary.

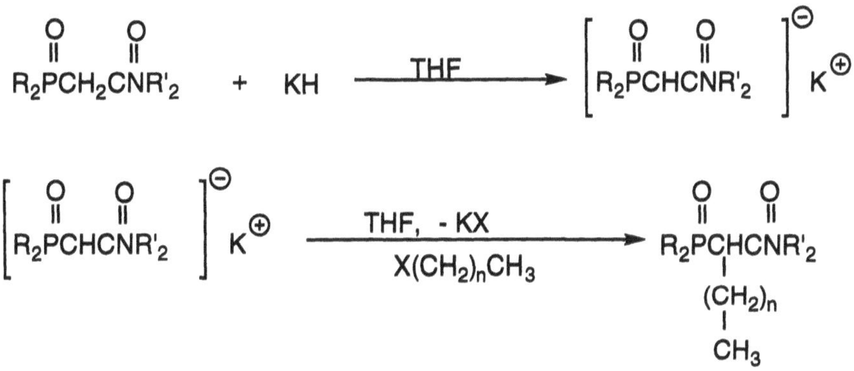


(1)

Figure 1 illustrates one significant difference between the product resulting from bridge-modification of CMPs compared to the products derived from bridge-modification of CMPOs. The asterisks in Figure 1 illustrate that, while the parent CMP possess no chiral center, the phosphorus atom of CMPO is chiral by virtue of having four different substituents attached to it in a tetrahedral arrangement. In each instance, modification of the central methylene results in the introduction of a chiral center at the methylene carbon, since the replacement of a hydrogen atom by an alkyl group means that the central, $sp^3$-hybridized carbon atom now possesses four different substituents. Bridge-modified CMPs, then, possess one chiral center while bridge-modified CMPOs possess two. The presence of two chiral centers allows for the presence of two sets of diastereomers and so, as has been previously noted, bridge modification of CMPO will yield two sets of diastereomers with differing physical properties.[12]

**B) Characterization - Chromatography**

Table 1 summarizes the results of GC and TLC measurements on the parent CMPs and CMPO as well as their bridge-modified derivatives. The GC measurements were obtained using a column, detection system, and programmed temperature profile essentially the same as one previously described for examination of TRUEX process solvent and its degradation products.[27] The column contains a stationary phase comprised of fused silica with a 95% dimethylsilicone/5% diphenylsilicone coating. Such a surface selects for nonpolar compounds and generally separates such nonpolar materials by molecular weight. This preference is reflected in the data of Table 1. In all cases, the retention time is increased by bridge modification of the parent compound and goes in the order parent < isobutyl modified bridge < octyl modified bridge, consistent with the increasing total molecular weight of the compounds. The parent series follows the order DEDECMP < DHDECMP < CMPO, again reflecting their relative molecular weights. This molecular weight correlation to retention times is only observed within a particular series as the alkyl group changes; this correlation is only approximate between series. For example, the molecular weight for the octyl-bridge modified DHDECMP, o-DHDECMP, is greater than CMPO by about 12 grams/mole, but the retention time for CMPO is greater than o-DHDECMP by approximately 2 minutes. Still, the difference in retention times between the parent compounds and all of the bridge-modified derivatives is substantial and allows ready quantification of the product's purity. Such quantification assumes that the peak ratios accurately reflect the parent versus bridge modified product's relative molar ratios, an assumption that has previously been shown to be valid for other, more disparate, neutral organophosphorus molecules.[28] Each individual, bridge-modified CMPO can be further resolved into two bands separated by somewhat less than 1 minute in retention time. The presence of two bands is consistent with the expected presence of diastereomers as previously noted, and their resolution further illustrates that a simple correlation of retention time to molecular weight is an insufficient description of the compound's column behavior. The ratios of the two bands present in the bridge-modified CMPO products are slightly different than the statistically expected ratios of 1:1. In each instance, the band with the lower retention time is slightly less abundant, with the ratio of peak area ranging from 1:1.4 (isobutyl-CMPO) to 1:1.8 (octyl-CMPO).

Thin layer chromatography on silica gel using ethyl acetate as the eluant was also performed on these compounds: the results are summarized in Table 1. Significant differences between the $R_F$ values of the parent CMPs and CMPO and their bridge-modified products are observed. Modification of the bridge results in an increased preference for the mobile phase, with $R_F$ values essentially doubling. Unlike the GC results, the $R_F$ values for DHDECMP and CMPO are identical, and the effects of bridge modification are similar and independent of the type of alkyl group introduced. The DEDECMP shows the same trends

**Table 1.** Chromatography Summary for CMPs and CMPOs

| Sample | GC retention times[a] (minutes) | TLC ($R_F$)[b] |
|---|---|---|
| DHDECMP | 18.76 | 0.34 |
| Isobutyl-DHDECMP | 19.16 | 0.64 |
| Octyl-DHDECMP | 24.04 | 0.64 |
| CMPO | 24.96 | 0.34 |
| Isobutyl-CMPO | 25.26 & 26.06 | 0.64 & 0.73 |
| Octyl-CMPO | 37.31 & 39.26 | 0.64 & 0.76 |
| DEDECMP | 12.35 | 0.21 |
| Octyl-DEDECMP | 17.3 | 0.30 |

as does DHDECMP and CMPO, but the $R_F$ values for both the parent as well as the magnitude of the $R_F$ change upon alkylation are somewhat different. As seen with the GC results, both sets of diastereomers in the CMPO series are resolved, which suggests that column chromatography could be used to separate the diastereomers. Such a separation was outside the scope of the present study and was not attempted.

**C) Characterization - Mass Spectroscopy**

Figures 2 and 3 summarize and illustrate the major observed and identified fragments in the mass spectra following 70EV electron impact ionization of the output after separation by GC. The mass spectra of CMPOs have been previously studied,[23b] and the major observed fragmentation patterns noted in that study are described by routes **1-3** in Figures 2 and 3 for the CMPs and CMPOs, respectively. Somewhat surprisingly, given the ionization method used, parent ion fragments are observed in many instances. Even for those cases where no parent ion is observed, both fragments resulting from a single bond breakage can be observed. Usually, the single bond breakage that shows the base peak (100%), as well as a fairly intense fragment for the other piece of the molecule, comes from cleavage of the

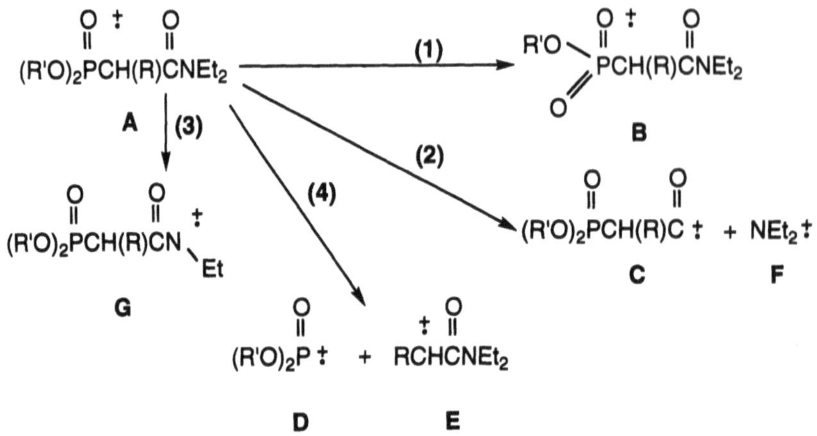


**Figure 2.** Mass spectra summary and suggested fragmentations for CMPs.

carbonyl carbon and nitrogen bond in the amide functionality. Previous work has shown that cleavage of the C-N amide bond is the major decomposition route for a variety of carbamoylmethyl-phosphine oxides.[23b] However, a fourth major decomposition route, route **4** in Figures 2 and 3, also can be discerned. This pathway, not observed for the parent CMPO, provides a new type of base ion for the bridge-modified CMPOs examined. This fourth decomposition route involves cleavage at the phosphorus-carbon bond, which connects the modified methylene to the phosphoryl group. Both fragments from this bond cleavage can be observed in good abundance for the bridge-modified CMPOs. The presence of either a parent ion, or both fragments resulting from a single bond cleavage, provides good evidence for the proposed composition of these bridge-modified CMPOs and CMPs. For the two bands observed in the GC of the bridge-modified CMPOs, the mass spectrum of each band appears essentially identical to the other, consistent with the assignment of these bands to two sets of diastereomers.

## D) Characterization - Infrared Spectroscopy

Little change is observed in the infrared spectra as a result of bridge modification. Figure 4 illustrates how the spectrum changes for DHDECMP upon replacing a bridge hydrogen with an octyl group. Similar features are observed for DEDECMP and CMPO substituted compounds. The pattern and relative intensities of the bands remain very similar. Assignments of the frequency for the carbonyl bond stretching frequency and the phosphoryl bond stretching frequency were made by analogy to previous literature assignments.[29,30] For DHDECMP, the ν(CO)[29,30a] band for the amides was measured at 1646 $cm^{-1}$ and the ν(PO)[29,30a] band at 1257 $cm^{-1}$. Bridge alkylation results in no detectable change in the

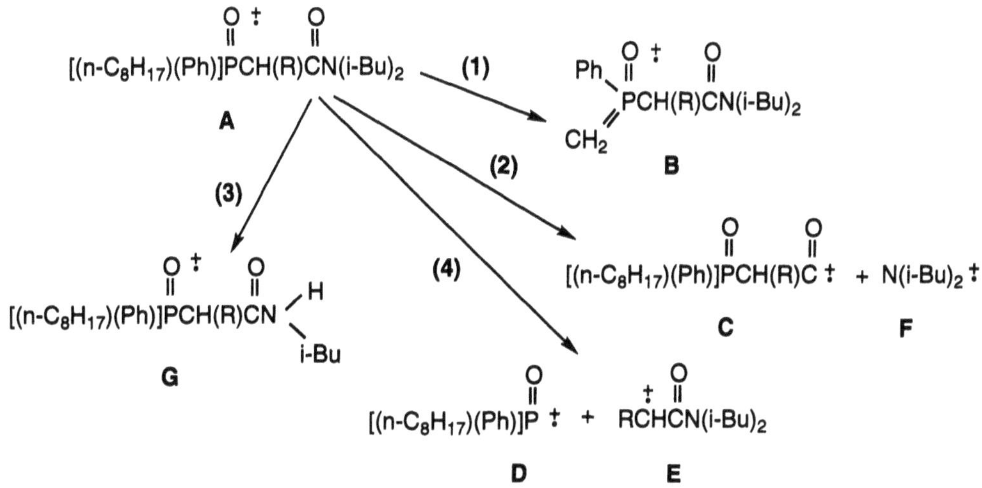


**Figure 3.** Mass spectra summary and suggested fragmentations for CMPOs.

energy of the carbonyl vibration of the amide, but the phosphoryl band decreases by 5 to 7 $cm^{-1}$. Such a decrease in ν(PO) suggests only a minor decrease, if any, in P-O bond order and an concomitant increase in the phosphoryl group's basicity as a result of alkylation.[30a] With the DEDECMP series, the ν(CO)[29] band for the amides was measured at 1642 $cm^{-1}$ and the ν(PO)[29] band at 1256 $cm^{-1}$. Bridge alkylation resulted in no significant change in the carbonyl band, and the phosphoryl band decreased by 5 $cm^{-1}$. For the CMPO series, the carbonyl band was measured at 1627 $cm^{-1}$, and the closest band to the expected phosphoryl band[30] at 1200 $cm^{-1}$ was seen at 1172 $cm^{-1}$. Bridge alkylation here appears to increase the amide carbonyl band by 7 $cm^{-1}$ and the phosphoryl band by 16 to 17 $cm^{-1}$. However, interpretation of the CMPO series is complicated by solvent effects. The CMPO is a solid and its IR spectrum was obtained as a chloroform solution. Solvent interference was present just above 1200 $cm^{-1}$, which is very close to literature values for the phosphoryl band in CMPO solutions.[30] Solvent effects were not important for the bridge-modified compounds, which are oils, and their IR spectra were obtained neat. For these bridge-modified CMPOs, the phosphoryl band was seen just below 1200 $cm^{-1}$, which suggests only a minor decrease, if any, in the phosphoryl band. Even with a phosphoryl band of 1172 $cm^{-1}$, the magnitude little change, if any, in basicity occurs as a result of bridge modification in CMPs and CMPO.

### E) Characterization - NMR Spectroscopy

$^{1}H$, $^{13}C\{^{1}H\}$, and $^{31}P\{^{1}H\}$ NMR spectra of the parent compounds and their bridge-modified products were obtained. The $^{1}H$ and $^{13}C\{^{1}H\}$ NMR spectra of the bridge-modified products were extremely complex compared to the parent spectra. For the CMPs,

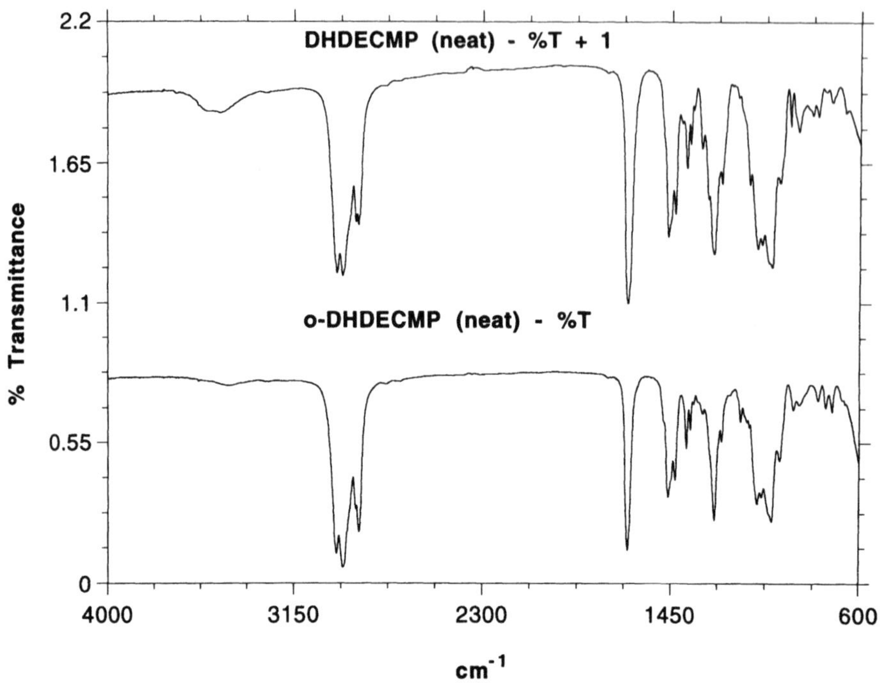


**Figure 4.** Infrared spectra for DHDECMP (top) and octyl, bridge-alkylated DHDECMP, o-DHDECMP (bottom). Spectra were recorded neat on NaCl plates.

the added complexity is due to the inequivalence of the alkoxy groups attached to the phosphorus atom that is present for the bridge-modified materials as opposed to the parent CMPs. For the bridge-modified CMPOs, the added complexity results from the presence of pairs of diastereomers as mentioned above. Where the resonance is simple and well separated from other resonances, interpretation of the changes that result from bridge modification is possible. The $^{13}C$ carbonyl resonances increase by about 3 ppm as a result of bridge-alkylation in the CMP series. For the bridge-alkylated CMPOs, two resonances in the carbonyl region are observed, again consistent with the presence of two sets of diastereomers. For the CMPO series, the increase in the carbonyl resonances is greater, around 7 ppm. Similar increases are observed in the $^{31}P\{^1H\}$ NMR spectra as a result of bridge alkylation. As with the carbonyl resonances in the $^{13}C$ NMR, for the CMPO series, bridge alkylation results in the presence of two separate signals. Two $^{31}P$ NMR signals were also observed in a previous study for the pentyl, bridge-alkylated, CMPO.[12] Although slight overlap prevented good integration of the two resonances, it appears that their relative intensities are in qualitative agreement with the GC integrations described above. In all cases, alkylation increases the chemical shift of the phosphorus signal: 2 to 3 ppm for the CMPO series, about 3 ppm for the DHDECMP series, and around 4 ppm for the DEDECMP case. The changes for the CMPO series are somewhat less than those reported for CMPO alkylated by a pentyl group; in that instance, an increase of 6 to 7 ppm was observed.[12] A previous study has shown a good correlation between the phosphoryl group's basicity (approximated by the extraction constant for nitric acid) and the $^{31}P$ NMR shift in $CHCl_3$.[23b]

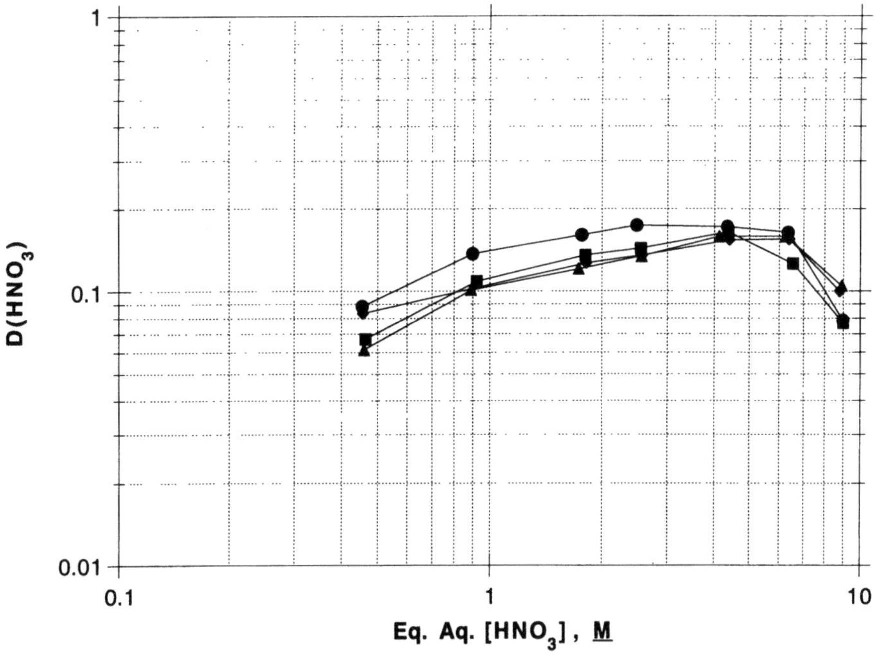


**Figure 5.** Plot of the $HNO_3$ distribution values for CMPs as a function of the equilibrium aqueous nitric acid concentration. Extractant solutions were 0.5M in tert-butyl benzene. Data for DHDECMP are shown as circles, data for o-DHDECMP as diamonds, data for i-DHDECMP as squares, and data for o-DEDECMP as triangles.

Therefore, this increase in the $^{31}P$ NMR shifts suggests an increase in the phosphoryl oxygen's basicity as a result of alkylation. However, the magnitude of the $^{31}P$ shifts is fairly small, and it is unlikely that the expected basicity difference could be discernable in a measured distribution constant for nitric acid.

### F) Distribution Measurements - $HNO_3$

Nitric acid distribution measurements for the DHDECMP, CMPO, and all bridge-modified derivatives were performed for 0.5M solutions of the extractant using tert-butyl benzene as the diluent. The results are shown in Figures 5 and 6, respectively. Tert-butyl benzene was chosen as the diluent to minimize potential third-phase formation.[31] As shown in Figures 5 and 6, bridge modification has essentially no impact on the extraction of nitric acid from 0.5M nitric acid to 10M initial aqueous nitric acid concentrations. The strong similarity of the shapes of these curves as a function of aqueous nitric acid concentration suggests that no changes in the nature of the nitric acid-extractant complex in the organic phase have occurred as a result of these bridge modifications. Furthermore, based on a comparison between o-DHDECMP and the octyl, bridge-alkylated, DEDECMP, o-DEDECMP, changing the alkoxy group on the phosphorus from ethyl to hexyl also has no influence on the magnitude of nitric acid extraction. A direct comparison of DHDECMP and DEDECMP was not possible due to the known water solubility of DEDECMP. The lack of change in distribution values for nitric acid, then, implies that the bridge modifications have

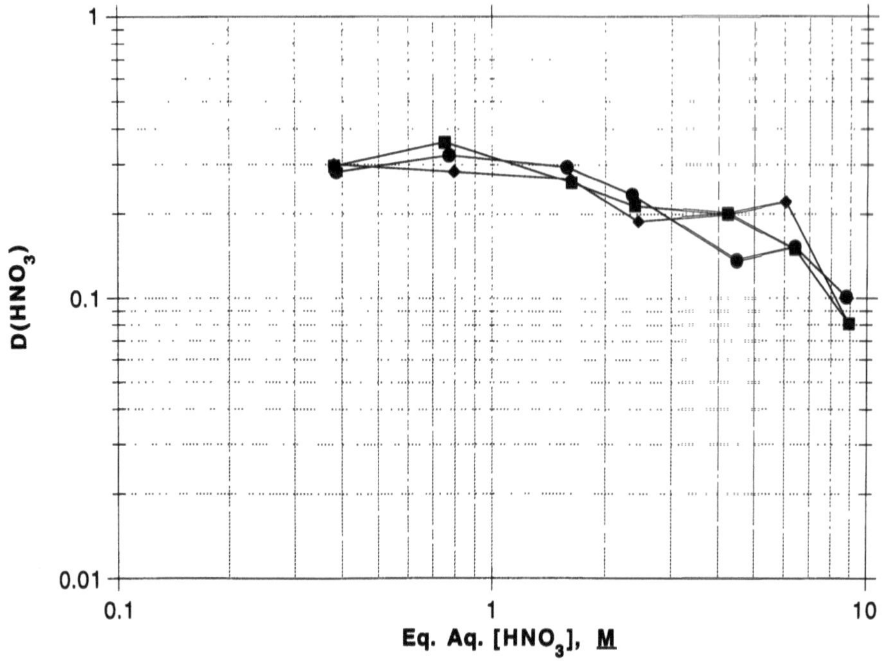


**Figure 6.** Plot of the $HNO_3$ distribution values for CMPOs as a function of the equilibrium aqueous nitric acid concentration. Extractant solutions were 0.5M in tert-butyl benzene. Data for CMPO are shown as circles, data for o-CMPO as diamonds, and data for i-CMPO as squares.

had no significant change on the phosphoryl group's basicity in these extractants (the phosphoryl group is generally accepted to be the site of binding to Lewis acids[5,7,23b,30a]). Based on the nitric acid distribution data, DHDECMP and DEDECMP appear to be of comparable basicity to each other as well as to their bridge-modified analogues. Furthermore, in the CMPO series, all of the bridge-modified compounds appear to be of comparable basicity to CMPO itself.

## G) Distribution Measurements - Am(III)

Figures 7 and 8 show the distribution of Am(III) from 0.5M solutions of CMPs and CMPOs, respectively, as a function of the equilibrium, aqueous, nitric acid concentration. The equilibrium, aqueous, nitric acid concentrations were obtained by measuring the aqueous solution directly following contact of the organic phase with a nitric acid solution. In all cases alkylation of the central methylene results in a marked decrease in the compound's ability to extract Am(III). The decrease from the parent extractants DHDECMP and CMPO is significant, often decreasing by an order of magnitude or more. However, the shapes of the curves as a function of nitric acid appear similar, which suggests that no change in extraction mechanism has taken place as a result of bridge modification of these organophosphorus extractants. The extractant's ability to remove Am(III) from acidic nitrate solutions decreases with the steric bulk of the alkyl group attached to the central methylene, as isobutyl shows a much greater effect than does n-octyl. A comparison of octyl-alkylated extractants o-DEDECMP and o-DHDECMP reveals a small but detectable increase in the ability of the

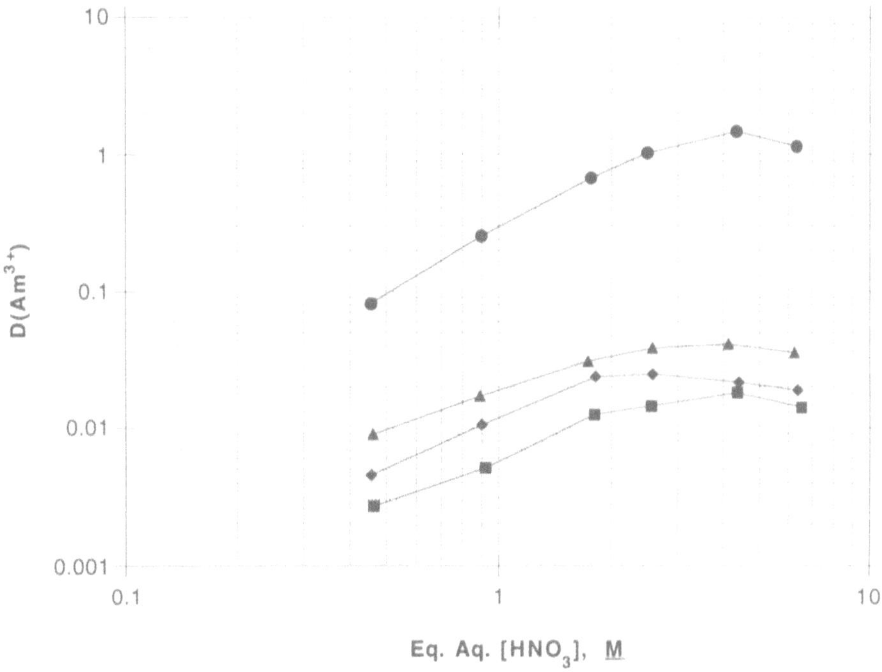


**Figure 7.** Plot of the $Am^{3+}$ distribution values for CMPs as a function of the equilibrium aqueous nitric acid concentration. Extractant solutions were 0.5M in tert-butyl benzene. Data for DHDECMP are shown as circles, data for o-DHDECMP as diamonds, data for i-DHDECMP as squares, and data for o-DEDECMP as triangles.

compound with the shorter alkyl group to extract Am. The shapes and magnitude of D(Am) for the unmodified extractants, DHDECMP and CMPO, as a function of nitric acid concentration, agree well with prior literature reports that used comparable extractant concentrations and diluents.[4,5,32-35]

**H) Distribution Measurements - Pu(IV)**

Figures 9 and 10 show the distribution of Pu(IV) from 0.1M solutions of CMPs and 0.025M solutions of CMPOs, respectively, as a function of the initial aqueous nitric acid concentration. Initial aqueous nitric acid concentrations were used in this instance since titration of nitric acid solutions showed an insignificant change in the aqueous nitric acid concentrations following contact with 0.1M CMP and 0.025M CMPO solutions. Use of 0.5M extractant solutions was initially attempted, but the measured activity in the aqueous solutions was very low, close to the background levels and lead to an undesirably high uncertainty in the distribution values. Therefore lower extractant concentrations were used, yielding more precise distribution values.

In the DHDECMP series, distribution values decreased as a result of bridge modification. The magnitude of these decreases is very much less than that of Am(III), with the greatest difference between DHDECMP and the isobutyl bridge-alkylated compound, i-DHDECMP. Here the difference in $D(Pu^{4+})$ is slightly greater than a factor of 2. As with the extraction of Am(III), the extractant is less able to remove $Pu^{4+}$ from acidic nitrate

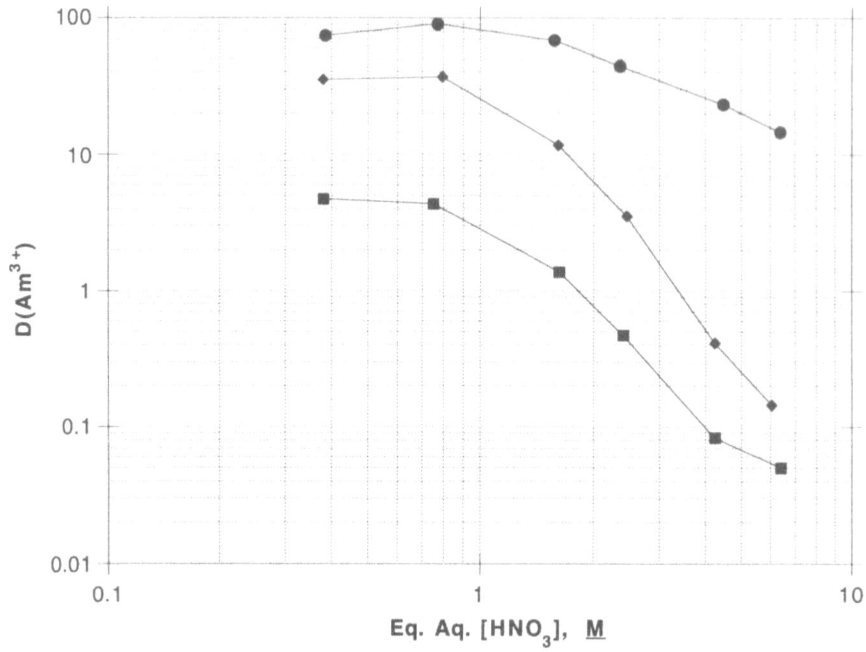


**Figure 8.** Plot of the $Am^{3+}$ distribution values for CMPOs as a function of the equilibrium aqueous nitric acid concentration. Extractant solutions were 0.5M in tert-butyl benzene. Data for CMPO are shown as circles, data for o-CMPO as diamonds, and data for i-CMPO as squares.

solutions as the steric bulk of the alkyl group attached to the central methylene increases, with isobutyl showing a greater effect than n-octyl. The shapes of the plots of $D(Pu^{4+})$ versus aqueous nitric acid concentration appear similar. A comparison of o-DEDECMP with o-DHDECMP shows a more marked flattening of the plot of $D(Pu^{4+})$ versus aqueous nitric acid concentration, so much so that, while $D(Pu^{4+})$ for o-DHDECMP is slightly greater at high nitric acid concentrations, $D(Pu^{4+})$ if o-DEDECMP is higher at low nitric acid concentrations, although the magnitude of these differences is small, less than a factor of 2.

The CMPO series behaves differently than with Am(III). A comparison of Figure 8 with Figure 10 shows that while $D(Am^{3+})$ is greatly changed by bridge modification, $D(Pu^{4+})$ is unaffected over the entire range of nitric acid concentrations examined. The similar shapes of the plots of distribution values versus nitric acid concentration indicate that no change in extraction mechanism takes place as a result of bridge modifications in the CMPO system. There is little literature concerning the extraction of $Pu^{4+}$ of well-purified DHDECMP or pure CMPO to compare with these results. Horwitz and coworkers have reported the extraction of $Th^{4+}$ with 0.05M DHDECMP in o-xylene as a function of nitric acid concentration: The shape and magnitude of the D versus nitric acid plot agrees with the results presented here.[36]

## I) Distribution Measurements - $UO_2^{2+}$

Figures 11 and 12 show the distribution of $UO_2^{2+}$ from 0.5M solutions of DHDECMPs and 0.025M solutions of CMPOs as a function of the equilibrium aqueous

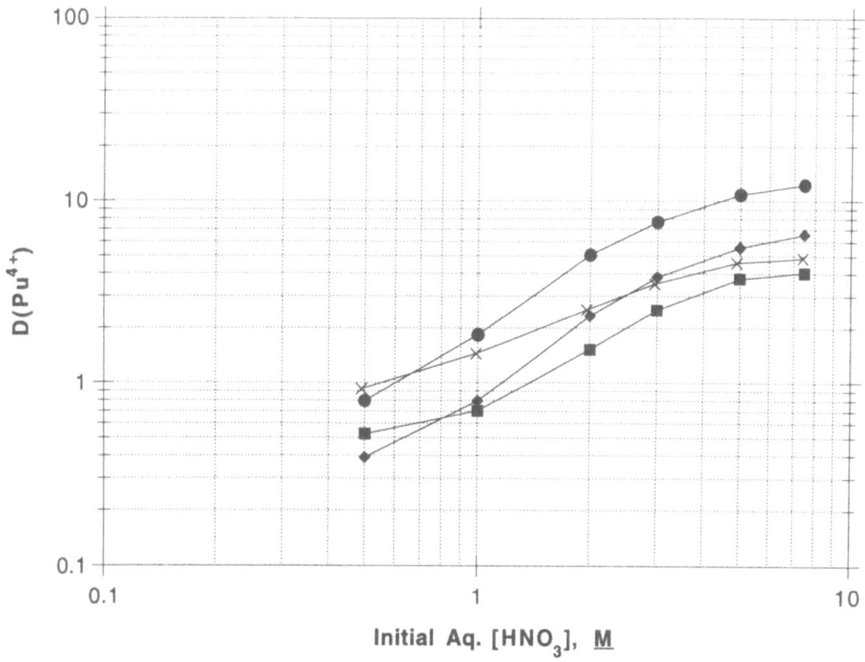


**Figure 9.** Plot of the $Pu^{4+}$ distribution values for CMPs as a function of the initial aqueous nitric acid concentration. Extractant solutions were 0.1M in tert-butyl benzene. Data for DHDECMP are shown as circles, data for o-DHDECMP as diamonds, data for i-DHDECMP as squares, and data for o-DEDECMP as crosses.

nitric acid concentration and initial aqueous nitric acid concentrations, respectively. As noted above, the low-extractant concentration for the CMPO system reduces the amount of nitric acid extracted so that the initial and equilibrium, aqueous, nitric acid concentrations differ little. In addition, the decreased uranyl ion distribution values allow for more precise measurements, although the large range and good sensitivity provided by laser fluorimetry reduces the need for such a requirement.

Little systematic change can be seen for either the DHDECMP or CMPO systems as a result of bridge modification. The plot shapes for $UO_2^{2+}$ distribution values versus nitric acid concentration are the same, which suggests that no change in extraction mechanism takes place as a result of these extractant modifications. The shapes and magnitude of $D(UO_2^{2+})$ for the unmodified extractant DHDECMP as a function of nitric acid concentration agrees with previous literature reports under comparable extractant concentrations and diluents.[34] Little equivalent data has been published concerning the extraction of $UO_2^{2+}$ from pure CMPO solutions.

## CONCLUSIONS

The purpose of this study was to examine, by systematic modifications of carbamoylmethyl phosphonate and phosphine oxide extractants, the effects of bridge

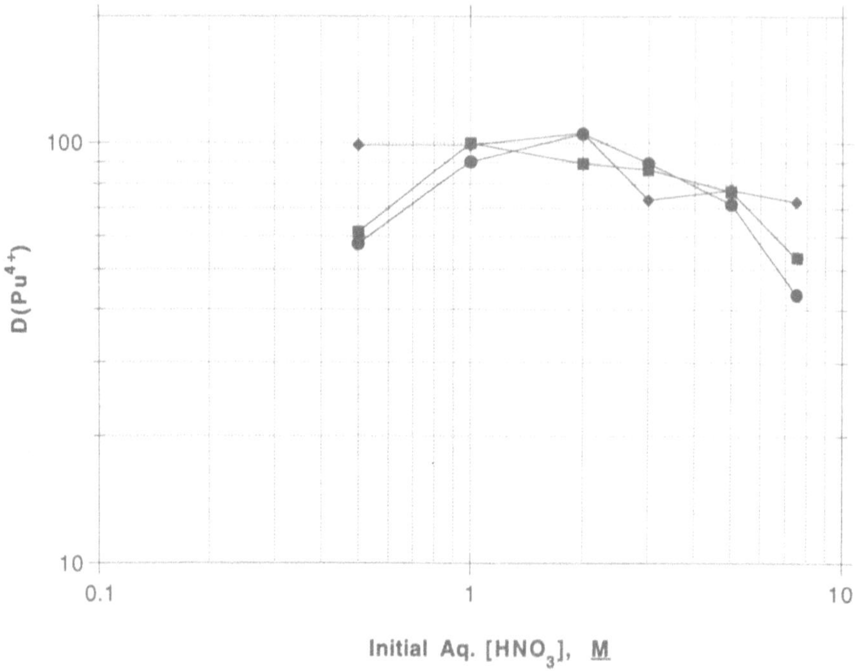


**Figure 10.** Plot of the $Pu^{4+}$ distribution values for CMPOs as a function of the equilibrium aqueous nitric acid concentration. Extractant solutions were 0.025M in tert-butyl benzene. Data for CMPO are shown as circles, data for o-CMPO as diamonds, and data for i-CMPO as squares.

alkylation on actinide extraction. Examination by IR and multinuclear NMR spectroscopies indicates that alkylation of the central methylene increases the basicity of the phosphoryl group slightly, if at all, but results in no increase in the basicity of the carbonyl. A more direct indicator of the phosphoryl group's basicity, the distribution value for nitric acid extraction, showed no significant change as a function of bridge alkylation and indicates that the electronic characteristics of the phosphoryl and carbonyl oxygens are unchanged as a result of these modifications. The extraction of the actinides Pu(IV), Am(III) and $UO_2^{2+}$ were examined under identical conditions for each of the bridge-modified extractants and their carbamoylmethyl-phosphonate or -phosphine oxide parent. The results varied and were different for each of the actinides examined. Americium(III) showed a marked decrease for both CMPs and CMPOs as a result of bridge alkylation, with the decrease in the measured distribution values often exceeding an order of magnitude. The magnitude of the effect strongly depended on the alkyl group involved; bridge-alkylation with an isobutyl group gave a much poorer extractant than did alkylation with n-octyl. In the CMP series, the length of the alkoxy group attached to the phosphorus also has an effect, with alkylated DEDECMP being a slightly better extractant than similarly alkylated DHDECMP.

These trends observed for Am(III) extraction are completely consistent with a simple correlation of the extractant effectiveness to the steric demands of the extractant molecule for the following reasons. First, an explanation based on differing electronic characteristics can be eliminated from the nitric acid extraction data. Second, for the CMP series, the smaller

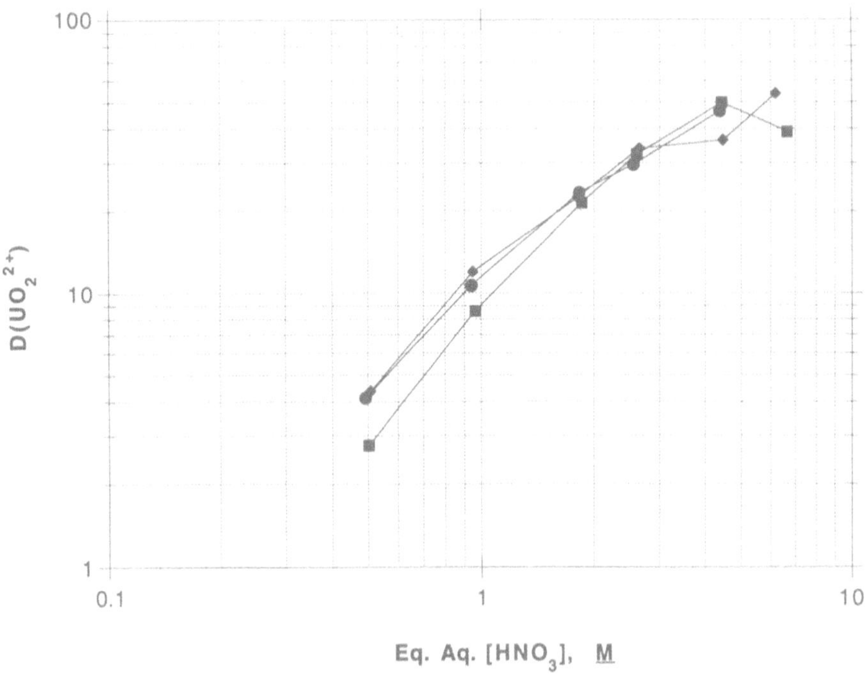


**Figure 11.** Plot of the $UO_2^{2+}$ distribution values for CMPs as a function of the initial aqueous nitric acid concentration. Extractant solutions were 0.5$\underline{M}$ in tert-butyl benzene. Data for DHDECMP are shown as circles, data for o-DHDECMP as diamonds, and data for i-DHDECMP as squares.

(EtO) was a slightly more effective extractant than (HxO), and third, attaching a primary alkyl group, n-octyl, gave a more effective extractant than attaching the more sterically demanding secondary alkyl group, isobutyl.

Extraction of Pu(IV) showed a difference in behavior between the CMPs and CMPOs. No change in distribution values as a function of nitric acid concentration was observed for bridge-alkylated CMPO versus CMPO itself. With the CMP series, the trends were essentially the same as those observed with Am(III), but the magnitude of the changes upon bridge alkylation was greatly decreased.

Extraction of uranyl ions showed a still different behavior. Modifying the extractant by bridge alkylation resulted in no change in the distribution values for $UO_2^{2+}$ for either the CMPs or CMPOs.

These differences in extractant behavior for the different actinides as a result of bridge alkylation can be explained considering the nature of the extracted species in the organic phase. With either DHDECMP or CMPO as the extractant (E), it has been well established that the nature of the extracted species in similar organic solvents is $Am(NO_3)_3{\cdot}3E$.[4,5,32,33,35,37,38,39] For uranyl ions, again it is known that for either DHDECMP or CMPO, the nature of the extracted species in organic solution is $UO_2(NO_3)_2{\cdot}2E$.[36,38,39] In the case Pu(IV), the situation is less clear. With CMPO as the extractant, the nature of the extracted species has been fairly well established as $Pu(NO_3)_4{\cdot}2(CMPO)$.[38,39] However, no information appears to exist on the nature of the

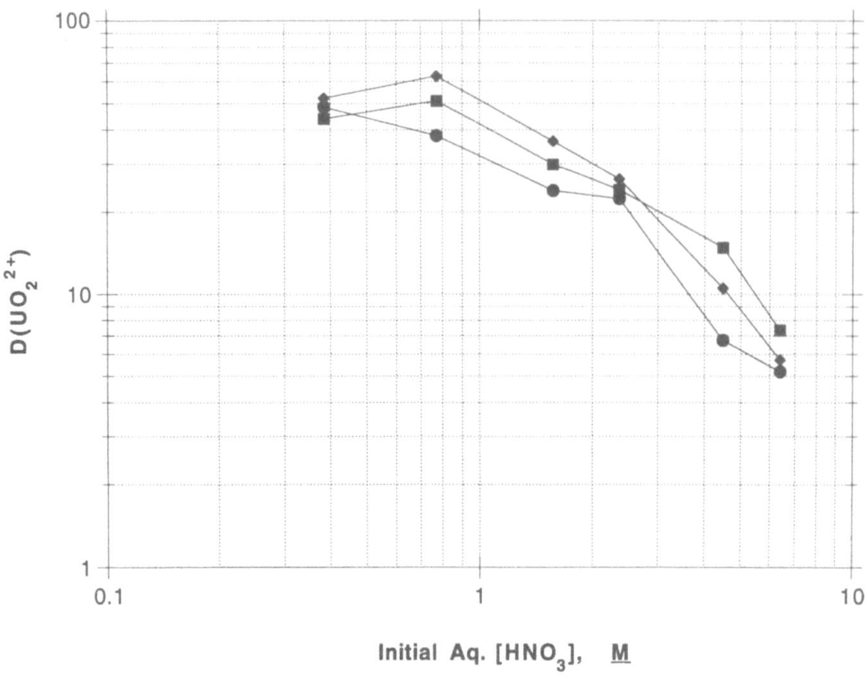


**Figure 12.** Plot of the $UO_2^{2+}$ distribution values for CMPOs as a function of the equilibrium aqueous nitric acid concentration. Extractant solutions were 0.025M in tert-butyl benzene. Data for CMPO are shown as circles, data for o-CMPO as diamonds, and data for i-CMPO as squares.

extracted species with well-purified DHDECMP as the extractant. However, previous research with Th(IV) indicates that the extracted species with DHDECMP is of a non-integral value, somewhat greater than 2 but less than 2.5.[36]

With the assumptions that the extracted species for Pu(IV) with DHDECMP is the same as that found for Th(IV) with DHDECMP and that the nature of the extracted species with actinides does not change from that observed for DHDECMP and CMPO upon bridge alkylation, then a steric-based explanation for the observed actinide extraction behavior is apparent: where the extractant to metal ratio in the extracted species is high, congestion around the metal center increases, and the steric demands of the extractant become more significant. For example, if the extracted species possesses a metal-to-extractant ratio of 3 (or more), the distribution values decrease strongly as the steric bulk of the extractant increases. This behavior is observed for Am(III) extraction with both CMPs and CMPOs. For extracted species with a metal to extractant ratio of 2 (or less), no dependence on the distribution values is observed as the steric demands of the extractant change. This behavior is observed for uranyl ion with both the DHDECMP- and CMPO-based extractants and with Pu(IV) for the CMPO based extractants. An intermediate case shows an intermediate effect. This behavior was observed with Pu(IV) for the CMP-based extractants.

To summarize, a series of extractants, modified from DHDECMP and CMPO by alkylation of the central methylene carbon, have been synthesized and characterized. The distribution of nitric acid, Am(III), Pu(IV) and $UO_2^{2+}$ have been measured as a function of nitric acid concentration. The changes in distribution values observed for these actinides can be explained based on the steric demands of the extractant and the ratio of extractant to metal in the extracted complex. However, extrapolation of the results to other systems should be done cautiously: other possible effects remain to be evaluated. For instance, understanding how these modified extractants interact with other diluents may prove important to fully integrate these results to previously studied systems.

## ACKNOWLEDGEMENTS

The author would like to thank Michael J. Wagner and Rebecca J. Barrington for technical assistance and Dr. Gregg J. Lumetta of Pacific Northwest Laboratory and Professor R. T. Paine of the University of New Mexico for their reviews of this manuscript. Pacific Northwest Laboratory is operated for the U.S. Department of Energy by Battelle Memorial Institute under Contract DE-AC06-76RLO 1830.

## REFERENCES

1) T.H. Siddall, *J. Inorg. Nucl. Chem.* 25:883-892 (1963).

2) T.H. Siddall, *J. Inorg. Nucl. Chem.* 26:1991-2003 (1964).

3) W.E. Steward and T.H. Siddall, *J. Inorg. Nucl. Chem.* 32:3599-3604 (1970).

4) D.G. Kalina, E.P. Horwitz, L.Kaplan, and A.C. Muscatello, *Sep. Sci. Technol.* 16:1127-1145 (1981).

5) E.P. Horwitz, D.G. Kalina, L.Kaplan, G.W. Mason, and H. Diamond, *Sep. Sci. Technol.* 17:1261-1279 (1982).

6) D.G. Kalina and E.P. Horwitz, *Solv. Extr. Ion Exch.* 3:235-250 (1985).

7) E.P. Horwitz, K.A. Martin, H. Diamond, and L. Kaplan, *Solv. Extr. Ion Exch.* 4:449-494 (1986).

8) R.C. Gatrone, M.L. Dietz, and E.P. Horwtiz, *Solv. Extr. Ion Exch.* 11:411-422 (1993).

9) E.P. Horwitz, D.G. Kalina, H. Diamond, G.F. Vandergrift, and W.W. Schultz *Solv. Extr. Ion Exch.* 3:75-109 (1985).

10) W.W. Schulz and E.P. Horwitz, *Sep. Sci. Technol.* 23:1191-1210 (1988).

11) S.L. Blaha, L.J. Caudle, G.C. Conary, and R.T. Paine, *Radiochimica Acta* 48:59-64 (1989).

12) R.C. Gatrone, E.P. Horwitz, P.G. Rickert, and H. Diamond, *Solv. Extr. Ion Exch.* 7:793-811 (1989).

13) N.E. Kochetkova, O.E. Koiro, N.P. Nesterova, T.Ya. Medved, M.K. Chmutova, B.F. Myasoedov, and M.I. Kabachnik, *Radiokhimiya* 28:338-345 (1986).

14) B.F. Myasoedov, M.K. Chmutova, N.E. Kochetkova, O.E. Koiro, G.A. Pribylova, N.P. Nesterova, T.Ya. Medved, and M.I. Kabachnik, *Solv. Extr. Ion Exch.* 4:61-81 (1986).

15) A.Yu. Shadrin, I.V. Smirnov, R.N. Kiseleva, N.P. Nesterova, Yu.M. Polikarpov, and M.I. Kabachnik, *Radiokhimiya* 35:50-56 (1993).

16) C. Musikas, *Sep. Sci. Technol.* 23:1211-1226 (1988).
17) C. Cuillerdier, C. Musikas, P. Hoel, L. Nigond, and X. Vitart, *Sep. Sci. Technol.* 26:1229-1244 (1991).
18) C. Cuillerdier, C. Musikas, and L. Nigond, *Sep. Sci. Technol.* 28:155-175 (1993).
19) L. Nigond, C. Musikas, and C. Cuillerdier, *Solv. Extr. Ion Exch.* 12:297-323 (1994).
20) B.M. Rapko and G.J. Lumetta, *Solv. Extr. Ion Exch.* in press.
21) G.J. Lumetta, B.M. Rapko, M.J. Wagner, C.D. Carlson, and R.J. Barrington, *Sludge Treatment and Extraction Technology Development: Results of FY 1993 Studies*, PNL-9387, Pacific Northwest Laboratory, Richland, Washington (March 1994).
22) G.J. Lumetta, *Pretreatment of Neutralized Cladding Removal Waste Sludge: Results of the Second Design Basis Experiment*, PNL-9747, Pacific Northwest Laboratory, Richland, Washington (May 1994).
23a) DHDECMP at δ 23.6. S.M. Bowen, E.N. Duesler, R.T. Paine, and C.F. Campana, *Inorg.Chim. Acta* 59:53-63 (1982). b) CMPO at δ 37.67. R.C. Gatrone and P.G. Rickert, *Solv. Extr. Ion Exch.* 5:1117-1139 (1987).
24) The distribution value (D) is defined as the concentration in the organic phase divided by the concentration in the aqueous phase.
25a) J.P Shulka, A. Kumar, and R.K. Singh, *Radiochimica Acta* 60:103-107 (1993). b) S.A. Pai, J.P. Shulka, and M.S. Subramanian, *J. Inorg. Nucl. Chem.* 43:1045-1047 (1981) and reference 10 therein.
26) G.S. Conary, D.J. McCabe, R.L. Meline, E.N. Duesler, and R.T. Paine, *Inorg. Chim. Acta* 203:11-19 (1993) and references therein.
27) R.C. Gatrone, P.G. Rickert, E.P. Horwitz, B.F. Smith, C.S. Bartholdi, and A.M. Martinez, *J. Chrom.* 516:395-404 (1990).
28) For example, mixed systems containing tributyl phosphate and CMPO demonstrate linear behavior with analogous detectors.[27]
29) For DHDECMP as a KBr pellet,[23a] the carbonyl band is found at 1641 $cm^{-1}$ and the phosphoryl band at 1262 $cm^{-1}$. For DEDECMP as a KBr pellet,[23a] the carbonyl band is found at 1651 $cm^{-1}$ and the phosphoryl band at 1264 $cm^{-1}$. For DHDECMP in tetrachloroethylene,[30a] the carbonyl band is found at 1657 $cm^{-1}$ and the phosphoryl band at 1254 $cm^{-1}$. For CMPO in solution,[30b] the carbonyl bands range from 1634 to 1643 $cm^{-1}$ and the phosphoryl bands from 1202 to 1209 $cm^{-1}$.
30a) K.A. Martin, E.P Horwitz and J.R. Ferraro, *Solv. Extr. Ion Exch.* 4:1149-1169 (1986). b) E.P. Horwitz, K.A. Martin, and H. Diamond, *Solv. Extr. Ion Exch.* 6:859-888 (1988).
31) W.W. Schulz and J.D. Navratil, *Solvent Extraction with Neutral Bidentate Organophosphorus Reagents*, in "CRC Recent Developments in Separation Science, Vol. VII", N.N. Li, ed., CRC Press, Boca Raton, FL (1982).
32) E.P Horwitz and D.G. Kalina, *Solv. Extr. Ion Exch.* 2:179-200 (1984).
33) J.N. Mathur, M.S. Murali, and P.R. Natarajan, *J. Radioanal. Chem.* 152:127-135 (1991).

34) J.N. Mathur, M.S. Murali, and P.R. Natarajan, *J. Radioanal. Chem.* 162:171-178 (1992).
35) E.P. Horwitz, A.C. Muscatello, D.G. Kalina, and L. Kaplan, *Sep. Sci. Technol.* 16:417-437 (1981).
36) E.P. Horwitz, D.G. Kalina, and A.C. Muscatello, *Sep. Sci. Technol.* 16:403-416 (1981).
37) B.J. Mincher, *Solv. Extr. Ion Exch.* 10:615-622 (1992).
38) J.N. Mathur, M.S. Murali, P.R. Natarajan, L.P. Badheka, and A. Banerji, *Talanta* 39:493-496 (1992).
39) M.L. Dietz and E.P. Horwitz, *LC-GC* 11:424-436 (1993).

# NEW WATER-SOLUBLE PHOSPHONATE AND POLYCARBOXYLATE COMPLEXANTS FOR ENHANCED f ELEMENT SEPARATIONS*

Kenneth L. Nash, Paul G. Rickert, Erik P. Lessmann**,
Michael D. Mendoza**, Janine F. Feil**, and James C. Sullivan

Chemistry Division
Argonne National Laboratory
9700 S. Cass Avenue
Argonne, IL 60439-4831

## ABSTRACT

While lipophilic extractant molecules and ion exchange polymeric materials are clearly essential to efficient separation of metal ions by solvent extraction or ion exchange, the most difficult separations often could not be accomplished without the use of water soluble complexants. This report focuses on recent developments in the design, synthesis and characterization of phosphonic acid and polycarboxylic acid ligands for enhanced f element separations. Emphasis is on the basic solution chemistry and crystal structures of complexes of the f elements with selected amino-derivatives of methanediphosphonic acid and with tetrahydrofuran-2,3,4,5-tetracarboxylic acid. The former series of compounds exhibit high affinity for lanthanides and actinides in acidic solutions. The latter ligand exhibits an unusual (and very useful) "anti-selectivity" for uranyl ion in a solvent extraction process, which permits efficient separation of uranyl from more radioactive components of nuclear wastes. Most of the observed effects can be explained through examination of the structure of the ligand, and comparison of the spectroscopic and thermodynamic parameters for complexation of various metal ions.

* Work performed under the auspices of the Office of Basic Energy Sciences, Division of Chemical Sciences, U. S. Department of Energy, under contract number W-31-109-ENG-38 and the Westinghouse Hanford Company

** Student Research Participant

## INTRODUCTION

It is appropriate to begin the discussion of aqueous complexation of f elements in separations with a brief summation of the relevant chemical parameters which lead to differences in the chemical behavior of these important metal ions in separations processes. Both series are characterized by valence electrons which occupy f orbitals: lanthanides the 4f, actinides the 5f. The 4f orbitals in the lanthanides are relatively deeply submerged and do not participate directly in bonding under normal conditions. The 5f orbitals of the actinides exhibit a slightly greater spatial extension, and some "hybridization" involving these orbitals is indicated for the light members of the series. This is particularly evidenced by the existence of the upper oxidation states of uranium, neptunium, plutonium, and americium as linear dioxocations ($AnO_2^+$, $AnO_2^{2+}$) in solution (1). The three principal chemical characteristics critical to their separations chemistry are the multiplicity of oxidation states of the light actinides, the steadily decreasing radii of the trivalent cations, and the slightly stronger interaction of actinides with soft-donor ligands.

The light actinides (U to Am) exist in aqueous solutions in the III, IV, V, and VI oxidation states. The multiplicity of oxidation states of the light actinides can be utilized to accomplish very efficient separation of these elements, both from the lanthanides and from each other. Except for actinium (only trivalent), the actinide ions through plutonium either exist predominantly in higher oxidation states (Th(IV), Pa(IV,V)) or can be interconverted with relative ease among any of four oxidation states (III, IV, V, VI). The relative strength of complexes formed by the actinide cations in these oxidation states is An(IV) > An(VI) > An(III) > An(V). This order also applies to most separations reactions involving these cations. The tri-, tetra-, and hexavalent oxidation states are most important in separations.

The separation of individual members of the trivalent lanthanide and transplutonium actinide series is based on the decrease in ionic radius which occurs with increasing atomic number (2). This characteristic is only useful for intragroup separations as the cation radii for the trivalent actinides overlap those of the lanthanide cations. Trivalent lanthanide/actinide cation radii decrease across the series because the valence f electrons compensate relatively poorly for the steadily increasing nuclear charge. Decreasing radii implies that the heavy members of both series should bind solute (and suitable solvent) molecules more tightly than the light members.

Numerous reports in both the separations literature and the classical inorganic chemistry of actinides indicate that actinides ions interact more strongly with soft donor ligands, particularly for the lighter members of the series (3). There is no clear agreement on whether this difference is attributable to a covalent contribution to bonding or a more subtle polarizability effect, particularly in solution. What is clear is that extractants and water-soluble complexants containing "soft" donor atoms are central to most ion exchange and solvent extraction separations of lanthanides from trivalent actinides.

Actinide separation processes based on exploiting differences in cation oxidation states typically achieve the separation based on a difference in interaction strength of the cation with a lipophilic complexant or resin. For the separation of trivalent ions, however, aqueous complexants are often critical, both for group (lanthanide-actinide) and intragroup (lanthanide-lanthanide or actinide-actinide) separations. Two examples of efficient group separations based on aqueous complexants are chloride based anion exchange (4) and the TALSPEAK process (5). Each process depends on the stronger interaction of trivalent actinides with soft donor ligands (Cl- in anion exchange, DTPA in TALSPEAK). The feature example of aqueous complexants for intragroup separations is α-hydroxy-isobutyric acid cation exchange separation of lanthanides or trivalent actinides (6).

Aside from these very specific examples of the use of aqueous complexants in f element separations, efficient complexing agents are often used in processing by solvent extraction or ion exchange to improve the efficiency of (for example) stripping of used process solvent (or resin) to allow its recycle. In this case, cation selectivity is often secondary in importance to complex strength. Occasionally, a stripping agent may be employed to partition actinides based on differences in complexation strength of actinide ions in different oxidation states (7).

A variety of different types of ligands are available when a complexant is required for improved process efficiency. The five general classes typically considered for f elements are: inorganic acids, polycarboxylic acids, aminopolycarboxylic acids, macrocycles, and phosphonic acids. The use of each has associated advantages and drawbacks. Anions of inorganic acids are inexpensive, but typically corrosive and with few exceptions ($PO_4^{3-}$ and $CO_3^{2-}$) form weak complexes. Polycarboxylic acids are probably the most studied of all complexants for f elements. Their f element complexes exhibit a wide variety of properties, but most do not form strong complexes below pH 3-4. Aminopolycarboxylic acids (e.g., EDTA) do form very strong complexes with f elements at pH 2 and above. However, they are poorly soluble at lower pH and are very resistant to degradation. The latter characteristic has been associated with a variety of safety and contamination problems (8). Macrocyclic complexants like crown ethers have not been extensively studied with f elements. They may be designed to exhibit good selectivity and would probably be generally benign with respect to waste problems, but are almost without exception moderately expensive. Phosphonic acids derivatives of methanediphosphonic acid, in particular, are unique in their ability to strongly complex f elements in acidic solutions (pH $<$ 2, including molar acid) (9-11). Recent research has also demonstrated that certain derivatives of this general ligand type are readily decomposable (7). The principal drawback to their application is the need to dispose of a waste solution containing phosphate, which is incompatible with present strategies for the disposal of high level radioactive wastes in glass waste forms.

Our research program has focussed on the chemistry of coordination complexes of f elements with derivatives of methanediphosphonic acid (I, 7, 9-11) and with tetrahydrofuran-2,3,4,5-tetracarboxylic acid (THFTCA, II, 12,13). These complexing agents are seen as very efficient general stripping agents,

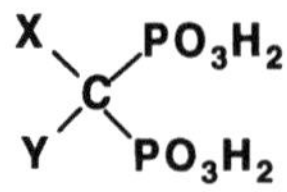


I

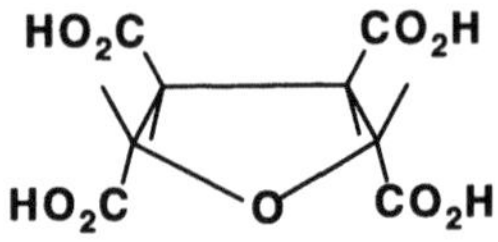


II

but also exhibit a variety of properties which could be exploited to perform unique separations, or improve upon reagents which are now used for such separations. We have found in our studies to date that in acidic media f element complexes with simple diphosphonic acid ligands are 2 to 6 orders of magnitude more stable than carboxylic acid ligands of comparable acidity (14). The dominant complexed species in solution are typically protonated (e.g., $MH_2L$, $MH_4L_2$). For thorium, results indicate that stable complexes may be formed without deprotonation of the phosphonate groups (10). Thermodynamic results and laser-induced fluorescence decay studies of europium complexes strongly suggest intramolecular hydrogen bonding contributes to the unusual stability of the f element complexes in acid solutions (15). Studies of the kinetics of uranyl complexation by simple diphosphonates indicate that differences in complex stability are derived from different complexation rates (16). For polycarboxylic and amino-polycarboxylic acid complexes of f elements, formation rates are generally diffusion controlled. Stability differences are manifested in relative complex dissociation rates (17).

Our earlier investigation of the stability of europium complexes with THFTCA suggested the importance of ligand backbone rigidity in the determination of relative complex stability (12). THFTCA was reported to form complexes five times stronger than the analog complexant diglycolic acid in dilute acid solution. More stable complexes are observed at lower pH partly because of the lower pK's of THFTCA and partly as a result of rigidity imparted to the oxydiacetate coordination zone by the tetrahydrofuran backbone. This ligand exhibits excellent solubility of both the ligand and its lanthanide complexes from pH 2 to 6.

The multiple purposes of the present report are to update our investigation of simple methanediphosphonic acid derivatives with additional insights, to report results on some ongoing investigations of amino-derivatives of methanediphosphonic acid, and to explore in greater detail the chemistry of THFTCA, its unique properties in actinide separations, and its f element complexes. The results of these investigations are reported within the context of potential applications of these chelating agents for enhanced f element separations.

## EXPERIMENTAL

The results described in this paper are derived from a variety of different experiments, some of which either have been or will soon be published independently elsewhere. The experimental methods will be described to the appropriate degree of detail in this section.

### Synthesis of Amino Methane/ethane Diphosphonic Acids

The aminomethane diphosphonate ligands in Table 1 were synthesized from the appropriate amides or nitriles using the approach of Rusinia (18) as adapted from Ploger et al (19). Aminomethanediphosphonic acid (AMDPA), dimethylaminomethanediphosphonic acid (DMAMDPA), morpholinomethanediphosphonic acid (MMDPA), and piperidinylmethanediphosphonic acid (PMDPA) were prepared by dropwise addition of the parent amide to a solution of $PCl_3 + H_3PO_3$. The basic reaction is described in equation 1:

$$PCl_3 + H_2PO_3 + R^1R^2NCH(O) \rightarrow R^1R^2CH(PO_3H_2)_2 \ . \qquad (1)$$

Chloride was hydrolyzed with ice water, and the liberated HCl removed by vacuum distillation. The respective acids were precipitated with methanol.

The 1-aminoethane-1,1-diphosphonic acid (AEDPA) was prepared by dropwise addition of acetonitrile to $PCl_3$. The mixture was stirred overnight. Glacial acetic acid was added to this solution and again stirred overnight. The top layer containing acetic acid was removed leaving a thick oil. Water was slowly added to the oil to hydrolyze chloride. Water and HCl were removed by vacuum distillation and the AEDPA was precipitated from 6 M HCl by adding ethanol.

The product aminophosphonic acids were characterized by melting point, $^{31}P$ and $^{1}H$ NMR spectroscopy, and potentiometric titration. The parent amides were selected to provide a variety of different geometries for the amine and simultaneously a range of basicities (Table 1). Previous reports have appeared in the literature for each of these ligands, most detailing the protonation equilibria or stability of transition metal complexes (20).

### Purification of THFTCA

THFTCA was further purified from the previous report by dissolution of the solid (purchased from Aldrich), filtration to remove insoluble residues, an acetone wash, and recrystallization from water. In contrast to earlier attempts, large, well-defined single crystals were precipitated from concentrated aqueous solutions. The crystals were characterized by potentiometric titration. The isomer was established as the trans, cis,trans isomer (see below) by single crystal x-ray diffraction courtesy of Professor Robin Rogers at Northern Illinois University.

## Potentiometric Titration of Ligands and Complexes

Potentiometric titrations of both free ligand solutions and metal-ligand mixtures were done using a Mettler DL21 titrator interfaced to a 386SX computer with titrator control provided by Mettler's $TS^2$ software package. All titrations were run under nitrogen atmosphere at 25.0 (±0.1) °C and constant ionic strength (0.1 or 0.2 M), maintained with $NaClO_4$. Fresh titrant either purchased from Aldrich or prepared in our laboratory from 99.999% pure solid NaOH were used for titrations. Titrant solutions were standardized by titration of primary standard (NIST certified) potassium hydrogen phthalate. An Orion-Ross micro combination electrode was used as the pH probe. The electrode/measurement system was calibrated daily by titration of $HClO_4$/$NaClO_4$ solutions to generate a pH-p[H] calibration curve which was used to convert measured pH to p[H]. The accuracy of volume addition of the automatic burette was determined gravimetrically, and a quadratic correction applied to the indicated volume of titrant delivered. Metal ion complexation reactions were investigated for $La^{3+}$, $Eu^{3+}$, and $Tm^{3+}$ in the aminodiphosphonic acid systems, and for $La^{3+}$, $Nd^{3+}$, $Eu^{3+}$, $Dy^{3+}$, and $Tm^{3+}$ in the THFTCA system. Potentiometric titration data for 1:1 and 1:2 metal:ligand solutions with the same standardized base used for the free ligand titrations was analyzed using the program BEST (21). Because of persistent problems with solubility for the lanthanide complexes with the other aminodiphosphonate ligands, we were able to determine metal complex stability constants only the MMDPA and PMDPA systems.

## Distribution of THFTCA into Combined Process Solvent

The distribution of THFTCA between various aqueous solutions and the Combined Process solvent (0.2 M di(t-butylcyclohexano)18-crown-6, 0.2 M octyl(phenyl)-N,N-diisobutylcarbamoylmethylphosphine oxide, 1.2 M diamyl(amyl)phosphonate, in Isopar L) was investigated by potentiometric titration and NMR spectroscopy. Solutions of THFTCA at several different concentrations were prepared and standardized by potentiometric titration using the DL21 titrator. The consecutive protonation constants of THFTCA are well enough separated to permit (via Gran analysis of the titration curves) the observation of individual endpoints for the second, third, and fourth ionizations under most conditions. It is also possible in this analysis to determine $[THFTCA]_t$ and total acidity from a single titration . Titrimetric analysis was performed before and after equilibration of the THFTCA solution with the Combined Process solvent. To asssess the contributions of the crown ether and DA(A)P to THFTCA extraction, parallel experiments were run with 0.2 M CMPO, 1.2 M DA(A)P in Isopar L, and with 1.2 M DA(A)P in Isopar L. Phases were equilibrated overnight with rotator mixing, separated by centrifugation, and the aqueous analyzed titrimetrically. In all cases, phase separation was clean and mass balance was assumed to calculate distribution ratios. In one experiment, mass balance was checked by extensive water washing of the organic phase and found to be greater than 95%. The relative importance of interaction of THFTCA with CMPO and DA(A)P was determined by $^{31}P$ NMR spectroscopy of the loaded organic phases.

## RESULTS AND DISCUSSION

### Lanthanide Diphosphonate Complexes

Earlier results from this laboratory indicate that europium complexes ($MH_2L^+$) with 1-hydroxyethane-1,1-diphosphonic acid (HEDPA) are 3 times stronger than the equivalent complex with ethane-1,1-diphosphonic acid (9). HEDPA complexes are formed with more exothermic heats and smaller positive entropies than those with methanediphosphonic acid or vinylidene-1,1-diphosphonic acid (which have no α-hydroxy group) (16). This difference has not been specifically attributed to an interaction between the metal ion and the α-hydroxy group, but it is tempting to do so. We have examined the inner coordination sphere of lanthanide complexes in the solid state and in aqueous solution to provide new insight into this question

A crystal structure of the $Nd(H)(H_2HEDP)_2 \cdot 6H_2O$ was recently determined for us by Professor Robin Rogers at Northern Illinois University. The complex crystallizes in the P1 point group with cell constants 9.318, 10.272, 11.766, 91.09, 111.24, 98.40. The crystal contains two formula units per unit cell. The resolved structure (Figure 1) indicates the existence of two bridging and two terminal (chelated) HEDP ligands per unit cell. The inner coordination sphere of Nd is populated exclusively by phosphonate oxygens in an 8-coordinate geometry. The six water molecules and the free hydronium ions are in inter layer spaces remote from the metal ion. This structure is consistent with one reported earlier for the analogous erbium complex (22) indicating no significant relationship between metal complex geometry and the radius of the lanthanide cation. The positioning of the α-hydroxy group suggests that it is not directly coordinated with the metal ion.

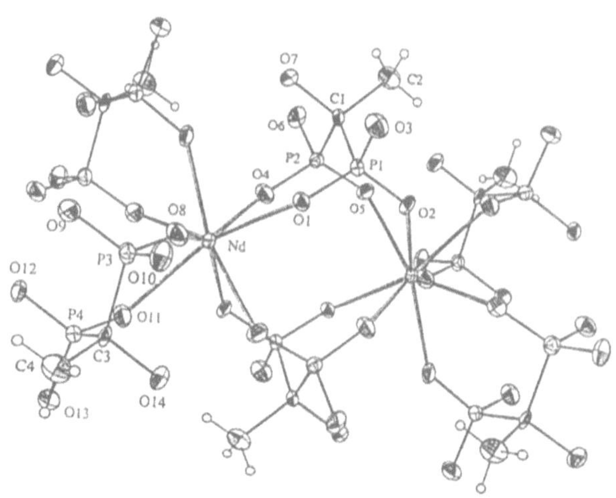


**Figure 1.** Crystal structure of $Nd(H)(H_2HEDP)_2 \cdot 6H_2O$.

In contrast, laser induced fluorescence studies of the 1:2 europium complexes ($Eu(H_3L)_2^+$) with diphosphonic acids indicate that in acidic aqueous solutions 3 water molecules are present in the inner coordination sphere of the metal ion (16). Based on a combination of metal complexation enthalpies, the fluorescence results, and molecular modeling calculations, the probable structure for this complex is as shown in Figure 2. Extensive intracomplex hydrogen bonding is suggested to explain the very low complexation entropies observed for the formation of a 1:2 complex despite the high degree of dehydration of the metal ion. The difference between the inner coordination sphere in the solid and solution states illustrates the danger of extrapolating structural data obtained from

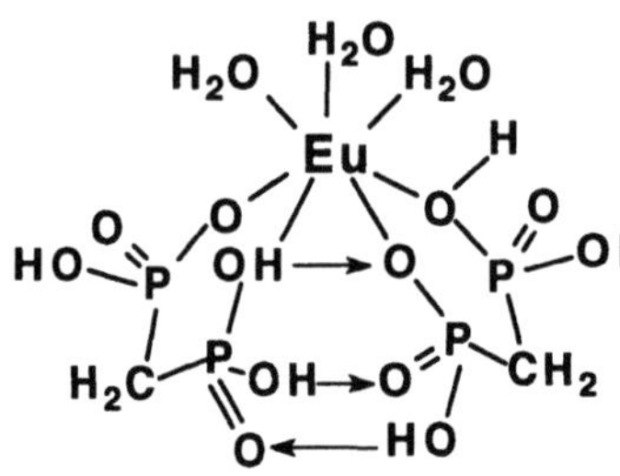


**Figure 2.** Proposed structure of $Eu(H_3MDP)_2^+$ in aqueous solution.

solid state crystal structure determination with solution structure. No consistent difference in cation hydration is indicated between the HEDPA and MDPA complexes, again suggesting that the α-hydroxy group is not coordinated directly with the metal ion.

## Protonation and Complexation Equilibria for Aminomethanediphosphonic Acids

For the parent methanediphosphonic acid, the protonation constants (Table 1) are $pK_1 = 1.51$, $pK_2 = 2.60$, $pK_3 = 6.85$, and $pK_4 = 10.05$. Addition of an amino functionality to methanediphosphonic acid fundamentally alters the pattern of protonation constants. In solution, the strongly basic amino group binds a hydrogen ion leaving the formally neutral ligand as a zwitterion. The presence of a formal positive charge on the nitrogen in $H_4L$ reduces the affinity of the phosphonate oxygens for hydrogen ion and produces the pattern of constants $pK_1 = 1.5$, $pK_2 = 4.7\text{-}5.4$, $pK_3 = 8.0\text{-}8.6$, and $pK_4 = 11.5\text{-}12.5$ (or more). Accurate determination of both $pK_1$ and $pK_4$ is difficult by direct potentiometry because the fourth proton associates with the neutral molecule very weakly and the first associates with the amino nitrogen extremely strongly. Except for the morpholino derivative, the relative order of the protonation constants for the amino nitrogens correlates with the basicity of the corresponding free amines. The amines in the diphosphonates bind $H^+$ about 100 times stronger than the free amine, reflecting the relative basicity of primary, secondary and tertiary amines.

**Table 1.** Protonation constants for aminodiphosphonic acids at 0.2 M ionic strength, 25.0 °C.

| Compound | $pK_1$ | $pK_2$ | $pK_3$ | $pK_4$ | Parent Amine |
|---|---|---|---|---|---|
| MDPA[a] | 1.51 | 2.60 | 6.85 | 10.05 | NA |
| AMDPA[b] | (1.4) | 5.30 | 8.07 | 11.47 | 9.25 ($NH_3$) |
| DMAMDPA[c] | (0.4) | 4.84 | 8.59 | (12.6) | 10.73 $(CH_3)_2NH$ |
| AEDPA[d] | 1.46 | 5.35 | 8.57 | 11.72 | 9.25 ($NH_3$) |
| MMDPA[e] | (0.8) | 4.68 | 8.35 | 12.21 | 8.33 (morpholine) |
| PMDPA[f] | 1.35 | 4.72 | 8.57 | (12.6) | 11.12 (piperidine) |

[a]Methanediphosphonic acid, FW = 176.02 g (I=0.1 M).

[b]Aminomethanediphosphonic acid, FW = 190.01 g.

[c]Dimethylaminomethanediphosphonic acid, FW = 219.07 g.

[d]1-Aminoethane-1,1-diphosphonic acid, FW = 204.04 g.

[e]Morpholinomethanediphosphonic acid, FW = 261.11 g.

[f]Piperidinylmethanediphosphonic acid, FW = 259.14 g.

**Table 2.** Lanthanide complex stoichiometries and stability constants for complexes with morpholinomethanediphosphonic acid (MMDPA) and piperidinylmethanediphosphonic acid (PMDPA) at 0.2 M ionic strength and 25.0 °C.

| Stoichiometry | $La^{3+}$ MMDP | $Eu^{3+}$ MMDP | $Tm^{3+}$ MMDP | $La^{3+}$ PMDP | $Eu^{3+}$ PMDP | $Tm^{3+}$ PMDP |
|---|---|---|---|---|---|---|
| $MH_2L^+$ | 26.0 | 28.4 | 27.6 | 27.5 | 27.5 | 26.7 |
| MHL | 23.6 | 25.5 | 24.9 | 24.4 | 24.4 | 24.3 |
| $ML^-$ | 15.0 | 16.7 | 16.0 | 14.0 | 14.4 | 14.7 |
| $M(OH)L^{2-}$ | 4.2 | 5.3 | 5.8 | — | — | 3.3 |
| $MH_3L_2^{2-}$ | — | — | — | 48.4 | — | — |
| $MH_2L_2^{3-}$ | — | 41.6 | 41.2 | 41.4 | 42.2 | 24.0 |
| $MHL^2$ | — | — | 31.8 | — | — | — |
| $ML_2^{5-}$ | — | 21.2 | — | — | 19.6 | 21.3 |

Figure 3 is a sample titration curve for the potentiometric analysis of the protonation of PMDPA and a 1:1 solution of PMDPA with $Eu^{3+}$. The free ligand titration curve shows the three distinct breaks expected for the well-separated protonation constants presented in Table 1. Resolved metal complexation equilibrium constants (I=0.1 M, T=25.0 °C) for $La^{3+}$, $Eu^{3+}$, and $Tm^{3+}$ are shown in Table 2. Stability constants are resolved for the 1:1 complexes in each system, but only a few parameters corresponding to the 1:2 complexes are determined. The apparent exclusion of the 1:2 complexes may be a result of the relative bulk of the ligand as compared with the parent MDPA, where several protonated complexes having both 1:1 and 1:2 metal:ligand stoichiometry are known (9). It is somewhat surprising that only the lanthanide complexes with the largest amino substituents are sufficiently soluble to permit analysis by potentiometry. Perhaps it is the bulk of the amine which interferes with formation of stable solid species. Solubility problems for DMAMDPA complexes with $Nd^{3+}$ and $Gd^{3+}$ were suggested in a previous report (23).

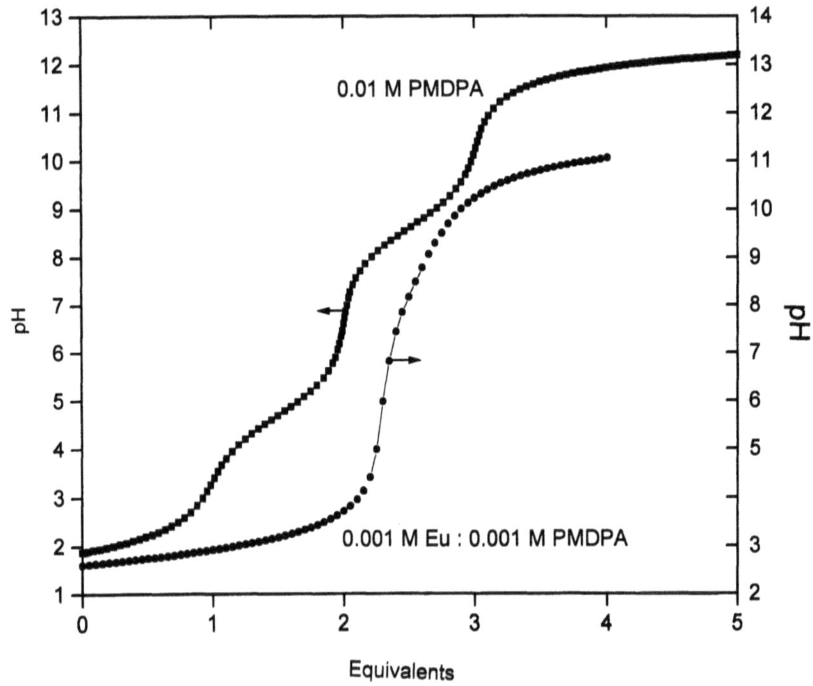


**Figure 3.** Titration results for 0.01 M piperidinylmethanediphosphonic acid (PMDPA) and 0.001 M $Eu^{3+}$, 0.001 M PMDPA at 25.0 °C, I = 0.2 M.

## Tetrahydrofuran-2,3,4,5-tetracarboxylic Acid Crystal Structure

Crystallographic analysis of duplicate samples of replicate preparations of single crystal samples of THFTCA were identified as the trans,cis,trans isomer, as shown in Figure 4. This crystallographic analysis also was conducted by Professor Rogers. THFTCA crystallizes in the P1 space group with indices 7.594, 8.137, 9.404, 107.08, 102.48, 101.30. There was no evidence in either sample for the presence of any other geometric isomer of THFTCA. This structure is consistent with previous reports of crystal structures of THFTCA complexes with $Cs^+$, $Ca^{2+}$, (24) and $Co(en)_3^{3+}$ (25). This isomer was previously indicated as one of the most probable species based on molecular mechanics calculations (13).

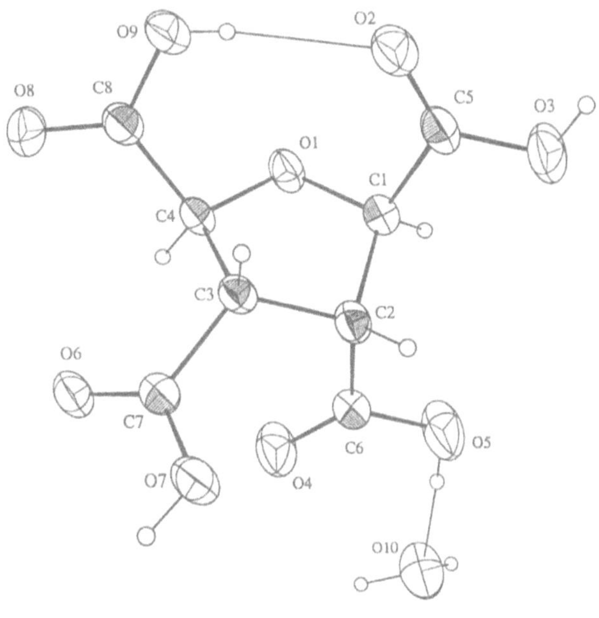


**Figure 4.** Crystal structure of THFTCA.

## THFTCA as a Stripping Agent in the Combined Process

The Combined Process is a solvent extraction based process designed to simultaneously remove actinides, lanthanides, and strontium from radioactive wastes (26). Solvent extraction processes for transuranic elements (TRUEX, 27) and strontium (SREX, 28) combined in a single process solvent permits the simultaneous removal of these radioactive metals from the bulk of the less-radioactive or nonradioactive components. The radioactive elements can then be selectively stripped from the loaded extractant phase through a combination of pH adjustments and selective chelating agents. Ongoing experiments indicate that THFTCA selectively strips trivalent and tetravalent actinides from the loaded solvent leaving uranyl and strontium in the extractant phase (29). Selective separation of uranyl from the trivalent and tetravalent actinides is important as the volume of wastes requiring disposal in a geologic repository could be significantly reduced if uranium is removed from other TRU's.

However, polycarboxylic acids are known to distribute significantly into extractant solutions containing octyl(phenyl)-N,N-diisobutylcarbamoylmethylphosphine oxide (CMPO) (30). Design of a process to use THFTCA for selective separation of uranium from the transuranium components of the waste must take into account the tendency of THFTCA to distribute to the CMPO phase. Successful design of THFTCA into the Combined Process requires that the distribution of THFTCA between the phases be known accurately under a variety of conditions and preferably reduced.

The results in Figure 5 indicate first order dependence of the distribution ratio of THFTCA into PS12 solvent. Saturation of the extractant is apparently reached at about 0.5 M $[THFTCA]_t$. The distribution at low concentrations of THFTCA clearly favors the organic phase. The $^{31}P$ chemical shifts for CMPO in these solutions are shown in Figure 6.

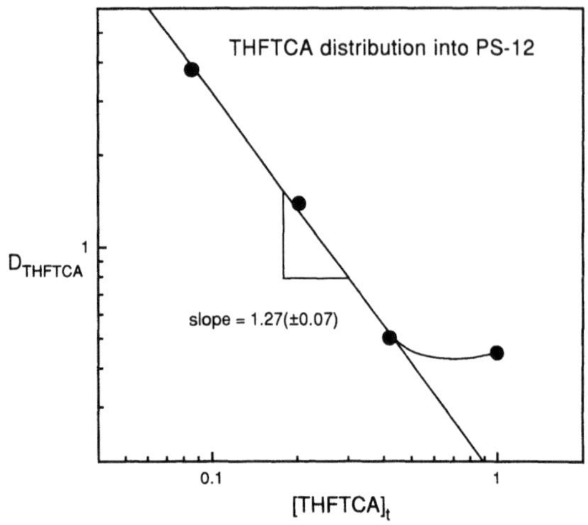


**Figure 5.** Distribution of tetrahydrofuran-2,3,4,5-tetracarboxylic acid (THFTCA) into PS12 Combined Process Solvent (0.2 M di(t-butylcyclohexano)18-crown-6, 0.2 M octyl(phenyl)-N,N-diisobutylcarbamoylmethylphosphine oxide, 1.2 M diamyl(amyl)phosphonate, in Isopar L).

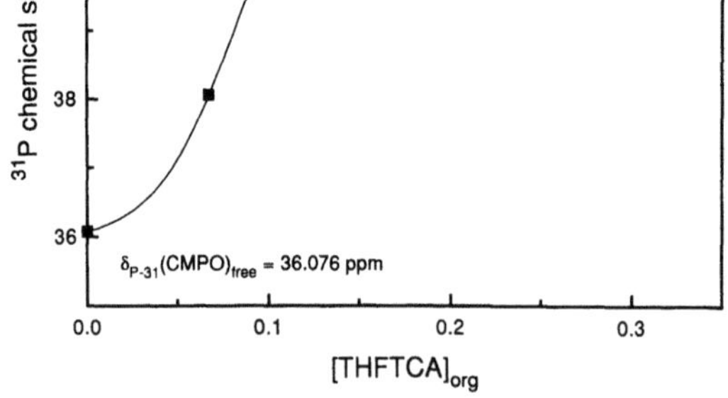


**Figure 6.** $^{31}P$ Chemical shift for CMPO in PS12 Combined Process Solvent as a function of the [THFTCA] extracted into the organic phase.

A significant downfield shift of nearly 7 ppm is indicated with saturation indicated at 1:1 stoichiometry (0.2 M). Excess THFTCA above 0.2 M is probably weakly associated with the phase modifier DA(A)P (small shift in $^{31}P$ resonance), but association of THFTCA with the crown ether cannot be ruled out. Complementary experiments run with 0.2 M CMPO, 1.2 M DA(A)P in Isopar L, and with 1.2 M DA(A)P in Isopar L suggest some contribution of each extractant to THFTCA extraction into the Combined Process Solvent. Addition of nitric acid to the aqueous solution effects the observed D's unpredictably. More recent work (in progress) indicates that the mono- and disodium salts of THFTCA are less extracted. In the pH 2.5 regime ($Na_{1.5}H_{2.5}THFTC$), THFTCA remains primarily in the aqueous phase while trivalent and tetravalent actinides are selectively stripped (leaving uranyl in the organic phase for subsequent isolation) (29).

## THFTCA Complexes with Lanthanides

If consistent trends are observed for a series of lanthanides, separations procedures analogous to α-hydroxyisobutyric acid are feasible. The relationship between metal ion separation factors and the stability of water soluble metal complexes is illustrated in equation 1.

$$S_B{}^A = \frac{D^A}{D^B} = \frac{D_O{}^A(1+\Sigma\beta_1{}^B(L)^\ell)}{D_O{}^B(1+\Sigma\beta_1{}^A(L)^\ell)} \qquad (2)$$

Note that while the separation factor ($S_B{}^A$) is directly proportional to the relative distribution coefficients of the metal in the absence of aqueous complexants, it is inversely dependent on the relative strength of their aqueous complexes. In the α-hydroxyisobutyric acid system, $D_A = D_B$ (Dowex 50 exhibits little inherent selectivity for individual lanthanides). Lanthanide separations in this system are based primarily on the relative strength of the lanthanide (or trivalent actinides) complexes with α-hydroxyisobutyric acid.

To assess whether THFTCA might improve the performance of the α-hydroxyisobutyric acid system, we determined the stoichiometry and stability of La, Nd, Eu, Dy, and Tm complexes. These metals were selected to represent a spread of light and heavy lanthanides. The resolved constants are given in Table 3. Most of the possible protonated complexes from $MH_2L$ to $ML_2$ are indicated for each metal ion. Hydrolysis is also indicated for the complexes at pH > 8. Solubility of the complexes was generally good from pH 2 to 10 (in the 0.001 to 0.01 M concentration range). Calculated speciation for DyTHFTCA at 0.001 M $Dy^{3+}$, 0.002 M THFTCA is shown in Figure 7.

For a viable separation scheme, equation 2 indicates the desirability of consistent trends in the stability constants of the complexes with members of the series. In Figure 8, the $\beta_{101}$ values for the lanthanide THFTCA complexes are compared with literature data for structurally similar ligands. Note that lanthanide complexes with α-hydroxyisobutyric acid

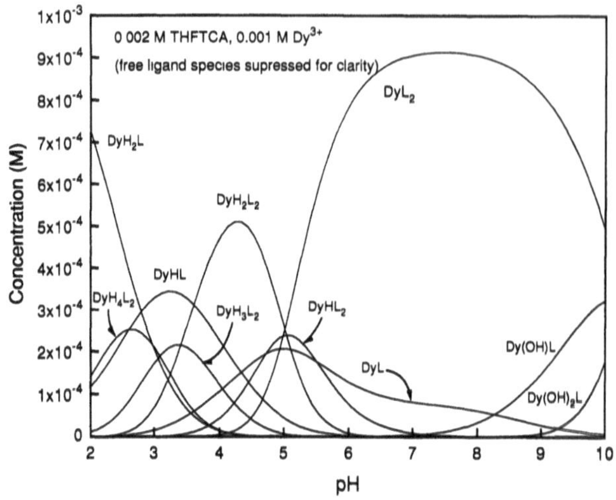


**Figure 7.** Calculated species distribution of $Dy^{3+}$-THFTCA at 0.001 M $[Dy^{3+}]_t$, 0.002 M $[THFTCA]_t$-from pH 2 to 10, T = 25.0 °C, I = 0.1 M.

**Table 3.** Selected lanthanide stability constants (log $\beta_{mhl}$) with THFTCA at 0.1 M ionic strength, 25.0 °C.

| Stoichiometry | $La^{3+}$ | $Nd^{3+}$ | $Eu^{3+}$ | $Dy^{3+}$ | $Tm^{3+}$ |
|---|---|---|---|---|---|
| $MH_2L^+$, log $\beta_{121}$ | 15.33 | 16.16 | 16.74 | 17.59 | 17.38 |
| MHL, log $\beta_{111}$ | 12.05 | 13.38 | 14.15 | 14.79 | 14.78 |
| $ML^-$, log $\beta_{101}$ | 8.45 | 9.47 | 10.22 | 10.46 | 10.37 |
| $M(OH)L^{-2}$, log $\beta_{1-11}$ | -1.97 | 1.00 | 2.10 | 2.19 | 2.62 |
| $M(OH)L^{-3}$, log $\beta_{1-21}$ | — | -9.76 | -8.34 | -8.07 | -7.12 |
| $MH_4L_2^-$, log $\beta_{142}$ | 30.02 | 32.04 | 31.80 | 32.20 | — |
| $MH_3L_2^{-2}$, log $\beta_{132}$ | 26.73 | 28.70 | 28.56 | 29.11 | — |
| $MH_2L_2^{-3}$, log $\beta_{122}$ | — | 25.36 | 25.44 | 25.70 | 24.91 |
| $MHL_2^{-4}$, log $\beta_{112}$ | — | 20.10 | 20.47 | 20.66 | 20.00 |
| $ML_2^{-5}$, log $\beta_{102}$ | 11.65 | 15.61 | 15.98 | 15.17 | 14.93 |

exhibit a consistent trend with decreasing cation radii, but the slope is relatively low. The slope is higher at higher concentrations due to the presence of 1:2 and 1:3 complexes. Separation factors for adjacent lanthanides using α-hydroxyisobutyric acid as the eluant

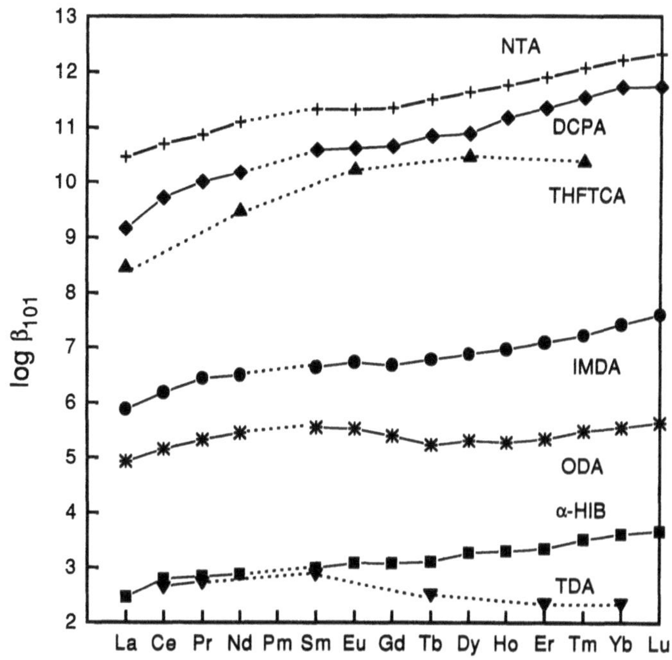


**Figure 8.** Stability constants for lanthanide complexes with thiodiacetic acid (TDA), α-hydroxyisobutyric acid (α-HIB), oxydiacetic acid (ODA), iminodiacetic acid (IMDA), tetrahydrofuran-2,3,4,5-tetracarboxylic acid (THFTCA) 2,6-dicarboxypiperidine-N-acetic acid (DCPA), and nitrilotriacetic acid (NTA). All data except the THFTCA taken from reference 31.

average 1.6 (±0.3). It is somewhat surprising that, despite the steady decrease of lanthanide cation radii across the series, the linear trend exhibited by the α-hydroxyisobutyrate complexes is not more generally observed. This apparent inconsistency is a manifestation of the complexity of overlapping effects which contribute to metal complex stability.

In the first half of the lanthanide series, a much steeper slope is observed for lanthanide-THFTCA complexes than for the 1:1 α-hydroxyisobutyric acid system. Similar relationships are noted for the various protonated complexes. The difference between oxydiacetic acid and THFTCA is particularly striking, as it indicates the impact of ligand preorganization on thermodynamic stability. The structural rigidity effect (largely entropy derived) is further supported by a comparison of DCPA with NTA (DCPA is NTA with a piperidine ring constraining the ligand binding groups). The slope for lanthanide stability constants with DCPA is measurably steeper than that for NTA, particularly for the lighter members of the series.We are presently examining the separation potential of this system for both lanthanides and trivalent actinides.

## CONCLUSION

Two classes of ligands which form strong complexes with f elements have been examined for their potential use in f element separations. Aminodiphosphonic acids form strong complexes with trivalent lanthanide cations, but it appears unlikely that the $\alpha$-amino group can be directly coordinated with the metal ion. Two derivatives, morpholinomethanediphosphonic acid (MMDPA) and piperidinylmethanediphosphonic acid (PMDPA), form complexes with $La^{3+}$, $Eu^{3+}$, and $Tm^{3+}$ which are soluble (at millimolar concentrations) from pH 2 to 10. However, thermodynamic data indicate little difference in complex stability across the lanthanide series. These ligands may have potential as general stripping agents as a result of the enhanced solubility of the complexes, but do not appear to have any significant potential for separation of individual members of either the lanthanide or actinide series.

The second category of f element complexants examined is a polycarboxylic acid, tetrahydrofuran-2,3,4,5-tetracarboxylic acid. Like the aminodiphosphonic acids, this complexant also promotes lanthanide solubility over a substantial range of pH. Investigations of the basic thermodynamics of lanthanide complexation suggests the potential for THFTCA as a viable reagent for enhanced separation of individual lanthanides. Investigations of its performance in a combined TRUEX-SREX solvent extraction process for treatment of radioactive wastes suggest that THFTCA may be a key reagent for selective partitioning of uranyl from transuranium actinides. This is an important and desirable objective as the low specific activity of natural uranium allows its disposal in relatively inexpensive near-surface disposal sites rather than a deep geological repository. Work continues on both classes of reagents to assess their usefulness in specific separations applications and the science underlying their behavior.

## REFERENCES


1. S.P. McGlynn and J.K. Smith, *J. Molecular Spectroscopy* 6:164 (1961).
2. R.D. Shannon, *Acta Cryst.* A32:751 (1976).
3. K.L. Nash, *Solv. Extr. Ion Exch.* 11:729 (1993).
4. K. Street, Jr. and G.T. Seaborg, *J. Am. Chem. Soc.* 72:2790 (1950).
5. B. Weaver and F.A. Kappelmann, *J. Inorg. Nucl. Chem.* 30:263 (1968).
6. G.R. Choppin and R.J. Silva, *J. Inorg. Nucl. Chem.* 3:153 (1956).
7. K.L. Nash and P.G. Rickert, *Sep. Science Techn.* 28:25 (1993).
8. C. Delegard, "Laboratory Studies of Complexed Waste Slurry Volume Growth in Tank 241-SY-101" Rockwell Hanford Informational Report RHO-LD-124 (1980).
9. K.L. Nash and E.P. Horwitz, *Inorg. Chim. Acta* 169:245 (1990).
10. K.L. Nash, *Radiochim. Acta* 54:171 (1991).
11. K.L. Nash, *Radiochim. Acta* 61:147 (1993).

12. K.L. Nash, E.P. Horwitz, R.C. Gatrone and P.G. Rickert, *J. Alloys Cpds.* 180:375 (1992).
13. J.F. Feil, R.D. Rogers and K.L. Nash, "Stoichiometry and stability constants of complexes of selected lanthanide with tetrahydrofuran-2,3,4,5-tetracarboxylic acid" submitted to *Inorg. Chim. Acta* (1994).
14. K.L. Nash, "Actinide Phosphonate Complexes in Aqueous Solutions" in *Proceedings of Actinides 93 International Conference*, D. L. Clark, D. E. Horbart, J. Fuger, Eds., Elsevier Science Publishers, (1994) pp 300-304.
15. K.L. Nash, L.-F. Rao, and G.R. Choppin, "A Calorimetric and LaserInduced Fluorescence Investigation of the Complexation Geometry of Selected Europium-(gem)diphos-phonate Complexes in Acidic Solutions," *Inorg. Chem.*, (1994) in press.
16. M.A. Hines, J.C. Sullivan, and K.L. Nash, *Inorg. Chem.* 32:1820 (1993) .
17. S.F. Lincoln, "Solvation, solvent exchange, and ligand substitution reactions of trivalent lanthanide ions" in *Advances in Inorganic and Bioinorganic Reaction Mechanisms;* Sykes, A. G., Ed.; Academic Press: London, (1986) pp 217-287.
18. M.N. Rusinia, T.M. Balashova, B.V. Zhadanov, A. Yu. Tsitrina, and I.A. Polyakova, *Zh. Obshch. Khim.* 47:1721 (1977).
19. W. Ploger, N. Schindler, K. Wollmann, and K.H. Worms, *Z. Anorg. Allg. Chem.* 389:119 (1972).
20. E.N. Rizkalla, *Reviews in Inorg. Chem.* 5:223 (1983).
21. A, E. Martell and R.J. Motekaitis, *Determination and Use of Stability Constants*, VCH Publishers, New York (1988).
22. L.M. Shkol'nikova, A.A. Masyuk, G.V. Polyanchuk, E.G. Afonin, A.L. Poznyak, and V.E. Zavodnik, *Koord. Khim.* 15:1424 (1989).
23. J.E. Bollinger and D.M. Roundhill, *Inorg. Chem.* 32:2821 (1993).
24. J.C. Barnes and J.D. Paton, *Acta Cryst.* C40:1809 (1984).
25. J.C. Barnes and J.D. Paton, *Acta Cryst.* B38:1588 (1982).
26. E.P. Horwitz, M.L. Dietz, H. Diamond, R.D. Rogers, and R. A. Leonard, *Combined TRU-Sr Extraction/Recovery Process* in Solvent Extraction in the Process Industries, Volume 3, Proceedings of ISEC 93, D. H. Logsdail, M.J. Slater Eds. Elsevier, London, pp 1805-1812.
27. W.W. Schulz and E.P. Horwitz, *Sep. Sci. Technol.* 23:1191 (1988).
28. E.P. Horwitz, M.L. Dietz, and D.E. Fisher, *Solv. Extr. Ion Exch.* 9:1 (1991).
29. E.P. Horwitz, H. Diamond, Personal Communication, 1994.
30. G.A. Clark, R.C. Gatrone, and E.P. Horwitz, *Solv. Extr. Ion Exch.* 5:471 (1987).
31. NIST Standard Reference Database 46, Critical Stability Constants of Metal Complexes Database Version 1.0, R. M. Smith, A. E. Martell.

# SEPARATION OF AMERICIUM FROM EUROPIUM BY SOLVENT EXTRACTION FROM AQUEOUS PHOSPHONATE MEDIA

Dale D. Ensor

Chemistry Department
Tennessee Technological Universitiy
Cookeville, TN 38505

Kenneth L. Nash

Chemistry Division
Argonne National Laboratory
Argonne, IL 60439

## ABSTRACT

The stoichiometry and stability constants for americium and europium complexes with phosphonoacetic acid ($HO_2CCH_2PO_3H_2$ - PAA) in dilute acid solutions (0.005 to 0.02 M) have been determined by distribution methods at 0.5 M ionic strength and 25.0°C. The identified Am and Eu complexes and log $\beta_{mhl}$ values are $Am(H_2L)^{2+}$ - 13.82(±0.04), $Am(HL)^{+}$ - 11.62(±0.04), $Am(H_2L)_2^{+}$ - 27.35(±0.01), $Eu(HL)^{+}$ - 12.10(±0.05), $Eu(H_2L)_2^{+}$ - 27.81(±0.01), $Eu(HL)_2^{-}$ - 23.08(±0.05). The Eu complexes are stronger than Am complexes of comparable stoichiometry, in accord with the electrostatic nature of the metal-ligand interaction. PAA was tested as a potential agent of enhanced Am/Eu separation in solvent extraction systems by examining Am and Eu extraction using different extractants. For extraction into bis(2-ethylhexyl)phosphoric acid (HDEHP), PAA slightly reduces the Am/Eu separation factor. With dinonylnapthalenesulfonic acid (HDNNS), or octyl(phenyl)-N,N-diisobutyl-carbamoylmethylphosphine oxide (CMPO)-nitrate extraction systems, $S_{Eu}^{Am}$ is slightly enhanced by PAA.

*Separations off Elements*, Edited by K.L. Nash
and G.R. Choppin, Plenum Press, New York, 1995

## INTRODUCTION

The intra- and intergroup separation of trivalent actinide and lanthanide elements remains one of the most challenging and important problems in separation science. These elements play a significant role in the development of methods for cleanup and decommissioning of nuclear facilities and the long-term storage of high-level waste. Advanced concepts for actinide burnup either in reactors or by accelerator transmutation demand efficient methods for separation of lanthanides from actinides. A recent review article on trivalent f-element separations (1) noted that most of the successful group separations were based on systems combining lipophilic extractants with aqueous complexing reagents.

Both polycarboxylic and aminocarboxylic acids have been used in f element separations. Recent work at Argonne has focussed on the thermodynamics of lanthanide and actinide complexes with phosphonic acid complexants. One of the objectives of this work has been to assess the potential of these reagents for use in lanthanide/actinide separations. Stability constants for Eu(III) (2), Th(IV) (3), and U(VI) (4) complexes with a wide variety of substituted diphosphonic acids have been measured. These reagents show a significant increase in complexing strength over their carboxylate analogs. The relative stability of the metal complexes is consistent with an electrostatic bonding model for these elements. It has been demonstrated that substituted methanediphosphonic acids are particularly efficient for stripping radioactivity from loaded organic phases in solvent extraction (5).

Little data is available on the complexation of the trivalent actinide ions with these phosphonate ligands. Elesin et al (6) report stability constants for the $M(H_2L)^{2+}$, $MHL^+$, and $M(HL)_2^-$ complexes of Pm(III), Am(III), and Cm(III) with phosphonoacetic acid (PAA). Their reported constants exhibit little difference in complex stability between the three metals. However, aspects of their methodology cast doubt on the absolute value of the constants.

In this work we report the stoichiometry and stability constants for Eu(III) and Am(III) complexes with PAA. Subsequently, the ability of PAA to enhance group separation of lanthanides and actinides was investigated by determining the distribution of Am(III) and Eu(III) from aqueous solutions of PAA into extractant solutions containing an acidic extractant (bis(2-ethylhexyl) phosphoric acid - HDEHP), a micellar extractant (dinonylnaphthalene sulfonic acid - HDNNS), and a neutral bifunctional extractant (octyl(phenyl)-N,N-diisobutylcarbamoylmethylphosphine oxide - CMPO) from nitrate media.

## EXPERIMENTAL

### Reagents

PAA, 99%, was obtained from the Aldrich Chemical Company and used without further purification. The extractants HDEHP, HDNNS, and CMPO were purified before use by well-known methods taken from the literature. Reagent grade toluene was used to prepare most of the extractant solutions. Diethylbenzene was the diluent for the CMPO solutions. All other reagents used were analytical reagent grade or better.

**Procedures**

**Potentiometry.** Stock solutions of PAA were prepared and standardized by potentiometric titration using NaOH. Ionic strength was controlled at 0.5 M using $NaNO_3$ as the supporting electrolyte. A Ross combination electrode was used as the pH probe. The electrode was calibrated by titrating 0.01 M $HNO_3$, 0.49 M $NaNO_3$ with 0.10 M NaOH, 0.40 M $NaNO_3$. The data provided a pH-p[H] calibration curve for conversion of the measured hydrogen ion activity to hydrogen ion concentration. Titrimetric analyses were performed using a Mettler DL21 titrator with associated $TS^2$ software for data acquisition. Gran plots were made to determine titration endpoints. Protonation constants for PAA were calculated as reported previously (2).

**Distribution Measurements.** The $^{241}$Am and $^{152\text{-}154}$Eu stock solutions were prepared from lab stocks by repeated evaporations to dryness in concentrated $HNO_3$ followed by final dissolution in 0.01M $HNO_3$, 0.49 M $NaNO_3$. Radiotracers were assayed using Packard Cobra Auto Gamma counter. The experiments for determination of Am-PAA stability constants were run in dual-label experiments with both $^{241}$Am and $^{152\text{-}4}$Eu present simultaneously using HDEHP as the extractant. The Am(III) counting channel covered the 60 kev $\gamma$-ray emission. The Eu channel included the Eu $\gamma$-rays between 100-700 kev. Each isotope's count rate was corrected for spill-over from the other based on prepared standards. Because the $D_0$(Eu) values were too high in the dual label experiments for accurate stability constant determination, Eu-PAA stability constants were determined in single label experiments. Separation factors with HDNNS and CMPO extraction systems were measured in dual-labeled experiments.

All distribution experiments were done at 25°C and 0.50 M ionic strength (adjusted with $HNO_3/NaNO_3$). For the determination of stability constants, PAA concentrations were varied between 0 and 0.40 M. This procedure was repeated at $[H^+] = 0.005$, 0.01, and 0.02 M. Under each set of experimental conditions at least two replicate experiments were performed with freshly prepared solutions. To maximize precision in the resolved stability constants, the concentration of HDEHP was varied with acidity to maintain the measured $D_0$ ([PAA]=0) between 5 and 20. All PAA solutions were prepared by volumetric dilution of a stock solution which had been standardized by potentiometric titration. Eu/Am separation factors with different extractants were measured at PAA concentrations between 0 and 0.34 M. The aqueous phase ionic strength was 0.5 M ($HNO_3/NaNO_3$) in the HDNNS experiments while $[NaNO_3]$ was maintained at 0.5 M in the CMPO experiments.

## RESULTS AND DISCUSSION

### Ligand Protonation and Metal Complex Stability

The protonation constants of PAA were calculated using the program PKAS (7). The results of the potentiometric titrations of free PAA shown in Table 1 are the average of five determinations at various PAA concentrations. Earlier reports at 2 M ionic strength (3) and infinite dilution (8) are consistent with the present results. The first and third protonation reactions describe association of hydrogen ions with the phosphonate group, the second hydrogen ion associates with the carboxylate. This sequence has been previously confirmed by Heubel and Popov (8) based on $^{31}P$ and $^{13}C$ NMR spectroscopy. Comparison of the acidity of the $-CO_2^-$ and $-PO_3^=$ groups in PAA with the pK's of analogous monoacids suggests that each group has minimal effect on the relative basicity of the other or perhaps that the sharing of hydrogen ions by the two functional groups offsets the inductive effect.

**Table 1.** Protonation Constants for PAA and related ligands at 25°C.

| Ligand | $pK_1$ | $pK_2$ | $pK_3$ | Conditions | Reference |
|---|---|---|---|---|---|
| PAA | 1.12 (0.15) | 4.66 (0.02) | 7.69 (0.02) | I=0.5 M, $NaNO_3$ | present work |
| PAA | 0.92 (0.10) | 4.57 (0.01) | 7.23 (0.01) | I=2.0 M $NaClO_4$ | 3 |
| PAA | 2.0 | 5.11 | 8.69 | I=0 | 8 |
| $CH_3CH_2PO_3^{2-}$ | 2.29 | 7.79 | | I=0.1 M | 9 |
| $CH_3CH_2CO_2^-$ | 4.69 | | | I=0.1 M | 9 |

To calculate the stability constants for the Am and Eu PAA complexes, it is first necessary to correct the distribution data for the presence of nitrate complexes. The stability constant for $EuNO_3^{2+}$ at 0.5 M ionic strength ($\log \beta_{EuNO_3} = 0.44$) was taken from Martell (9). Appropriate stability constants for $AmNO_3^{2+}$ ($\log \beta_{AmNO_3} = 0.38$) were extrapolated from existing data at higher ionic strength by comparing the Am ß's with more extensive data on $EuNO_3^{2+}$. The distribution data as a function of [PAA] and $[H^+]$ were reduced to the form $D_o/D-1-\beta_{MNO_3}[NO_3^-]$. Averaged distribution data used to determine stoichiometry and stability of the Am-PAA and Eu-PAA complexes are shown in Figures 1 and 2 respectively. In Table 2 we have included the complete data set for Am.

Preliminary analysis of the data consisted of plots of $D_o/D-1-\beta_{MNO_3}[NO_3^-]$ ($D_o/D-1-C$ for simplicity) vs. $[PAA^{3-}]$ at constant acidity. These plots were fit by a quadratic expression, indicating M:L stoichiometries of 1:1 and 1:2. The resolved conditional constants varied with

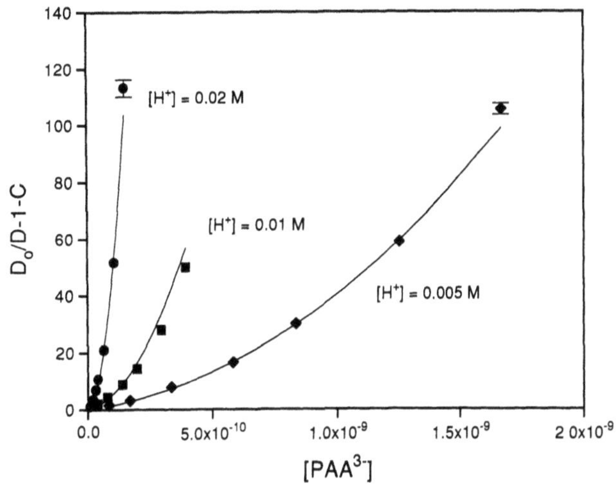


Figure 1 Distribution data and least-squares fit for Am-PAA system. Extraction into various concentrations of HDEHP in toluene. The term C on the y axis label represents $\beta_{MNO_3}[NO_3^-]$.

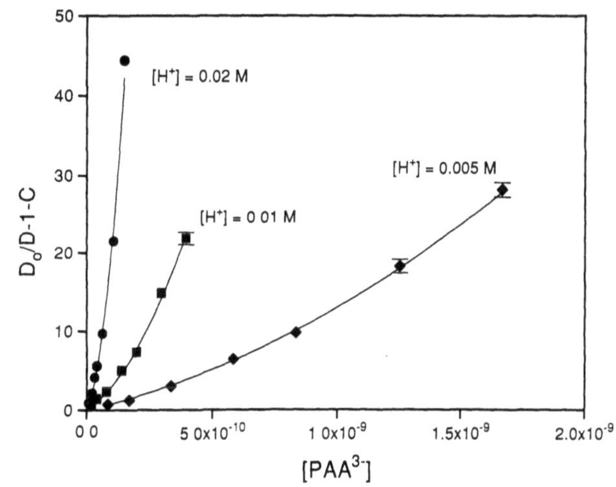


Figure 2 Distribution data and least-squares fit for Eu-PAA system. Extraction into various concentrations of HDEHP in toluene. The term C on the y axis label represents $\beta_{MNO_3}[NO_3^-]$.

**Table 2.** Distribution data for $Am^{3+}$ + PAA at 25°C, I=0.5 M ($NaNO_3$).

| $[PAA]_t$ | $[H^+]$ | $D_o$ | D | $D_o/D-1-C$ | $\sigma(D_o/D-1-C)$ |
|---|---|---|---|---|---|
| 0.01 | 0.005 | 6.942 | 2.320 | 0.7923 | 0.0658 |
| 0.01 | 0.005 | 6.942 | 2.426 | 0.6615 | 0.0540 |
| 0.02 | 0.005 | 6.942 | 1.973 | 1.319 | 0.109 |
| 0.02 | 0.005 | 6.942 | 2.118 | 1.078 | 0.089 |
| 0.04 | 0.005 | 6.942 | 1.354 | 2.926 | 0.235 |
| 0.04 | 0.005 | 6.942 | 1.332 | 3.013 | 0.268 |
| 0.07 | 0.005 | 6.942 | 0.793 | 6.558 | 0.538 |
| 0.07 | 0.005 | 6.942 | 0.815 | 6.317 | 0.518 |
| 0.1 | 0.005 | 6.942 | 0.574 | 9.900 | 0.780 |
| 0.1 | 0.005 | 6.942 | 0.586 | 9.643 | 0.867 |
| 0.15 | 0.005 | 6.942 | 0.330 | 18.86 | 1.49 |
| 0.15 | 0.005 | 6.942 | 0.350 | 17.63 | 1.42 |
| 0.2 | 0.005 | 6.942 | 0.233 | 27.58 | 2.26 |
| 0.2 | 0.005 | 6.942 | 0.223 | 28.88 | 2.34 |
| 0.01 | 0.01 | 10.17 | 3.518 | 0.6902 | 0.0550 |
| 0.01 | 0.01 | 10.17 | 3.935 | 0.3840 | 0.0339 |
| 0.02 | 0.01 | 10.17 | 2.354 | 2.120 | 0.808 |
| 0.02 | 0.01 | 10.17 | 3.502 | 0.7038 | 0.0600 |
| 0.04 | 0.01 | 10.17 | 2.168 | 2.489 | 0.213 |
| 0.04 | 0.01 | 10.17 | 2.329 | 2.167 | 0.175 |
| 0.07 | 0.01 | 10.17 | 1.399 | 5.070 | 0.440 |
| 0.07 | 0.01 | 10.17 | 1.438 | 4.872 | 0.404 |
| 0.1 | 0.01 | 10.17 | 1.085 | 7.170 | 0.759 |
| 0.1 | 0.01 | 10.17 | 1.051 | 7.471 | 0.619 |
| 0.15 | 0.01 | 10.17 | 0.597 | 14.83 | 1.29 |
| 0.15 | 0.01 | 10.17 | 0.599 | 14.76 | 1.25 |
| 0.2 | 0.01 | 10.17 | 0.412 | 22.47 | 1.78 |
| 0.2 | 0.01 | 10.17 | 0.433 | 21.29 | 1.76 |
| 0.0194 | 0.02 | 19.78 | 6.592 | 0.8001 | 0.0474 |
| 0.0194 | 0.02 | 19.78 | 6.302 | 0.9384 | 0.0530 |
| 0.0485 | 0.02 | 19.78 | 4.598 | 2.102 | 0.135 |
| 0.0485 | 0.02 | 19.78 | 4.618 | 2.082 | 0.137 |
| 0.0727 | 0.02 | 19.78 | 3.220 | 3.942 | 0.223 |
| 0.0727 | 0.02 | 19.78 | 3.088 | 4.204 | 0.254 |
| 0.0921 | 0.02 | 19.78 | 2.560 | 5.527 | 0.283 |
| 0.0921 | 0.02 | 19.78 | 2.553 | 5.547 | 0.360 |
| 0.145 | 0.02 | 19.78 | 1.672 | 9.632 | 0.654 |
| 0.145 | 0.02 | 19.78 | 1.662 | 9.700 | 0.617 |
| 0.242 | 0.02 | 19.78 | 0.830 | 21.64 | 1.252 |
| 0.242 | 0.02 | 19.78 | 0.841 | 21.31 | 1.192 |
| 0.339 | 0.02 | 19.78 | 0.422 | 44.61 | 2.309 |
| 0.339 | 0.02 | 19.78 | 0.427 | 44.13 | 2.716 |

the acidity to indicate the stoichiometry of the 1:1 and 1:2 complexes with respect to $[H^+]$. The hydrogen ion dependence for the 1:1 and 1:2 complexes of Am were 1.66(±0.01) and 3.92(±0.01) respectively, suggesting the presence of the species $MH_2L^{2+}$, $MHL^+$, and $M(H_2L)_2^+$. The corresponding values for Eu were 0.92(±0.54) and 3.59(±0.03) indicating the dominant species were $MHL^+$, $M(H_2L)_2^+$, and $M(HL)_2^-$. Data files like those represented in Table 2 were fit using a nonlinear least-squares program described previously (2). The following fit equation was used to adjust the data:

$$D_o/D-1-C = \beta_{121}[H^+]^2[L^{3-}]+\beta_{111}[H^+][L^{3-}]+\beta_{142}[H^+]^4[L^{3-}]^2+\beta_{122}[H^+]^2[L^{3-}]^2 \quad (1)$$

The σ's in Table 2 are based primarily on radioactivity counting statistics for the determination of distribution ratios D and $D_0$. All other sources of error were considered to contribute equally to the uncertainty of the individual points. The data were fit using $1/\sigma$ as the weighting factor.

Neither set of data supported all four of the model parameters in equation 1 ($\beta_{121}$, $\beta_{111}$, $\beta_{142}$, $\beta_{122}$), but each system was best described by a different three parameter equation. The best fit parameters are given in Table 3. The solid lines in Figures 1 and 2 are the least squares fit lines calculated using the stability constants in Table 3. Stepwise constants for $Am(H_2L)/Am(H_2L)_2$ ($\log K_{121} = 13.82(\pm0.04)$, $\log K_{142} = 13.53(\pm0.04)$) show only a factor of 2 decrease for addition of the second $H_2L^-$ ligand to the 1:1 complex implying that the strength of interaction of the second ligand with $Am^{3+}$ is effected little by the presence of the first. For europium, the step-wise constants for $Eu(HL)^+/Eu(HL)_2^-$ are 12.10(±0.05) and 10.98(±0.07) respectively. Here, the 1:2 complex is an order of magnitude weaker than the 1:1 complex, perhaps reflecting the greater perturbation of the inner coordination sphere of Eu by the doubly ionized $H(PAA)^{2-}$ anion. The relatively greater importance of $H(PAA)^{2-}$ in the Eu system reflects stronger electrostatic attraction of the smaller $Eu^{3+}$ cation for the ligand.

**Table 3.** Stability constants for americium and europium complexes with phosphonoacetic acid at 25° C[a].

| Metal | log $\beta_{121}$ | log $\beta_{111}$ | log $\beta_{142}$ | log $\beta_{122}$ | I, ref |
|---|---|---|---|---|---|
| $Am^{3+}$ | 13.82(0.04) | 11.62(0.04) | 27.35(0.01) | - | 0.5 M, p.w. |
| $Eu^{3+}$ | - | 12.10(0.05) | 27.81(0.01) | 23.08(0.05) | 0.5 M, p.w. |
| $UO_2^{2+}$ | 15.42 | 13.91 | 29.87 | 26.50 | 0.1 M, 4 |
| $Th^{4+}$ | 15.73 | - | 30.51 | - | 2.0 M, 3 |

Besides the present results, earlier data on $UO_2^{2+}$ (4) and $Th^{4+}$ (3) are included in Table 3. Only the $M(H_2L)_2$ complex is common to all four systems. The relative order of complex stability for this series is $Am^{3+} < Eu^{3+} < UO_2^{2+} < Th^{4+}$. This order is consistent with expectation, though the uranyl complexes appear to be exceptionally strong, perhaps reflecting a favorable contribution to overall complex stability by hydrogen bonding interactions between the protonated phosphonate group and the uranyl oxygens. The stepwise constants for both uranyl and thorium show an order-of-magnitude dropoff for addition of the second ligand, similar to the Eu-PAA system.

**Am/Eu Separation Factors**

The net separation factor for metal ions in a solvent extraction process which uses aqueous complexes is described by equation 2:

$$S_B^A = \frac{D^A}{D^B} = \frac{D_o^A(1+\Sigma\, \beta_i^B\, [L]^i)}{D_o^B(1+\Sigma\, \beta_i^A\, [L]^i)} \qquad (2)$$

where $D_0$ represents the distribution of the respective metal ions in the absence of aqueous complexes and β's are the stability constants of the respective metal ions. It is clear from this equation that the separation factors of A from B are directly dependent on their relative extraction but inversely dependent on the relative stability of their aqueous complexes. For the most efficient separation, both the extractant and the aqueous complexant must have appropriate properties.

In the case of HDEHP, literature reports and the $D_0$ values determined herein indicate that Eu is extracted about 30 times more efficiently than Am (log $K_{ex}^{Eu} = -1.06$, log $K_{ex}^{Am} = -2.61$, after correction for nitrate complexes). Since the Eu-PAA complexes are more stable than those of Am, the extractant and complexant ligands in this system work in the opposite sense resulting in lower separation factors when the complexant is present than when it is absent. At 0.01 M acid and [HDEHP] = 0.15 M, the distribution ratios and separation factors are $D_{Am}$ = 4.49, $D_{Eu}$ = 140, and $S_{Eu}^{Am}$ = 0.032. For extraction from 0.2 M PAA solutions under the same conditions, the corresponding values are $D_{Am}$ = 0.42, $D_{Eu}$ = 8.22, and $S_{Eu}^{Am}$ = 0.051. Calculations based on the resolved stability constants as applied in equation 2 reproduce the latter results. Similar results were obtained at 0.005 M acid.

Sulfonic acid extractants like HDNNS are known to exhibit little selectivity when extracting metal ions of the same charge (1). Because the $D_0^A/D_0^B$ term in equation 2 is near 1, the aqueous complexant should enhance Am/Eu separation efficiency in this system. In dual-label experiments using 0.0112 M HDNNS in toluene as the extractant, the following results were obtained at 0.02 M acid, 0.5 M $NaNO_3$: $D_{Am}$ = 16.1, $D_{Eu}$ = 19.4, and $S_{Eu}^{Am}$ = 0.83. For extraction from 0.34 M PAA at $[H^+]$ = 0.02 M and 0.5 M $NaNO_3$, the observed distribution ratios and separation factors are: $D_{Am}$ = 1.70, $D_{Eu}$ = 0.94, and $S_{Eu}^{Am}$ = 1.81. Corresponding values calculated from the experimental $D_0$ values and the resolved stability constants are: $D_{Am}$ = 0.79, $D_{Eu}$ = 0.42, and $S_{Eu}^{Am}$ = 1.88. While the separation factor is consistent, the distribution ratios are lower than the experimental values suggesting the possible partial extraction of both the Eu and Am-PAA complexes by HDNNS. Subsequent experiments confirmed the partial extraction of $Eu^{3+}$ and $Am^{3+}$ - PAA complexes by HDNNS.

Neutral bifunctional extractants like CMPO also exhibit little lanthanide/actinide selectivity when extracting these metal ions from nitrate media (1). The distribution ratios and separation factors for Am and Eu extraction by 0.5 M CMPO-diethylbenzene from 0.5 M $NaNO_3$ at pH 1.88 are $D_{Am}$ = 110, $D_{Eu}$ = 107, and $S_{Eu}^{Am}$ = 1.03. When 0.2 M PAA is introduced into the aqueous phase, the corresponding values are $D_{Am}$ = 4.57, $D_{Eu}$ = 2.21, and $S_{Eu}^{Am}$ = 2.07. Distribution ratios calculated from the experimental $D_o$ values (corrected for the mononitrate complex in the aqueous phase) and the stability constants reported in Table 3 are $D_{Am}$ = 10.7 and $D_{Eu}$ = 5.00 ($S_{Eu}^{Am}$ = 2.14). The experimental separation factor is well described by the calculation, but the distribution ratios are lower than the calculated values. At 0.3 M PAA, the discrepancy between experimental and calculated distribution ratios is larger while the separation factors are still consistent. We cannot at present explain this observation, but it may be related to the procedures used to preequilibrate the extractant phase or perhaps to the extraction of mixed ligand complexes. In this work, we have examined only a few selected sets of conditions.Because solvent extraction systems based on neutral extractants offer multiple parameters for adjustment of the relative D's, these conditions do not necessarily represent the optimized configuration for the system.

## CONCLUSION

Complexes between $Am^{3+}$ or $Eu^{3+}$ and phosphonoacetic acid differ in relative stability in accord with the electrostatic model of cation binding. The smaller $Eu^{3+}$ cation forms stronger complexes with PAA than the larger $Am^{3+}$ cation. The observed metal complexes in the acid range from 0.005 M to 0.02 M (at I= 0.5 M) are $Eu(HL)^+$, $Eu(H_2L)_2^+$, $Eu(HL)_2^-$ and $Am(H_2L)^{2+}$, $Am(HL)^+$, $Am(H_2L)_2^+$. When used as a holdback reagent, PAA slightly enhances the separation of Am/Eu when used with sulfonic acids or CMPO/nitrate, but reduces separation efficiency with HDEHP. The calculated stability constants can be used to explain the separation factors, but do not always accurately predict metal distribution ratios in the CMPO systems, implying that there are details of this system which have not been fully elucidated.

## ACKNOWLEDGMENT

Work performed under the auspices of the Office of Basic Energy Sciences, Division of Chemical Sciences, U. S. Department of Energy under contract No. W-31-109-ENG-38.

## REFERENCES

1. K. L. Nash *Solv. Extr. Ion Exch. 11* 729 (1993).
2. K. L. Nash *Inorg. Chim. Acta 169* 245 (1990).
3. K.,L. Nash *Radiochim. Acta 54* 171 (1991).

4. K. L. Nash *Radiochim. Acta 61* 147 (1993).
5. E. P. Horwitz, H. Diamond, R. C. Gatrone, K. L. Nash, P. G. Rickert "TUCS: A new class of aqueous complexing agents for use in solvent extraction processes" in *Solvent Extraction 1990* (ed. T. Sekine ) Elsevier Science Publishers (1992) pp. 357-362.
6. A. A. Elesin, A. A. Zaitzeva, S.S. Kazakova, G. N. Yakovlev *Radiokhimiya 14* 541 (1972).
7. A. E. Martell and R. J. Motekaitis *Determination and Use of Stability Constants* VCH Publishers New York (1988).
8. P. C. Heubel and A. I. Popov *J. Soln. Chem. 8* 615 (1979).
9. NIST Database 46. Critical Stability Constants of Metal Complexes Database, Version 1.0, R. M. Smith, A. E. Martell, R. J. Motekaitis, U. S. Department of Commerce, Gaithersburg, Md. September, 1993.

# EUROPIUM(III) INTERACTION WITH A WATER-SOLUBLE COMPLEXANT: FROM SPECIATION TO PHOTODESTRUCTION

James V. Beitz

Chemistry Division
Argonne National Laboratory
Argonne, IL 60439 USA

## ABSTRACT

Speciation of trivalent europium ions in solutions containing tetrahydrofuran-1,2,3,4-tetracarboxylic acid (THFTCA) has been carried out using time- and wavelength-resolved laser-induced fluorescence. THFTCA is a candidate stripping agent in solvent extraction processing of nuclear waste that enables removal of trivalent metal ions while suppressing removal of uranyl. Laser-induced fluorescence provides evidence for the predominance of a complex containing 1 metal ion and 2 THFTCA molecules in a solution whose composition, 0.3 M THFTCA + 0.4 M NaOH, is typical of that likely to be used in processing nuclear waste. Direct photochemical destruction of THFTCA in aqueous solution has been demonstrated for the first time using broad band ultraviolet photolysis and is a promising method to facilitate recovery of stripped metal ions for subsequent processing.

## INTRODUCTION

The complexant studied in the present work is tetrahydrofuran-1,2,3,4-tetracarboxylic acid (THFTCA). The unusual ability of THFTCA in aqueous solution to act as a stripping agent for removal of 3+ lanthanide and actinide ions, such as $Eu^{3+}$ and $Am^{3+}$, from organic phases while minimizing removal of uranyl is reported by Nash and coworkers[1] in this volume. The chemical structure of THFTCA and its expected 1:1 complex[1,2] with a trivalent metal ion are shown schematically in Figure 1.

Application of THFTCA as a stripping agent for separating components of nuclear waste via solvent extraction processing[1] will require developing a means of degrading or destroying THFTCA-metal ion complexes that is environmentally acceptable and cost effective. The considerable resistance of THFTCA to oxidation by materials commonly employed in solvent extraction processing of actinides makes waste minimization in destroying THFTCA complexes challenging.[3] The work reported here provides new insight

*Separations of f Elements*, Edited by K.L. Nash
and G.R. Choppin, Plenum Press, New York, 1995

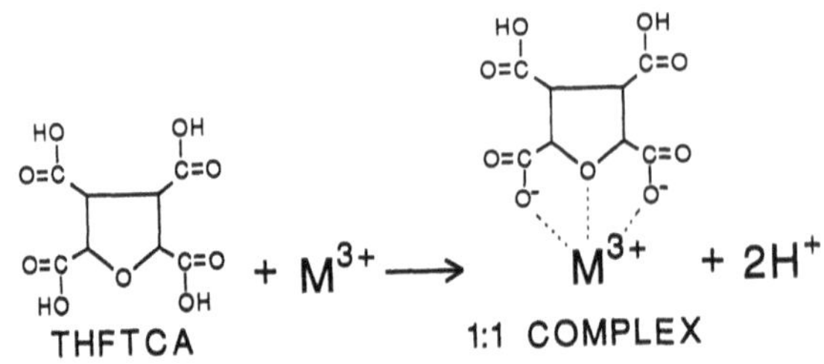


**Figure 1.** The complexant studied, tetrahydrofuran-1,2,3,4-tetracarboxylic acid (THFTCA), and its expected coordination[1,2] as a 1:1 complex with a 3+ lanthanide or actinide metal ion ($M^{3+}$) are shown.

into THFTCA-metal ion binding in solution and demonstrates photochemistry as a means of destroying the coordinating power of THFTCA in aqueous solution.

The favorable spectroscopic, photophysical, and photochemical properties of $Eu^{3+}$ make it the trivalent lanthanide ion of choice for investigation of THFTCA-metal ion complexes. In addition, europium is a fission product commonly present in Am-contaminated nuclear waste[4] that may be treated using THFTCA. Some spectroscopic properties and photophysical processes of 4f electron states of aquated $Eu^{3+}$ are shown in Figure 2. The optical spectrum of $Eu^{3+}$ in dilute acid solution is presented; the spectral region over which strong solvent bands prevent solute absorbance measurements is shown shaded with diagonal lines. For the wavenumber range shown, the strongest observed absorption bands occur in the near-ultraviolet. Because lanthanide ion 4f-4f transition energies shift by only a small amount on complexation of the metal ion,[5,6] the spectroscopic

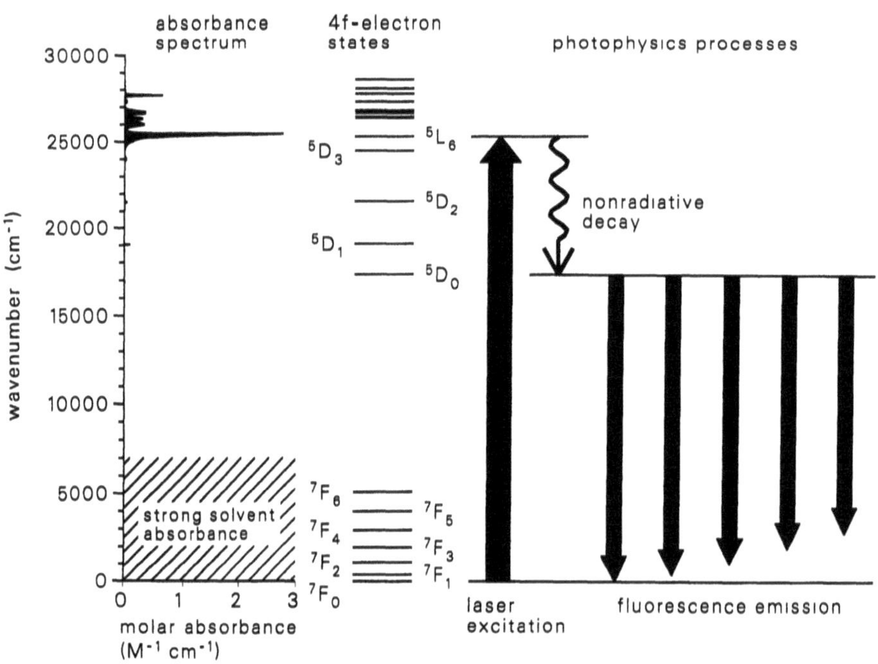


**Figure 2.** Comparison of the optical absorbance spectrum of aquated $Eu^{3+}$, calculated[10] "free" ion 4f electron states of $Eu^{3+}$, and some of the photophysical processes observed in the present study.

properties of the aquated ion shown in Figure 2 provide the basis for assigning the excitation and emission bands observed in the present work.

Our work identifies spectroscopic and photophysical properties of $Eu^{3+}$-THFTCA complexes that are of significance for fluorescence-based real-time monitoring of stripping solutions as well as for speciation studies. Detailed time- and wavelength-resolved laser-induced fluorescence methods have been employed to probe the interaction of $Eu^{3+}$ with THFTCA in aqueous solutions similar to those expected when THFTCA is used as a stripping agent in nuclear waste processing. Due to the acidity of organic phases in such processing, and to suppress solubility of THFTCA in the organic phase, the stripping solution of choice is partially neutralized THFTCA.[1] In consequence, our work has centered on investigation of solutions containing 0.3 M THFTCA and 0.4 M NaOH. For comparison purposes, 0.5 M THFTCA was studied. Our speciation results are in good in agreement with expectations based on speciation modeling using stability constants derived from potentiometric data on THFTCA at lower concentration[1] and molecular modeling.[2]

Because photons have no mass or charge, light can be a particularly attractive "reagent" for chemical processing that is designed to minimize waste generated during degradation or destruction of metal ion complexes. The studies reported in the present work on photodestruction of THFTCA and THFTCA-metal ion complexes in solution are preliminary in nature. Our work is the first photochemical investigation of THFTCA and has provided insight into its photodestruction. Known direct photochemical processes for simple carboxylates include decarboxylation (loss of $CO_2$) and creation of radical species.[7] Photochemistry can also occur via ligand-to-metal charge transfer (LMCT) that is spectroscopically observable in many $Eu^{3+}$ complexes with organic ligands.[6] The most relevant $Eu^{3+}$ LMCT work is that of Matsumoto and Azuma.[8] These workers reported photocatalytic destruction of two simple organic carboxylates, acetate and formate, in aqueous solution based on excitation of a LMCT band of complexed $Eu^{3+}$. Our work provides evidence of similar behavior for $Eu^{3+}$ in THFTCA solutions.

## EXPERIMENTAL APPARATUS AND METHODS

A schematic diagram of the laser-induced fluorescence apparatus used in this work is shown in Figure 3. A tunable dye laser (National Research Group model DL-0.03, PBBO dye in toluene + ethanol solvent), pumped by a pulsed nitrogen laser (Molectron UV-14), provided 10 ns duration excitation pulses of typically 100 μJ energy and 0.03 nm spectral bandwidth at 20 pulses per second. Dielectrically-coated bandpass optical filters, a cooled photomultiplier (RCA 31034), and a digital storage oscilloscope (LeCroy 9310) were used to acquire fluorescence decay rate data. The $Eu^{3+}$ $^5D_0$ state fluorescence decay data were recorded while monitoring $^5D_0 \rightarrow {}^7F_2$ emission and were well fit by nonlinear least squares adjustment to a single exponential decay model. The fluorescence decay of $^5D_1$ was recorded using a bandpass filter centered at 560 nm. A small amount of light leakage from longer wavelengths ($Eu^{3+}$ $^5D_0$ emission) contributed to slightly curved residuals when fitting $^5D_1$ data.

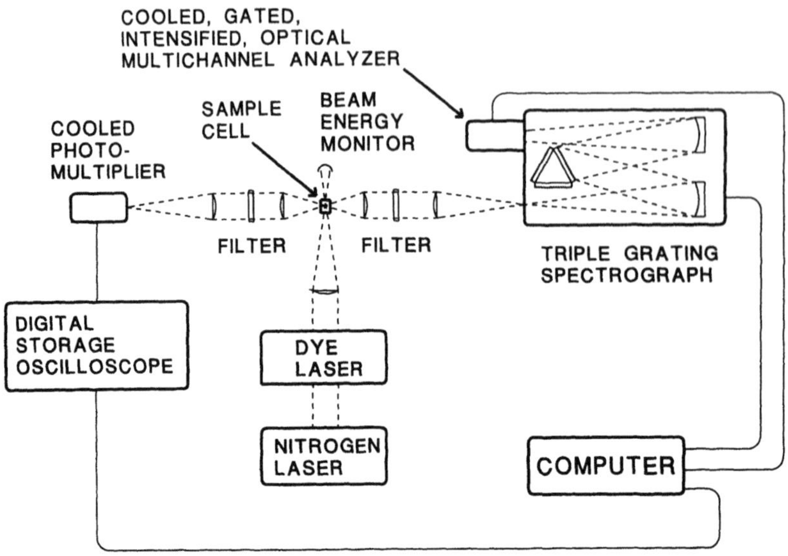


**Figure 3.** Schematic diagram of the laser-induced fluorescence apparatus. Optical beam paths shown as dashed lines. The digital storage oscilloscope and computer provide add-to-memory signal averaging for fluorescence decay and time-gated optical multichannel analyzer signals, respectively.

A red-enhanced, intensified, gated optical multichannel analyzer (Princeton Instruments model RE/IRY-700 with ST120 controller) equipped with a photocathode cooler (CFC-100, 253 K coolant) and a high voltage pulse generator (PG-10) together with a 27.5 cm focal length grating spectrograph (Acton model 275 equipped with 150, 1200, and 1800 groove per millimeter gratings, 75 micron effective slit width) were used to acquire time-gated emission spectra. In most cases, on-array averaging within the RE/IRY-700 was used together with add-to-memory averaging of the photodiode array readout data that were transferred from the ST120 to a personal computer. Fixed pattern photodiode array noise was removed by subtracting out data runs recorded with no light falling on the photocathode. Care was taken to ensure that the array values decayed to their dark level between spectral data acquisition runs.

Emission lines from a low pressure neon lamp were used to calibrate the optical multichannel analyzer system. In most cases, the optical multichannel analyzer was gated on after a fixed time delay following the laser pulse. When emission spectra were recorded during and immediately following the laser pulse, a digital pulse generator (Stanford Research Systems DG535) turned on the optical multichannel analyzer prior to externally triggering the nitrogen laser. Due to the $\pm 1$ μs trigger jitter of the nitrogen laser, the optical multichannel analyzer was gated on for 7 μs with the gate turn on beginning 2 μs prior to triggering the nitrogen laser. Fluorescence data were recorded at 295 $\pm 2$ K.

Ultraviolet photolysis was carried out using an unfiltered 450 watt medium pressure mercury arc lamp placed within a closed bottom Suprasil fused silica jacket. This assembly was placed within a temperature controlled recirculating bath filled with deionized water held at 288 K The photolysis cells were fabricated from mandrel shrunk 8x8 mm inner diameter Suprasil tubing with 6 mm tubulation attached. The cells were placed symmetrically about the photolysis lamp and immersed in the bath. During photolysis, gas was evolved from the THFTCA-containing solutions in the cells. The gas was conducted away to water-filled bubblers using polytetrafluoroethylene fittings and tubing.

THFTCA was purchased from Tokyo Kasei as high purity material and was used as received. A slight turbidity was present when the THFTCA was dissolved to make stock solutions. The stock solutions were allowed to stand and aliquots were withdrawn by pipetting or decanting. The pH of solutions was measured using an Orion Ross temperature compensated pH electrode and model 290A pH meter that were periodically standardized against pH 4.00 and 7.00 buffers. Optical absorption spectra were recorded using a Cary 17I dual beam spectrophotometer. Aliquots of a 0.37 M stock solution of $Eu(ClO_4)_3$ in dilute $HClO_4$ were used to prepare Eu-containing sample solutions that typically contained 0.01 M $Eu(ClO_4)_3$. The stock solution was prepared by dissolving $Eu_2O_3$ (Apache, 99.999%) in $HClO_4$. The $D_2O$ solutions were prepared by evaporating an aliquot of an $H_2O$ solution to dryness at 358 K under a stream of dry nitrogen gas and then adding an equal volume of $D_2O$ (Aldrich, >99.96 D-atom%). This evaporation-addition cycle was repeated a minimum of 4 times.

## EXPERIMENTAL RESULTS

In our initial work, we recorded the ultraviolet absorption spectrum of THFTCA and found it to be typical of a saturated organic carboxylic acid[7] with a peak at 223 nm. We measured the fluorescence emission spectra and lifetimes of $Eu^{3+}$ complexed by THFTCA in several solutions (see Figures 4 and 5, respectively, for some typical examples). As might be expected, complexation increased the fluorescence lifetime of $Eu^{3+}$ and altered its emission spectrum. As is evident in Figure 4, emission spectrum changes for the $^5D_0 \rightarrow {}^7F_{1\text{-}4}$ bands are rather modest. Sufficient spectral resolution and sensitivity were achieved to enable recording of the emission spectrum of the strongly disallowed $Eu^{3+}$ $^5D_0 \rightarrow {}^7F_0$ transition under a variety of solution compositions (see Figure 6).

Unexpectedly, we observed short-lived, very broadband emission whose intensity was influenced by ultraviolet photolysis and the presence or absence of $Eu^{3+}$ during

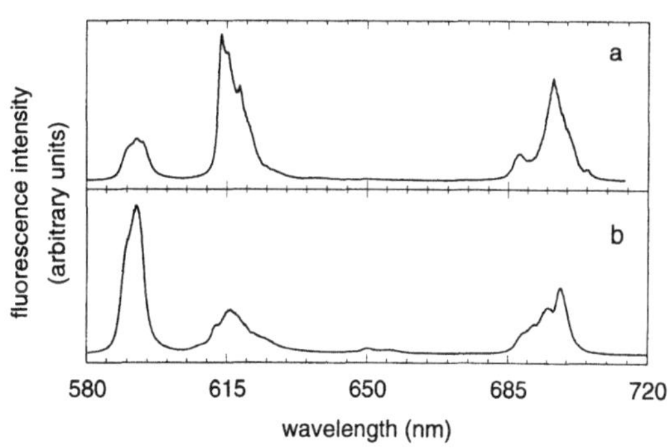


**Figure 4.** Comparison of observed emission spectra of the $Eu^{3+}$ $^5D_0 \rightarrow {}^7F_{1\text{-}4}$ transitions, recorded using a time-gated optical multichannel analyzer following $^7F_0 \rightarrow {}^5L_6$ excitation, for $Eu^{3+}$ as a complexed ion in 0.5 M THFTCA at pH = 1.1 (panel **a**) and an aquated ion (panel **b**).

photolysis (see Figure 7). Near-ultraviolet absorption spectrum of THFTCA solutions with and without added $Eu^{3+}$ provided evidence of formation of a $Eu^{3+}$ LMCT band as is frequently found[6] for organic complexes of $Eu^{3+}$. Prompted by this result and past

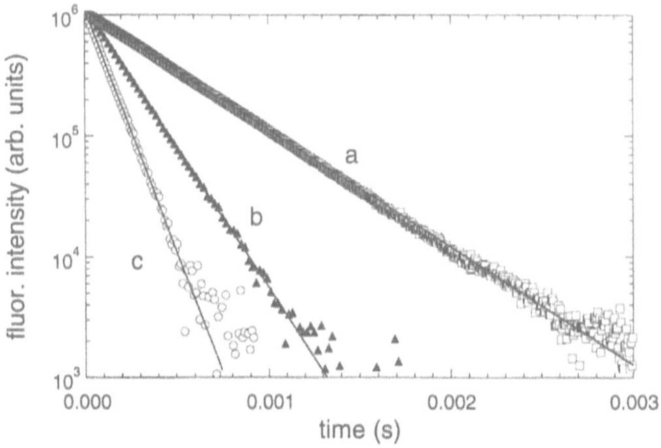


**Figure 5.** Examples of the observed (symbols) and calculated (solid line) fluorescence decay of the $^5D_0$ state of $Eu^{3+}$ following excitation into $^5L_6$. Curve **a** is $Eu^{3+}$ in 0.3 M THFTCA + 0.4 NaOH before photolysis, curve **b** is $Eu^{3+}$ added to 0.3 M THFTCA + 0.4 M NaOH that undergone 4 hours of ultraviolet photolysis, and curve **c** is aquated $Eu^{3+}$.

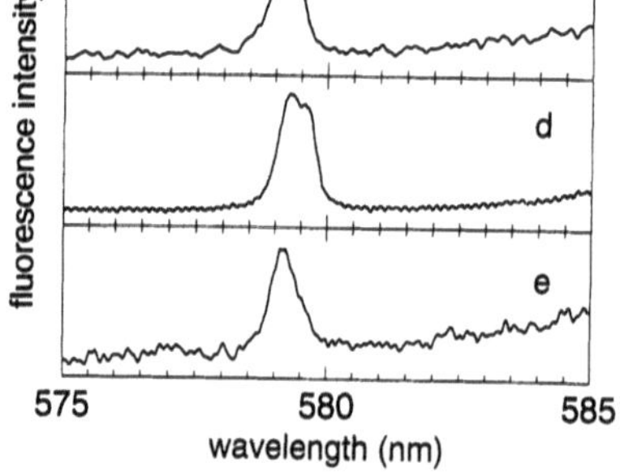


**Figure 6.** Time-gated optical multichannel analyzer emission spectra of the $Eu^{3+}$ $^5D_0 \rightarrow {}^7F_0$ band in dilute $HClO_4$ solution (panel **a**), in 0.3 M THFTCA + 0.4 M NaOH before photolysis (panel **b**) and after 4 hours of ultraviolet photolysis (panel **c**), in 0.5 M THFTCA before photolysis (panel **d**) and after 4 hours of ultraviolet photolysis (panel **e**). Data were recorded using $^7F_0 \rightarrow {}^5L_6$ excitation.

photochemical studies on $Eu^{3+}$ complexes with simple carboxylates,[8] we carried out ultraviolet photolysis on selected solutions and used laser-induced fluorescence on $Eu^{3+}$ to obtain spectroscopic and photophysical evidence as to ensuing changes in solution composition. The spectroscopic data are presented in Figures 6 and 7. Observed fluorescence lifetimes for the $^5D_0$ and $^5D_1$ states of $Eu^{3+}$ in THFTCA-containing solutions are given in Table 1 (estimated uncertainty is ±3 μs for $^5D_0$ lifetime values in $H_2O$ solution, ±60 μs for $D_2O$ solutions, and ±0.2 μs for $^5D_1$ lifetime values).

A broadband, fast decaying emission was observed to strongly increase on ultraviolet photolysis of THFTCA solutions (see Figure 7, data recorded using the 150 groove per mm grating). The observed decay time, 20 ns, of this emission following 395 nm excitation was that of our measurement apparatus. Decreasing sensitivity of the apparatus, including the short wavelength cutoff of the optical filter used to block scattered laser light, contributes to the rapidly diminishing signal intensity at short wavelengths. A linear abscissa is used in the upper panel to facilitate intensity comparisons (prominent $Eu^{3+}$ emission bands are marked with downward pointing arrows); a logarithmic abscissa in the lower panel aids recognition of an $H_2O$ solvent Raman peak (marked with an upward pointing arrow) at 458 nm. This Raman peak is obscured by shot noise in curves a and b of Figure 7 due the high intensity of the broadband emission. Ultraviolet photolysis of THFTCA strongly increases the intensity of the broadband emission from following 395 nm excitation. Ultraviolet photolysis after addition of $Eu^{3+}$ results in a lesser increase in the intensity of the broadband emission.

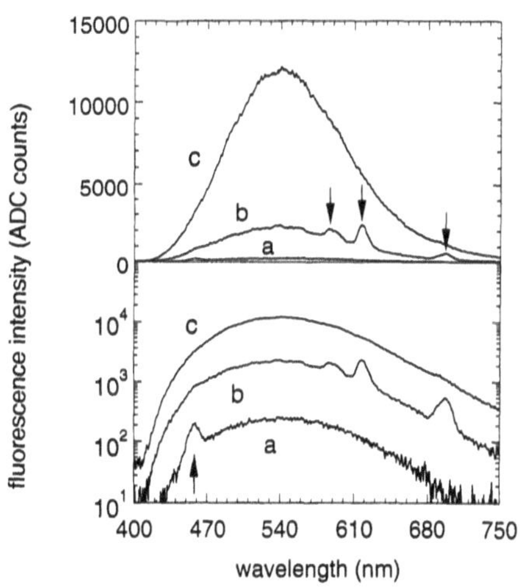


**Figure 7.** Linear and logarithmic plots of the observed broadband emission recorded during and immediately following pulsed 395 nm laser excitation of an aliquot of 0.5 M THFTCA before photolysis (curve **a**) and after 4 hours of ultraviolet photolysis (curve **c**) as well as for a solution containing 0.01 M $Eu^{3+}$ + 0.5 M THFTCA after 4 hours of ultraviolet photolysis (curve **b**).

## DISCUSSION

Investigation of the $Eu^{3+}$ $^5D_0 \rightarrow {}^7F_0$ transition is particularly useful because such J=0 to J=0 transitions can not be split by the electrostatic field of the surrounding ligands. For example, if 2 bands are observed in the $^5D_0 \rightarrow {}^7F_0$ spectral region, it is clear that at least 2 $Eu^{3+}$ species are present. In addition, Albin and Horrocks [9] have established a quantitative correlation between total charge, $p$, on coordinated ligands and the $^7F_0 \rightarrow {}^5D_0$ transition frequency $\nu$ in wavenumber units, $cm^{-1}$ ($1\ cm^{-1} = 1.2398 \times 10^{-4}$ eV), of the complexed $Eu^{3+}$:

$$\nu = -0.76\,p^2 + 2.29p + 17273. \tag{1}$$

The estimated uncertainty in the value of $p$ calculated using Eqn 1 is ±1 charge.[9] The total charge on coordinated ligands in $Eu^{3+}$ complexes can be calculated using Eqn 1 if the appropriate $Eu^{3+}$ transition frequency is measured.

**Table 1.** Comparison of observed $Eu^{3+}$ emission lifetimes (in μs) in the present work and the influence of 4 hour ultraviolet photolysis.

| emitting $Eu^{3+}$ state | aquated $Eu^{3+}$ | | 0.3 M THFTCA + 0.4 M NaOH before photolysis | | 0.3 M THFTCA + 0.4 M NaOH after photolysis[a] | 0.5 M THFTCA before photolysis | | 0.5 M THFTCA after photolysis[a] |
|---|---|---|---|---|---|---|---|---|
| | $H_2O$ | $D_2O$ | $H_2O$ | $D_2O$ | $H_2O$ | $H_2O$ | $D_2O$ | $H_2O$ |
| $^5D_0$ | 108 | 2150 | 446 | 2340 | 183 | 224 | 2280 | 185 |
| $^5D_1$ | 4.2 | | 1.6 | | | 2.0 | | |

[a] Europium added after 4 hour photolysis with a medium pressure mercury arc lamp.

The "free" ion 4f electron states of $Eu^{3+}$ as calculated by Carnall and coworkers[10] are shown in Figure 2. The largest gap between 4f states of $Eu^{3+}$ occurs between $^5D_0$ and $^7F_6$. Because trivalent lanthanide ion 4f state nonradiative decay in aqueous solution exhibits an approximately exponential dependence on the energy gap between the emitting and next lower f-state[5], $Eu^{3+}$ emission in aqueous solutions occurs primarily from $^5D_0$. Some of the photophysical processes important in the present work are shown as bold (radiative process) or wavy (nonradiative process) arrows in Figure 2. Fluorescence lifetime studies on the $^5D_0$ state of $Eu^{3+}$ enable determination of the number of inner sphere coordinated water molecules, $q$, via a relationship established by Horrocks and Sudnick:[11]

$$q = 1.05 \times 10^{-3}\,(\tau_{H_2O}^{-1} - \tau_{D_2O}^{-1}) \tag{2}$$

where $\tau_{H_2O}$ is the measured fluorescence lifetime in seconds of the $^5D_0$ state of $Eu^{3+}$ in $H_2O$ and $\tau_{D_2O}$ is the measured fluorescence lifetime in seconds of the $^5D_0$ state of $Eu^{3+}$ in $D_2O$ solution. The estimated uncertainty in $q$ is ±0.5 water molecule.[11] The $D_2O$ solution measurement provides a basis for assessing nonradiative decay of the $^5D_0$ state arising from interaction of $Eu^{3+}$ with its inner sphere coordinated ligands (other than water) and also accounts for the finite radiative lifetime of the $^5D_0$ state. The value of $\tau_{D_2O}$ for aquated $Eu^{3+}$ is reported to be 4.1 ms when observed values are extrapolated to zero $H_2O$ concentration.[12] Experimentally measured $\tau_{D_2O}$ values[11] typically are 20% smaller in solutions made using high isotopic purity $D_2O$.

Analysis of the data shown in Figure 6 began with conversion to vacuum wavenumber units[13] to enable calculation, via Eqn 1, of the total charge on coordinated ligands before and after photolysis (see Table 2). Note that $Eu^{3+}$ in 0.5 M THFTCA before photolysis contained two spectroscopically distinguishable $Eu^{3+}$ complexes, based on the splitting of the $^5D_0$ to $^7F_0$ band evident in panel d of Figure 6. Analysis of the lifetime data in Table 1 via Eqn 2, enables calculation of the number of inner sphere coordinated water molecules (see Table 2). We necessarily have assumed that the value of $\tau_{D_2O}$ is the same in THFTCA solutions after photolysis as the value measured before photolysis (photolysis of THFTCA in $D_2O$ is likely to lead to H-D exchange and so reduce the isotopic purity of the solvent). The strong influence of THFTCA on the emission lifetime of $Eu^{3+}$ suggests use of laser-induced fluorescence lifetime analysis as a near-real time monitor of metal-ion loading during stripping stages in solvent extraction processing of nuclear waste as well as of photodestruction of THFTCA in spent stripping solutions. A previous study[14] on complexed $Eu^{3+}$ provides evidence that laser-induced fluorescence also can be applied to metal ion complexes in organic phases typical of those used in solvent extraction processing.

Based on stability constants derived from potentiometric titration data taken at lower THFTCA concentration and lower ionic strength,[1] the predominant complex expected in 0.3 M THFTCA + 0.4 M NaOH contains a single $Eu^{3+}$ ion and 2 THFTCA ligands each with a net negative charge of -2. Assuming that the overall coordination number of $Eu^{3+}$ is 8 and that $THFTCA^{2-}$ is tridentate in its coordination (as expected from molecular modeling[2] and the crystal structure of THFTCA[1]), $Eu^{3+}$ in 0.3 M THFTCA + 0.4 M NaOH will have 2 inner sphere coordinated water molecules and a total charge on coordinated ligands of -4. Our laser-induced fluorescence measurements (see Table 2 and panel b of Figure 6) provides evidence that a single $Eu^{3+}$ complex predominates with a total charge on coordinated ligands of (-3.5 ± 1) and (1.9 ± 0.5) inner sphere coordinated water molecules. Increase in the number of coordinated water molecules to (5.3 ± 0.5) with an increase in pH and decrease in total ligand charge to (-1.5 ± 1) after ultraviolet photolysis provides evidence of photodestruction of most THFTCA such that the predominate solution species becomes a 1:1 complex between $Eu^{3+}$ and a presumed THFTCA fragment retaining carboxylate functionality. Increased pH is attributed to photodecarboxylation (loss of $CO_2$) of THFTCA and is consistent with observed gas formation during photolysis.

Before photolysis, the observed fluorescence decay of $Eu^{3+}$ in 0.5 M THFTCA was single exponential. This combined with the $^5D_0 \rightarrow {}^7F_0$ spectral evidence of predominantly 2 $Eu^{3+}$ species in this solution (see Figure 6, panel d), renders it probable that the 2 major Eu-THFTCA species present interconvert rapidly compared to the observed fluorescence decay rate. Consistent with this conclusion, time-gated optical multichannel analyzer spectra (not shown) recorded with differing gate delays and widths revealed no changes in spectral features as might occur if emission from two non-interconverting $Eu^{3+}$ complexes was present. The presence of a Eu-THFTCA complex containing a total charge on coordinated ligands of (-2.2 ± 1), see Table 2, can be understood on the basis that the high proton concentration in 0.5 M THFTCA shifts the $Eu^{3+}$ speciation to favor either reduction in denticity from 3 to 2 per coordinated THFTCA (due to increased $THFTCA^{-}$ concentration) with two such THFTCA coordinated per $Eu^{3+}$ or that a reduction in the number of coordinated $THFTCA^{2-}$ from 2 to 1 per $Eu^{3+}$ has occurred due to reduced $THFTCA^{2-}$ concentration. Due to the single exponential character of the $Eu^{3+}$ emission, no definitive information as to the number of inner sphere coordinated water molecules per complexed $Eu^{3+}$ was obtained in 0.5 M THFTCA.

**Table 2.** Influence of 4 hour ultraviolet photolysis on THFTCA solutions as probed by studies on $Eu^{3+}$ coordination.

| | | $Eu^{3+}$ coordination | |
|---|---|---|---|
| Solution | Solution pH | Number of inner sphere water molecules (±0.5)[a] | Total charge on coordinated ligands (±1)[b] |
| 0.3 M THFTCA + 0.4 M NaOH, before photolysis | 2.7 | 1.9 | -3.5 |
| 0.3 M THFTCA + 0.4 M NaOH, after photolysis[c] | 5.0 | 5.3 | -1.5 |
| 0.5 M THFTCA, before photolysis | 1.1 | 4.2[d] | -2.2 and -3.5[e] |
| 0.5 M THFTCA, after photolysis[c] | 1.5 | 5.2 | -1.5 |

[a] Calculated using Eqn 2 and lifetime values from Table 1.
[b] Calculated using Eqn 1 and spectral bands shown in Figure 6.
[c] Europium added after 4 hour photolysis with a medium pressure mercury arc lamp.
[d] 2 or more $Eu^{3+}$ species present. Value shown is average for all species present.
[e] Spectral evidence indicates predominantly 2 complexed $Eu^{3+}$ species (see Figure 3, panel d).

Our observation of a decrease in $Eu^{3+}$ $^5D_1$ state fluorescence lifetime in THFTCA containing solutions (see Table 1), in comparison with aquated $Eu^{3+}$, is unexpected in comparison with $^5D_0$ state behavior on complexation. Past studies[12] have provided evidence that about 50% of aquated $Eu^{3+}$ ions optically pumped into the $^5D_1$ state relax to the $^5D_0$ state. The remaining $^5D_1$ state ions evidently relax directly to the lower lying $^7F$ states via transfer of energy to vibrational modes of coordinated $H_2O$ molecules. The high frequency of these modes (~3400 $cm^{-1}$, based on infrared and Raman spectra[15] of $H_2O$) apparently prevents them from effectively relaxing $^5D_1$ population into the $^5D_0$ state. From Carnall and coworkers calculation,[10] the $^5D_1$-to-$^5D_0$ energy gap is 1758 $cm^{-1}$. The bending modes of liquid water occur at ~1640 $cm^{-1}$ based on infrared and Raman spectra[15] of liquid $H_2O$ whereas the C=O stretch of saturated carboxylic acids is reported[16] to occur at ~1760 $cm^{-1}$. In consequence, there is a diminished mismatch between the $^5D_1$-to-$^5D_0$

energy gap of $Eu^{3+}$ and the energy of the C=O stretching vibration in THFTCA, in comparison with the mismatch for bending or stretching vibrational modes of water molecules. This reduced energy mismatch and the character of the interaction of THFTCA with $Eu^{3+}$ evidently opens a new and more effective nonradiative decay path for the $^5D_1$ state of $Eu^{3+}$, namely conversion of 4f state electronic energy into excitation of vibrational modes of THFTCA.

The intensity, broadband character, and very short emission lifetime of the species whose concentration increases upon ultraviolet photolysis of THFTCA (see Figure 7) are consistent with an aldehyde or ketone.[7] An observed increase in near-ultraviolet absorbance after photolysis is also consistent with this supposition. Suppression of the growth of this species via addition of $Eu^{3+}$ prior to ultraviolet photolysis suggests that metal-mediated photochemical destruction of THFTCA (or its initial photoproducts) has occurred. Verification of this hypothesis will require further investigation.

Because photons are massless and carry no electrical charge, light is potentially a very attractive "reagent" for minimizing waste generated during degradation or destruction of metal ion complexes in solution. The primary disadvantage of photodestruction of such complexes is the cost of producing ultraviolet light that generally is required to assure ligand dissociation. In consequence, economic considerations strongly favor highly efficient photoprocesses. Additional studies will be needed to identify the most efficient photodestruction methods that are applicable to THFTCA.

## CONCLUSIONS

The first spectroscopic study of metal ion-THFTCA coordination is reported. Our luminescence studies provide evidence for the predominance of a complex containing 1 $Eu^{3+}$ ion and 2 THFTCA molecules in an aqueous solution typical of that likely to be used in processing nuclear waste. Evidence that THFTCA coordinates, at least in part, via carboxylate groups was found in analysis of the observed $Eu^{3+}$ $^5D_1$ state fluorescence decay rates. These conclusions are consistent with expectations based on potentiometric titration of metal ion-THFTCA solutions at lower THFTCA concentration and lower ionic strength[1] as well as molecular modeling work.[2] Direct photochemical destruction of THFTCA has been demonstrated for the first time in our preliminary investigation on the influence of ultraviolet photolysis on THFTCA solutions. Photodestruction of THFTCA is a promising method for minimizing waste generated during solvent extraction processing of nuclear waste. Laser-induced fluorescence is shown to provide a basis for near-real time monitoring of metal ion coordination in THFTCA solutions and photodestruction of THFTCA and its metal ion complexes.

## ACKNOWLEDGMENTS

I thank K. Nash for furnishing a preprint of his THFTCA work, H. Diamond and K. Nash for preparing some of the stock solutions used in this study, and W. T. Carnall for supplying the absorbance spectrum of aquated $Eu^{3+}$ in digital form. This work was performed under the auspices of the Office of Basic Energy Sciences, Division of Chemical Sciences, U. S. Department of Energy, under contract W-31-109-ENG-38.

## REFERENCES


1. K. L. Nash, P. G. Rickert, E. P. Lessmann, M. D. Mendoza, J. F. Feil, J. C. Sullivan, New water soluble phosphonate and polycarboxylate complexants for enhanced f element separations, this volume.
2. K. L. Nash, E. P. Horwitz, R. C. Gatrone, and P. G. Rickert, The effect of ligand rigidity on the stability of europium(III) complexes with substituted diglycolic acids, *J. Alloys and Compounds*, 180:375 (1992).
3. R. Gatrone and E. P. Horwitz, private communication.
4. D. C. Stewart, "Data for Radioactive Waste Management and Nuclear Applications," John Wiley and Sons, New York (1985).
5. W. T. Carnall, The absorption and fluorescence spectra of rare earth ions in solution, *in:* "Handbook on the Physics and Chemistry of Rare Earths," Vol. 3, K. A. Gschneidner, Jr. and L. Eyring, eds., North-Holland Publishing Co., Amsterdam (1979).
6. F. S. Richardson, Terbium(III) and europium(III) ions as luminescent probes and stains for biomolecular systems, *Chem. Rev.* 82:541 (1982).
7. J. G. Calvert and J. N. Pitts, Jr. "Photochemistry," John Wiley and Sons, New York (1966).
8. A. Matsumoto and N. Azuma, Photodecomposition of europium(III) acetate and formate in aqueous solution, *J. Phys. Chem.*, 92:1830 (1988).
9. M. Albin and W. DeW. Horrocks, Jr., Europium(III) luminescence excitation spectroscopy. Quantitative correlation between the total charge on ligands and the $^7F_0 \rightarrow {}^5D_0$ transition frequency in europium(III) complexes, *Inorg. Chem.* 24: 895 (1985).
10. W. T. Carnall, P. R. Fields, and K. Rajnak, Electronic energy levels of the trivalent lanthanide aquo ions. IV. $Eu^{3+}$, *J. Chem. Phys.* 49:4450 (1968).
11. W. DeW. Horrocks, Jr. and D. R. Sudnick, Lanthanide ion probes of structure in biology. Laser-induced luminescence decay constants provide a direct measure of the number of metal-coordinated water molecules, *J. Am. Chem. Soc.* 101:334 (1979).
12. Y. Hass and G. Stein, Pathways of radiative and radiationless transitions in europium(III) solutions: role of solvents and anions, *J. Phys. Chem.* 75:3668 (1971).
13. D. G. Johnson, Rapid calculation of vacuum wavenumbers, *Appl. Spec.* 33:183 (1979).
14. J. V. Beitz and J. C. Sullivan, Laser-induced fluorescence studies of europium-extractant complexes in organic phases, *J. Less-Common Metals*, 148:159 (1989).
15. B. Schrader and W. Meier, "Raman/IR Atlas of Organic Compounds", Verlag Chemie, Weinheim, (1975).
16. T. J. Bruno and P. D. N. Svoronos. "CRC Handbook of Basic Tables for Chemical Analysis," CRC Press, Boca Raton (1989).

# USE OF SELECTIVE INORGANIC ION EXCHANGERS FOR THE SEPARATION OF RARE EARTHS

R. Cahill, B. Shpeizer, G.-Z. Peng, L. Bortun, and A. Clearfield*
Texas A&M University, Department of Chemistry, College Station, TX 77843

## ABSTRACT

Presented is a preliminary study of the adsorption of lanthanides on several types of inorganic ion exchangers. The exchangers used are a proton form of $Na_4Ti_9O_{20}$, $Zr(C_6H_4P_2O_6)_x(HPO_4)_{2-2x}{\cdot}xH_2O$ (ZrMPP), and $Zr(C_{12}H_8P_2O_6)_x(HPO_4)_{2-2x}{\cdot}xH_2O$ (ZrBPP). The distribution coefficient for some of the lanthanide ions on these materials suggest that they can be used to remove lanthanide ions from solutions containing alkali and alkaline earth metal ions, and can also be used for separation of the individual lanthanide ions by column experiments.

## INTRODUCTION

Recently, inorganic ion exchangers have come under increased study due to the problems arising from the remediation of nuclear waste.[1] Unlike organic ion exchange resins, the inorganic materials are typically radiation stable and can have higher selectivity performance in extremely complex matrices such as those found in nuclear waste tanks and sea water. It was for the above reasons that we chose to examine the exchange properties of some inorganic ion exchangers developed in our laboratory. Previously, the separation of the lanthanides has been carried out by methods such as fractional crystallization, fractional precipitation (which takes advantage of the solubility products of the hydroxide species), and lastly, ion exchange procedures.[2] The ion exchange method for separation of the lanthanides consists of three different types: cation elution chromatography, anion elution

* To whom all correspondence should be addressed.

*Separations of f Elements*, Edited by K.L. Nash
and G.R. Choppin, Plenum Press, New York, 1995 

chromatography, and displacement chromatography. All three methods can be tedious and require the addition of complexing agents such as α-hydroxyisobutyrate and ethylenediamine- N,N,N',N'-tetraacetate systems to achieve adequate separation.[1] These complexing agents are required in order to distinguish between the individual lanthanide ions because without such methods the similarities in ionic radii (Table 1) and other physical properties are insufficient to effect separation. It is hoped that the development of new inorganic materials will help to simplify the ion exchange procedures or to effect special separations in certain hostile environments.

**Table 1.** The ionic radii of the lanthanide metals.

| Element | Radius (Å) | Element | Radius (Å) |
|---|---|---|---|
| La | 1.06 | Tb | 0.923 |
| Ce | 1.03 | Dy | 0.908 |
| Pr | 1.01 | Ho | 0.894 |
| Nd | 0.995 | Er | 0.881 |
| Pm | 0.979 | Tm | 0.869 |
| Sm | 0.964 | Yb | 0.858 |
| Eu | 0.950 | Lu | 0.848 |
| Gd | 0.938 | | |

Previous studies with the sodium nonatitanate have shown this compound to strongly favor a selectivity for the alkaline earth metals over the alkali metals. It was because of this characteristic that investigation of its possible preference for lanthanide or trivalent ions was undertaken. Lehto and co-workers[3] first discovered that this material was selective for $Sr^{2+}$ in the presence of alkali metals in 1981. However, at this time not much was known about the structural characteristics or composition of this particular titanate. Clearfield and Lehto[4] later determined that the material was a layered compound with the formula $Na_4Ti_9O_{20}{\cdot}xH_2O$ and could be made by several synthesis routes.

The other systems of interest are the derivatized $\alpha$-$Zr(HPO_4)_2{\cdot}H_2O$ (ZrP), whose structure was solved by Clearfield *et al.*[5] ZrP is an ion exchanger with an ion exchange capacity of 6.6 meq/g. In 1978, Alberti *et al.*[6,7] prepared an organic derivative of ZrP of composition $Zr(C_6H_5PO_3)_2$. This discovery initiated studies in several laboratories, including ours, in the area of combining both inorganic phosphate or phosphite with organic phosphonic and diphosphonic acids to make mixed-component materials of a single phase.[8,9] When using the organic diphosphonic acid together with inorganic phosphate, the new product contains pores formed by the phosphonate groups bridging the inorganic $ZrO_3P$-layers to produce a 3 dimensional structure with open cavities. By controlling the ratio of phosphonate to phosphate and other experimental conditions, it was hoped to alter the porosity in a systematic manner. This in turn would lead to ion exchangers with a range of selectivities as well as potential catalysts.

## EXPERIMENTAL

The $Na_4Ti_9O_{20}{\cdot}xH_2O$ was made by hydrothermal synthesis using titanium isopropoxide (Aldrich) and sodium hydroxide (Fisher) as has been described previously.[4] The $H_4Ti_9O_{20}{\cdot}xH_2O$ was prepared by reacting $Na_4Ti_9O_{20}{\cdot}xH_2O$ twice with 0.1 M HCl at room temperature for a period of 18 h. This product was collected by filtration and washed with distilled, deionized water and subsequently, dried overnight at 65 °C. This solid was then used for all ion exchange reactions.

The synthesis of the zirconium materials was achieved by reacting $ZrOCl_2{\cdot}8H_2O$, which was obtained as a commercial product from Magnesium Elektron, Inc. and recrystallized from 8 M HCl twice, with a mixture of phosphoric acid (Fisher, reagent grade) and 1,4-phenyldiphosphonic, or 4,4'-biphenyldiphosphonic acid. The phosphonic acids were made using a known procedure[10] with some modifications.[11] The phosphonate:phosphate reaction ratios were 1:1, 1:2, 1:4, 1:6, and 1:8. The mixtures were then placed into plastic bottles, sealed, and reacted for 6 days in a constant temperature oven at 60 °C. The solids were then collected by filtration and washed with water and dried at 60 °C.

Ion exchange reactions were carried out by placing 50-100 mg of sample in 20 mL of a 0.001 M solution of the metal ion of choice and shaken for 20 h at ambient temperature. The solid was separated from the solution by filtration and the resulting filtrate was analyzed for the ion concentration. The lanthanide solutions for the $H_4Ti_9O_{20}{\cdot}xH_2O$ were made using a 0.01 M sodium acetate/acetic acid buffer solution.

## INSTRUMENTATION AND ELEMENTAL ANALYSIS

X-ray powder patterns (XRD) were obtained on a Rigaku RU 200 rotating anode X-ray diffractometer using monochromated CuK$\alpha$ radiation. Thermogravimetric analysis were done using a DuPont Thermal Analyst 2000 under a flowing $O_2$ atmosphere at a heating rate of 10 °C/min. CHN analysis were carried out by Desert Analytics, Tucson, AZ and other elemental analyses were performed by means of an ARL DCP-AES spectrophotometer or a Varian 250 plus AA unit.

## RESULTS AND DISCUSSION

*Nonatitantate Exchanger*: It must first be pointed out that the sodium nonatitanate could not be used for the separation of lanthanides due to its ease of hydrolysis in water. This reaction is shown below.

$$Na_4Ti_9O_{20} + xH_2O \longrightarrow Na_{4-x}H_xTi_9O_{20} + xNaOH \qquad (1)$$

The production of hydroxide ion results in a pH high enough to cause precipitation of the lanthanide ion onto the solid and therefore, prevents the ion exchange reaction from occurring. The pH of the exchange solutions increased above the precipitation pH of the lanthanides even in the presence of 0.01 M sodium acetate/acetic acid and 0.001 M HCl solutions. For this reason, the $Na_4Ti_9O_{20}{\cdot}xH_2O$ was converted to $H_4Ti_9O_{20}{\cdot}xH_2O$ so that more control over the pH could be maintained. The XRD powder pattern of this material is shown in Figure 1. The layered nature of the H-nonatitanate has been adequately described.[4] TGA analysis, followed by examination of the materials obtained at corresponding temperatures, shows that the $Na_4Ti_9O_{20}{\cdot}xH_2O$ is the more stable phase with decomposition of the compound to other sodium titanates[4] occurring at temperatures above 350 °C but with complete water loss occurring at 200 °C (Figure 2). The $H_4Ti_9O_{20}{\cdot}xH_2O$ decomposes to form phases of $TiO_2$ at 200 °C.

The distribution coefficients ($K_D$) are defined as the concentration of cation in the exchanger divided by the concentration of cation in solution with units of mL/g. The

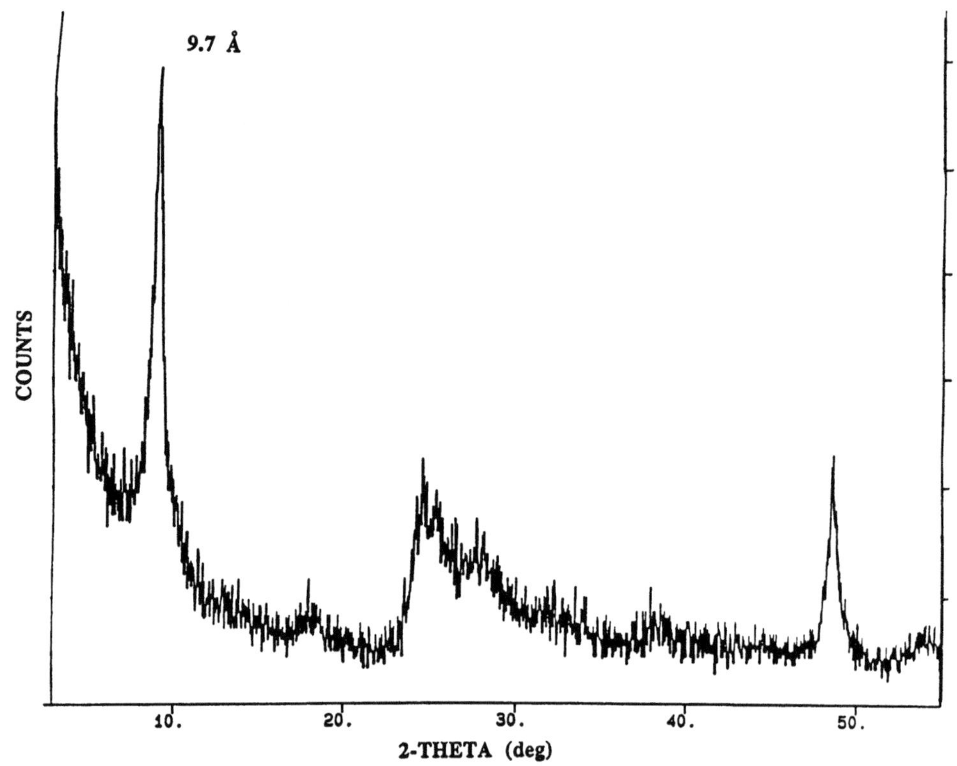


Figure 1. X-ray powder diffraction pattern of $H_4Ti_9O_{20}{\cdot}xH_2O$.

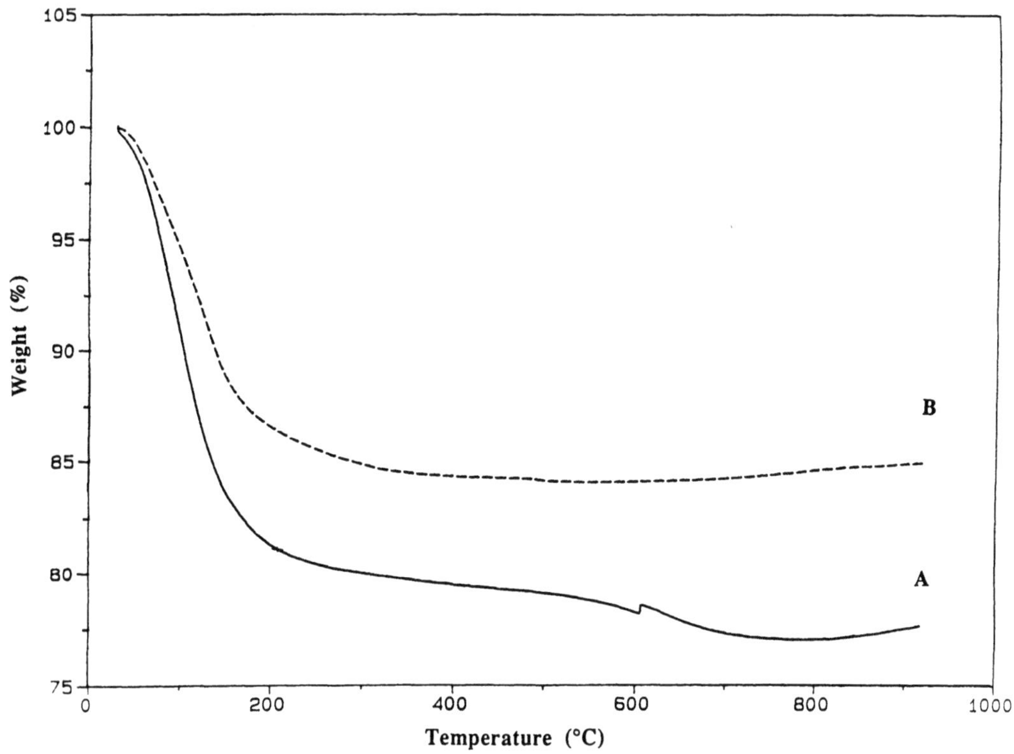


**Figure 2.** Thermogravimetric analysis curves of (A) $Na_4Ti_9O_{20}{\cdot}xH_2O$ (solid line) and (B) $H_4Ti_9O_{20}{\cdot}xH_2O$ (dashed line).

lanthanide $K_D$'s for the $H_4Ti_9O_{20}{\cdot}xH_2O$ exchanger are shown in Table 2. At the low pH range, the distribution coefficients are relatively low for all the lanthanides examined. However, by increasing the pH from 3.0 to 4.2, a definite increase in $K_D$ is observed as well

**Table 2.** $K_D$ values for lanthanides on the $H_4Ti_9O_{20}{\cdot}xH_2O$ exchanger along with the α values and order of elution of the ions.

| Sample | $La^{3+}$ | $Nd^{3+}$ | $Eu^{3+}$ | $Dy^{3+}$ | $Yb^{3+}$ |
|---|---|---|---|---|---|
| 3-75 in natural pH | 355 | 352 | 386 | 370 | 141 |
| Initial pH | 3.50 | 3.95 | 3.97 | 3.99 | 2.88 |
| Final pH | 2.96 | 2.87 | 2.98 | 2.94 | 2.70 |
| 3-75 in 0.01M NaAc /HAc | 2570 | 1280 | 3240 | 2800 | 1380 |
| Initial pH | 4.43 | 4.35 | 4.38 | 4.39 | 4.23 |
| Final pH | 4.32 | 4.36 | 4.22 | 4.25 | 4.11 |

| Order of Elution | Nd-Yb | Yb-La | La-Dy | Dy-Eu |
|---|---|---|---|---|
| α | 1.1 | 1.9 | 1.1 | 1.2 |

an increase in the separation factors for the lanthanide elements. The separation factor (α), which is defined as the $K_{D(A)}/K_{D(B)}$, is also shown in Table 2. From the $K_D$ values for the acetic acid/sodium acetate buffer solution, it can be seen that $Eu^{3+}$ is the most tightly held ion. The actual trend is $Nd^{3+}<La^{3+}<Eu^{3+}>Dy^{3+}>Yb^{3+}$ which suggests that an ionic radius approaching 0.950 Å is optimal for the exchanger. As the radius increases or decreases the $K_D$ value decreases. This suggests that the nonatitanate could only be used to separate particular regions of the lanthanide series via column techniques at the pH investigated. One of the interesting characteristics of the nonatitanate is the fact that because of the relatively low magnitude $K_D$ values, the separation for column experiments can be carried out using just a sodium acetate/acetic acid buffered solution for the elution of the lanthanides by pH variation. The solid can then be easily regenerated by washing with a 0.1 M HCl solution. Column experiments using $H_4Ti_9O_{20}{\cdot}xH_2O$ as the stationary phase are currently under investigation after first assessing the affect of pH on $K_D$.

*Zirconium 1,4-Phenyldiphosphonate/Phosphate Exchangers*: Unlike the nonatitanate whose structure is not known, the structure of the α-zirconium 1,4-phenyldiphosphonate/phosphate (ZrMPP) can be given a generalized description based on the known structure of α-zirconium phosphate[5] which forms the backbone of ZrMPP and zirconium phenylphosphonate, its closest analogue.[12] A more detailed description of these mixed derivatives of zirconium phosphate has been reported in the literature.[8,9,13] An idealized structure of ZrMPP is shown in Figure 3. The Zr atoms are bridged together by P-O bonds to form infinite 2

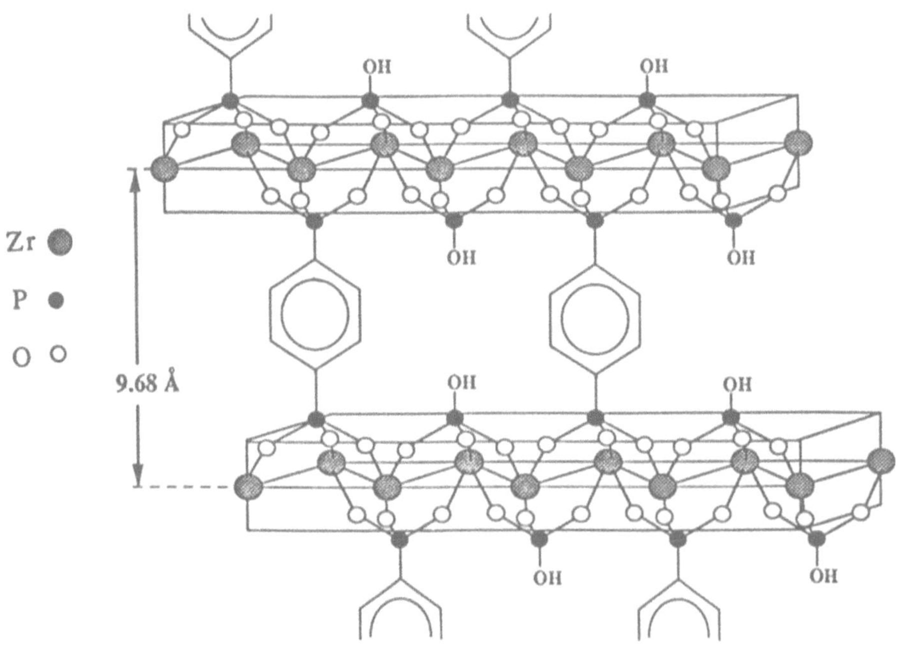


**Figure 3.** Idealized structure of ZrMPP.

dimensional sheets which are then cross-linked by the diphosphonate groups. The phosphate P-OH groups point towards the next layer (into the gallery region). Interestingly enough, these materials can be made to be either semi-crystalline or amorphous depending on whether or not the reaction is carried out in the presence of HF. This study will concentrate only on the amorphous samples. Empirical formulae of the monophenyl derivatized ZrP are shown in Table 3. A characteristic XRD powder pattern and TGA curve is shown in Figure 4A and 5A, respectively. The material is fairly amorphous with a very broad reflection at 10.6 Å

**Table 3.** Empirical formulas derived from the elemental analysis for the zirconium phosphonate/phosphate materials.

| Sample | Ratio | Formula |
|---|---|---|
| 1P-1 | 1:1 | $Zr(C_6H_4P_2O_6)_{0.885}(HPO_4)_{0.229} \cdot 2.2H_2O$ |
| 1P-2 | 1:2 | $Zr(C_6H_4P_2O_6)_{0.60}(HPO_4)_{0.80} \cdot 2.25H_2O$ |
| 1P-3 | 1:4 | $Zr(C_6H_4P_2O_6)_{0.33}(HPO_4)_{1.34} \cdot 2.3H_2O$ |
| 1P-4 | 1:6 | $Zr(C_6H_4P_2O_6)_{0.21}(HPO_4)_{1.58} \cdot 2.6H_2O$ |
| 1P-5 | 1:8 | $Zr(C_6H_4P_2O_6)_{0.16}(HPO_4)_{1.69} \cdot 3.1H_2O$ |
| 2P-1 | 1:1 | $Zr(C_{12}H_8P_2O_6)_{0.64}(HPO_4)_{0.71} \cdot 2H_2O$ |
| 2P-2 | 1:2 | $Zr(C_{12}H_8P_2O_6)_{0.53}(HPO_4)_{0.94} \cdot 0.28H_2O$ |
| 2P-3 | 1:4 | $Zr(C_{12}H_8P_2O_6)_{0.32}(HPO_4)_{1.36} \cdot 2.5H_2O$ |
| 2P-4 | 1:6 | $Zr(C_{12}H_8P_2O_6)_{0.24}(HPO_4)_{1.51} \cdot 2.5H_2O$ |
| 2P-5 | 1:8 | $Zr(C_{12}H_8P_2O_6)_{0.14}(HPO_4)_{1.72} \cdot 2.5H_2O$ |

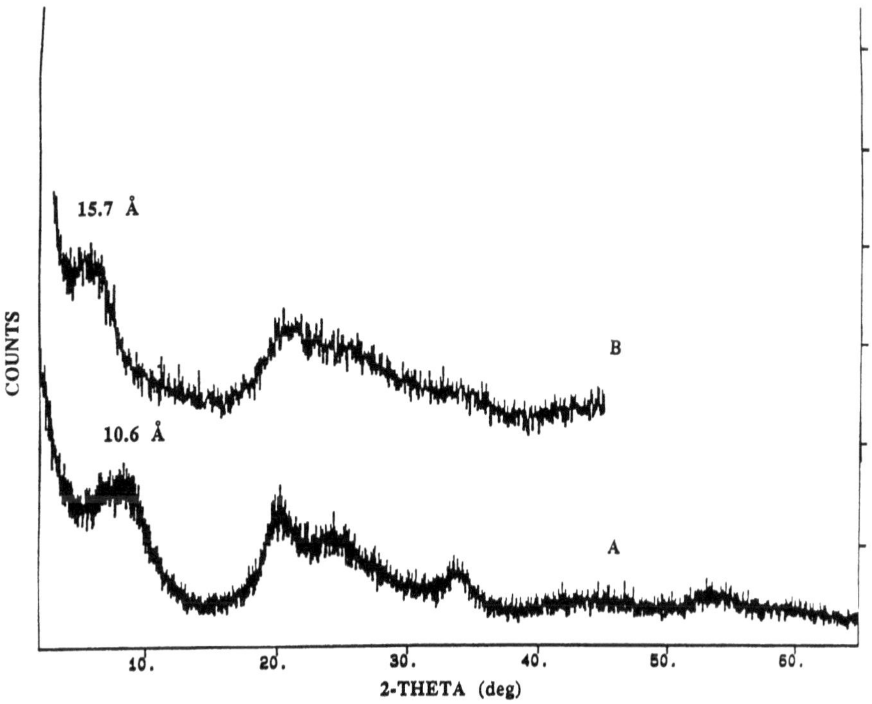


**Figure 4.** X-ray powder diffraction patterns of (A) ZrMPP 1P-1 and (B) ZrBPP 2P-2.

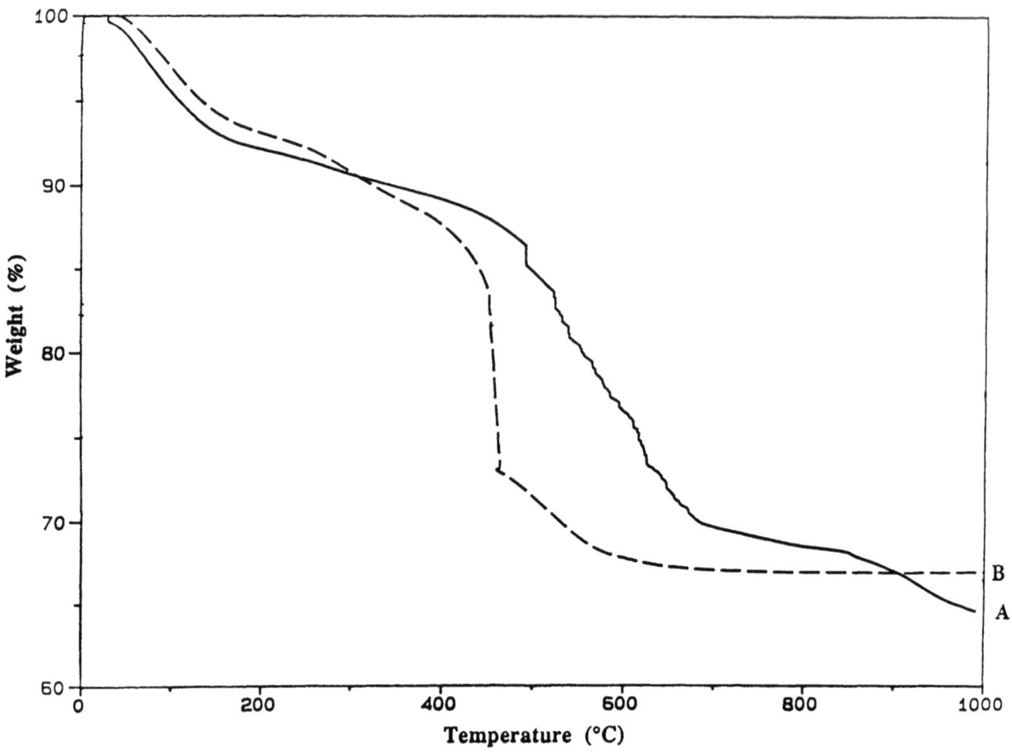


**Figure 5.** TGA curves of (A) ZrMPP 1P-1 (solid line) and (B) ZrBPP 2P-2 (dashed line).

which is larger than the reflection observed for the semi-crystalline phase (9.7 Å). This is attributed to the nonuniformity of the Zr layers. The TGA trace shows the compound to be stable up to 400 °C with no significant weight loss occurring until a temperature of 470 °C is reached which corresponds to the decomposition of the bridging phenyl groups.

The selectivity of these materials for the alkali and alkaline earth metals is considerable lower than that of the lanthanides as seen from Tables 4 and 5. As would be expected, samples with higher phosphate to phosphonate ratios generally have higher selectivities. This could be due to the increased number of exchange sites created by the increased phosphate ligand concentration in the samples. Another important fact to notice is that as the size of the cation increases for the alkali and alkaline earth metals, a corresponding increase in the magnitude of $K_D$ is observed. However, this does not hold true for the lanthanide ions. The $K_D$ typically increases with decreasing size of the ionic radii. A possible explanation for this phenomena can be the hydration sphere of these cations in solution. The hydrated ions, e.g. Li, Na, Mg, and Ca, are the largest in solution and may not have easy access into the interior of the ZrMPP without first losing their hydration shell. The lanthanide hydration shells for each individual lanthanide are of similar size, thus the exception to the previous

**Table 4.** $K_D$ values for the alkali and alkaline earth metal ions on the ZrMPP and ZrBPP samples at a pH=3.

| Sample | Li | Na | K | Rb | Cs | Mg | Ca | Sr | Ba |
|---|---|---|---|---|---|---|---|---|---|
| 1P-1 | <1 | <1 | <1 | 38 | 65 | 27 | 22 | 208 | 154 |
| 1P-2 | <1 | 2.4 | 11 | 140 | 277 | 44 | 27 | 140 | 72 |
| 1P-3 | 5.8 | 5.0 | 97 | 288 | 2090 | 188 | 457 | 595 | 292 |
| 1P-4 | 1.9 | 2.4 | 123 | 522 | 3350 | 320 | 589 | 613 | 420 |
| 1P-5 | <1 | <1 | 133 | 220 | 4500 | 187 | 361 | 461 | 285 |
| 2P-1 | <1 | <1 | 6.6 | <1 | 9.1 | <1 | <1 | <1 | <1 |
| 2P-2 | <1 | <1 | 12.3 | <1 | 20 | <1 | <1 | <1 | <1 |
| 2P-3 | 10.1 | 5.5 | 43.7 | 51 | 103 | 35.7 | 37.0 | 20.4 | 51.3 |
| 2P-4 | 0.8 | 5.9 | 50.9 | 10 | 212 | 126 | 123 | 97.8 | 219 |
| 2P-5 | 10.9 | 11.3 | 126 | 34.8 | 1561 | 207 | 337 | 258 | 687 |

**Table 5.** $K_D$ values for some lanthanide ions on the ZrMPP and ZrBPP samples

| Sample | La | Eu | Gd | Dy |
|---|---|---|---|---|
| 1P-1 | 12700 | 17600 | 93200 | 78500 |
| 1P-2 | 4940 | 23200 | 36400 | 35400 |
| 1P-3 | 23000 | 158000 | 119000 | 132000 |
| 1P-4 | 17400 | 63600 | 54100 | 46300 |
| 1P-5 | 33000 | 313000 | 231000 | - |
| 2P-1 | 411 | 587 | 922 | 1230 |
| 2P-2 | 666 | 2870 | 3940 | 3110 |
| 2P-3 | 3700 | 9950 | 10100 | 13200 |
| 2P-4 | 3200 | 7810 | 6430 | 15700 |
| 2P-5 | 6460 | 9710 | 17800 | 17600 |

class of elements. One of the more interesting properties observed is that the $K_D$'s for the lanthanides are several orders of magnitude greater than those of the 1+ and 2+ ions. This may make these materials ideal for separation of the lanthanides from more complex matrices containing other metals in lower pH regimes (all reactions were carried out at a pH in the neighborhood of 3).The order of elution as well as the separation factors are shown in Table 6. From this data one can conclude that for column separation of the lanthanides examined,

**Table 6.** Elution and $\alpha$ values for the lanthanides on ZrMPP and ZrBPP.

| Sample | Order of Elution and $\alpha$ | | |
|---|---|---|---|
| 1P-1 | La-Eu, 1.4 | Eu-Dy, 4.5 | Dy-Gd, 1.2 |
| 1P-2 | La-Eu, 4.7 | Eu-Dy, 1.5 | Dy-Gd, 1.03 |
| 1P-3 | La-Gd, 5.2 | Gd-Dy, 1.1 | Dy-Eu, 1.2 |
| 1P-4 | La-Dy, 2.7 | Dy-Gd, 1.2 | Gd-Eu, 1.2 |
| 1P-5 | La-Gd, 7.0 | Gd-Eu, 1.4 | Eu-Dy |
| 2P-1 | La-Eu, 1.4 | Eu-Gd, 1.6 | Gd-Dy, 1.4 |
| 2P-2 | La-Eu, 4.3 | Eu-Dy, 1.1 | Dy-Gd, 1.3 |
| 2P-3 | La-Eu, 2.7 | Eu-Gd, 1.01 | Gd-Dy, 1.3 |
| 2P-4 | La-Gd, 2.0 | Gd-Eu, 1.2 | Eu-Dy, 2.0 |
| 2P-5 | La-Eu, 1.5 | Eu-Dy, 1.8 | Dy-Gd, 1.01 |

1P-1 and 1P-3 may be the best of these materials. However, column experiments are needed to determine actual performance.

*Zirconium 4,4'-Biphenyldiphosphonate/Phosphate Exchangers*: The distance between the zirconium phosphate layers can be increased by increasing the size or length of the bridging ligands. The ZrBPP's idealized structure (Figure 6) is similar to that of ZrMPP but with an increase in the distance between the zirconium phosphate layers. A typical XRD powder pattern of these materials is shown in Figure 4B. Again, the materials are fairly amorphous with broad reflections around 15.7 Å. The TGA trace (Figure 5B) shows that the biphenyl

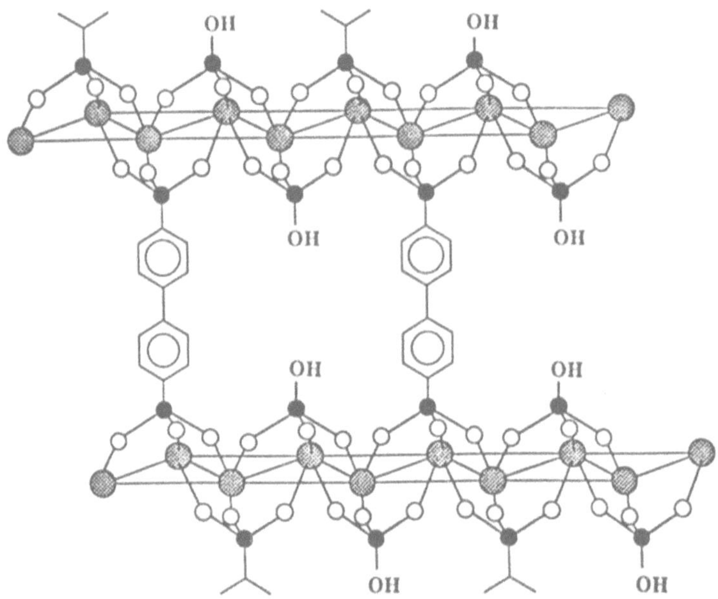

**Figure 6.** Idealized structure of ZrBPP.

derivative has a lower thermal stability than that of the monophenyl materials. Loss of the bridging phenyl groups occurs at a temperature of 420 °C which results in layer collapse of these solids. Empirical formulas for these materials can be found in Table 3 (2P-1 through 2P-5).

The selectivity coefficients for alkali and alkaline earth metals can be seen in Table 4. The $K_D$ values are smaller than those of the ZrMPP compounds and the same holds true for the selectivity of the lanthanide ions. The $K_D$'s overall exhibit an increase with increasing phosphate groups in the materials. This was seen previously in the monophenyl derivative samples. Like the ZrMPP, the ZrBPP samples have selectivity coefficients several orders of magnitude higher for the lanthanide ions investigated (Table 5) than those of the uni and divalent ions. The lower $K_D$ values compared to those of the monophenyl derivatives make these materials more ideal for column experiments. The order of elution of the lanthanide ions is shown in Table 6 with the order generally following La-Eu-Gd-Dy going from the largest (most loosely held ion) to the smallest (most tightly held ion). Sample 2P-1 and 2P-4 would appear to give the overall best separation factors.

## CONCLUSION

It has been shown that inorganic ion exchangers have potential for the separation of individual lanthanide ions as well as to remove them from solutions containing alkali and alkaline earth metal ions. The zirconium phenyldiphosphonate/phosphates appear to be promising materials to be used for removing lanthanides from acid waste solutions including nuclear waste solutions. The zirconium phosphate backbone enhances the radiation stability of the organic pillars. Both the monophenyl and biphenyl derivatives are stable in strong acid media but hydrolyze in strong basic solutions. The sodium nonatitanate, on the other hand, is quite stable in strong base but undergoes decomposition in solutions whose $pH \leq 0$. The proton form of the nonatitanate must be used for the separation of lanthanides to prevent precipitation. The $K_D$ values are much lower than those of organically pillared ZrP, but still show promise for column separation experiments. All of these materials warrant further study for specific separation situations.

## ACKNOWLEDGEMENTS

The authors would like to express appreciation for the funding provided by the National Science Foundation (Grant DM9107715) and DOE (Grant EE1268883-01) which was sponsered by Allied Signal Research Center and Battelle PNL.

## REFERENCES

1. Inorganic ion exchangers and adsorbents for chemical processing in the nuclear fuel cycle, *Proc. Int. Atomic Energy Agency Conf.*, Vienna, June 12-15, 1984. IAEA-TECDOC337, Vienna (1985).
2. J.E. Powell, Separation chemistry, "Handbook on the physics and chemistry of rare earths", K.A. Gschneidner Jr. and L. Eyring, eds., North-Holland Publishing Co., New York, 3 (1979).
3. O.J. Heinonen, J. Lehto, and J.K. Miettinen, Sorption properties of sodium titanate, *Radiochem. Radiaanal. Letters*, 46:381 (1981).

4. A. Clearfield and J. Lehto, Preparation, structure and ion-exchange properties of $Na_4Ti_9O_{20}\cdot xH_2O$, *J. Solid State Chem.*,73:98 (1988).
5. A. Clearfield and G.D. Smith, The crystallography and structure of α-zirconium bis(monohydrogen orthophosphate) monohydrate, *Inorg. Chem.*, 8:431 (1969).
6. G. Alberti, S. Allulli, U. Costantino, and N. Tomassini, Crystalline $Zr(R\text{-}PO_3)_2$ and $Zr(R\text{-}OPO_3)_2$ compounds: a new class of materials having layered structure of the zirconium phosphate type, *J. Inorg. Nucl. Chem.*, 40:1113 (1978).
7. G. Alberti and U. Costantino, J. Kornyei, and M.L. Luciani Giovagnotti, Derivatives of α-zirconium phosphate with two different functional groups, *React. Polym.*, 4:1 (1985).
8. P.M. DiGiacomo and M.B. Dines, Derivatized lamellar phosphates and phosphonates of M(IV) ions, *Inorg. Chem.*, 20:92 (1981).
9. K.P. Callahan, R.E. Cooksey, P.M. DiGiacomo, M.B. Dines, P.C. Griffith, and R.H. Lane, Catalysts supported on layered M(IV) phosphonates, *ACS Symp. Ser.*, 192 (1982).
10. P. Tavs, Reaction of aryl halides with trialkyl phosphites and dialkyl benenephosphonites to aromatic phosphonates and phosphinates by nickel salt catalysed arylation,, *Chem. Ber.*, 103:2428 (1970).
11. G.-Z. Peng, unpublished results, Texas A&M University (1991).
12. F.L. Campbell, A. Clearfield, H.-L. Hu, and M.D. Poojary, Determination of crystal structures from limited powder data sets: the crystal structure of zirconium phenylphosphonate, *Acta Crystallogr.*, B49:996 (1993).
13. A. Clearfield and J.D. Wang, Preparation of layered zirconium phosphonate/phosphate, zirconium phosphonate/phosphite and related compounds, *Mat. Chem. Phys.*, 35:208 (1993).

# FACTORS INFLUENCING THE EFFICIENCIES OF MULTISTAGE SEPARATIONS OF LANTHANIDES

S. Muralidharan, G. Ma, and H. Freiser

Strategic Metals Recovery Research Facility
Department of Chemistry
University of Arizona
Tucson, AZ 85721

## ABSTRACT

Centrifugal partition chromatography (CPC) a multistage liquid-liquid countercurrent distribution technique has been demonstrated to be valuable for the complete separation of adjacent tervalent lanthanide metal ions. Using bis(2,4,4-trimethylpentyl)phosphinic acid (Cyanex 272) and 1-phenyl-3-methyl-4-benzoyl-5-pyrazolone (HPMBP) as the extractants in the heptane-$H_2O$ and toluene-$H_2O$ phase pairs respectively, the base line separations of light and heavy lanthanides at 25 °C were achieved. It was also shown that a mixture of light and heavy lanthanides could be separated in a single run by the use of a pH gradient in the aqueous mobile phase. These separation studies clearly indicated that the main factor limiting the CPC separation efficiencies was the slow kinetics of back-extraction of the lanthanide complexes. The mechanism of the complex dissociation reactions could be discerned from the dependency of the width of the CPC bands on the extractant concentration and pH. These mechanism were independently verified by stopped-flow kinetic studies in micelles formed by the neutral surfactant Triton X-100. Further a linear correlation was found between CETP (channel equivalent of a theoretical plate) determined from the CPC band widths and the half-lives of the dissociation reactions determined by stopped-flow. The importance of CPC not only as a tool for the separation of tervalent lanthanides but also for gaining fundamental understanding of their complex formation and dissociation kinetics was established in this study.

## INTRODUCTION

The complete separation of tervalent lanthanides continues to be a challenging problem in the area of f-element separation. We have examined a variety of ligands, both by themselves and in the presence of auxiliary ligands, to achieve the separation of the tervalent lanthanides.[1] Even the ligands that exhibit the best selectivities for the adjacent

*Separations of f Elements*, Edited by K.L. Nash
and G.R. Choppin, Plenum Press, New York, 1995

lanthanides as indicated by the separation factors (separation factor for a pair of lanthanides = difference in their log $K_{ex}$ values; $K_{ex}$ = extraction equilibrium constant) are incapable of achieving their complete separation in a single stage, necessitating the use of multistage methods. The traditional approach of extraction chromatography wherein a solid support like silica is derivatized or coated with a suitable ligand has several shortcomings, the major ones being the low capacity of the columns and their deterioration.[2] Better approaches are obviously needed to achieve the complete separation of the tervalent lanthanides and one such approach is a relatively new technique that goes by the trade name of centrifugal partition chromatography (CPC), a multistage countercurrent liquid-liquid distribution technique, which has been shown to be effective for a variety of organic and biochemical separations.[3]

## CENTRIFUGAL PARTITION CHROMATOGRAPHY

The CPC apparatus manufactured by Sanki Engineering Company, Japan consists of a series of cartridges, with each cartridge containing 40 - 400 channels depending upon the desired internal volume.[3] These channels serve as a stage in the separation experiment and the total number of channels is 400 - 4800 depending upon the number of cartridges employed. These cartridges are arranged in rotor which is rotated at a certain speed (700 - 1200 rpm) and the centrifugal force generated keeps one the two phases (usually the organic phase) stationary while the other phase (usually the aqueous phase) is moved through it at a constant flow rate. The injected analyte mixture is carried by the aqueous mobile phase into the cartridges where they are extracted into the organic stationary phase by simple distribution if they are organic, or by complexation with a suitable ligand if they are metals. When the mobile phase is depleted of the analytes, further flow of the mobile phase causes the back-extraction (elution) of the analytes which can be detected by a suitable detection method such as uv-vis spectrophotometry. The back-extraction can be conducted by isocratic or gradient elution. If the analytes are completely separated, they appear as discrete peaks very much like in conventional chromatographic methods like HPLC, and hence it is called centrifugal partition chromatography. As such the traditional chromatographic methods of analysis are applicable for the CPC chromatograms. CPC is ideally suited for the complete separation of adjacent lanthanides because of the following advantages: large number of stages (400 - 4800 depending upon the operational volume chosen); high load capacity for extractants and analytes; negligible loss of extractants from the organic phase as bleeding of the stationary phase is minimal; flexible organic-aqueous phase volume ratios; and ready adaptability to process scale. The ratio of the volumes of the stationary and mobile phases in CPC is much bigger than in conventional LC.

The basic parameters employed in the analysis of the CPC chromatograms are : the retention volume $V_r$ which is related to the stationary phase and mobile phase volumes $V_s$ and $V_m$ respectively and the distribution ratio of the analyte D, equation 1; the chromatographic efficiency as measured by the number of theoretical plates N which is calculated from the retention volume $V_r$ and the width of the chromatogram w, equation 2; the chromatographic inefficiency by a quantity analogous to reduced plate height, namely the channel equivalent of a theoretical plate CETP, which is the ratio of the number of channels CH (2400 in our experiments) to N, equation 3; and the selectivity $\alpha$ achieved in the given separation between two analytes 1 and 2 is the ratio of their distribution ratios $D_1$ and $D_2$, equation 4.

Retention:

$$V_r = V_m + DV_s \tag{1}$$

Efficiency:

$$N = 16\left(\frac{V_r}{w}\right)^2 \tag{2}$$

Inefficiency:

$$CETP_{obs} = \frac{CH}{N} \tag{3}$$

Selectivity:

$$\alpha = \frac{D_2}{D_1} \quad (D_2 > D_1) \tag{4}$$

## RESEARCH OBJECTIVES

Our objective in this research is not only to develop CPC for the separation of lanthanides but also to understand the factors that affect the efficiencies of their separations. Such a knowledge would enable us to look for ways to improve their separation efficiencies in order to achieve their separations in the fewest possible number of stages. The parameter that helps us to elucidate these factors is $CETP_{obs}$, which consists of contributions from mass transfer and diffusion as well as other factors such as slow chemical kinetics (vide infra). The dependence of $CETP_{obs}$ on the concentrations of species in the aqueous and organic phases provides insights into kinetic factors limiting the efficiencies of CPC separations. This is a very distinct and major advantage of CPC over extraction chromatography, as in the latter case the presence of a solid support makes the separation of mass transfer and diffusion components, from other components, of the reduced plate height very difficult. CPC is not only a useful tool for the separation of metals but also for obtaining fundamental information on their kinetics. Both bulk and interfacial kinetic information can be derived from CPC chromatograms. The significant results of our research in the separations of lanthanides and their complex formation and dissociation kinetics using CPC are discussed here.

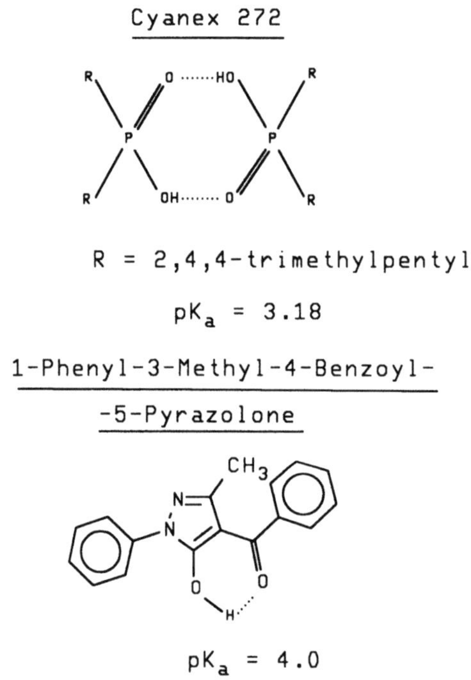


**Figure 1.** Structures of Cyanex 272 and 1-phenyl-3-methyl-4-benzoyl-5-pyrazolone.

## EXTRACTANTS INVESTIGATED

The extractants that we have investigated to achieve the separations of the lanthanides are an organophosphinic acid, bis(2,4,4-trimethylpentyl)phosphinic acid

(Cyanex 272) and an acylpyrazolone, 1-phenyl-3-methyl-4-benzoyl-5-pyrazolone (HPMBP), Figure 1. The extractions of lanthanides need to be performed at pH values ≤ 4 to avoid the formation of hydroxide species of these metals. In this regard, the $pK_a$ values of Cyanex 272 (3.18)[4] and HPMBP (4.0)[5] make them ideally suited for the extraction of lanthanides. The fundamental molecular properties of the two ligands, namely, the dimeric nature of Cyanex 272 in organic solvents which makes it a dimeric chelating agent and HPMBP, an acylpyrazolone existing predominantly in its enol form due to inter/intramolecular hydrogen bonding, make them excellent choices for the investigation of the factors affecting the multistage separations of lanthanides, which is one of the objectives of our research. Further, earlier work in our laboratory involving batch extractions indicated that Cyanex 272 has better selectivity over HPMBP for the lanthanides,[6,7] which in a multistage separation technique like CPC could lead to better resolutions of the adjacent lanthanides with Caynex 272 than with HPMBP.

## SEPARATIONS WITH CYANEX 272 AND HPMBP

The base line separation of lanthanides have been achieved with both the extractants.[8,9] Typical examples of separations with 0.1 M HPMBP using the toluene-$H_2O$ phase pair at pH 2.32 and 0.1 M Cyanex 272 using the heptane-$H_2O$ phase pair at pH = 2.10 are shown in Figures 2 and 3 respectively. The lanthanides were detected at 654 nm by post column derivatization of the lanthanides with arsenazo III.

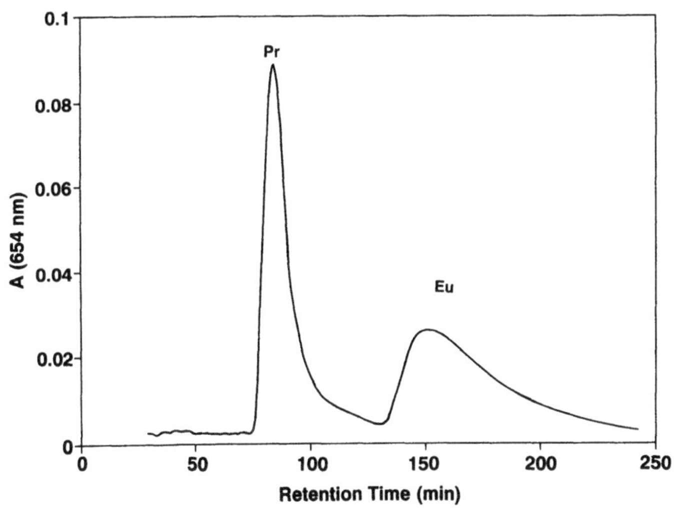


**Figure 2.** Separation of Pr and Eu with 0.1 M HPMBP using the toluene-$H_2O$ phase pair at pH = 2.32 and $V_s/V_m$ = 0.32.

The extraction equilibria for Cyanex 272 and HPMBP are equations 5 and 6 where $(HL)_2$ and HL represent the dimeric Cyanex 272 and monomeric HPMBP respectively, $M^{3+}$ the tervalent lanthanide, and the subscript o the species in the organic phase.

$$M^{3+} + 3(HL)_{2(o)} \overset{K_{ex}}{\rightleftharpoons} M(HL_2)_{3(o)} + 3H^+ \tag{5}$$

$$M^{3+} + 3HL_o \overset{K_{ex}}{\rightleftharpoons} ML_{3(o)} + 3H^+ \tag{6}$$

The corresponding extraction equilibrium constants in terms of the concentrations of the species and the different equilibrium constants are equations 7 and 8, where $K_{DC}$ and $K_{DR}$

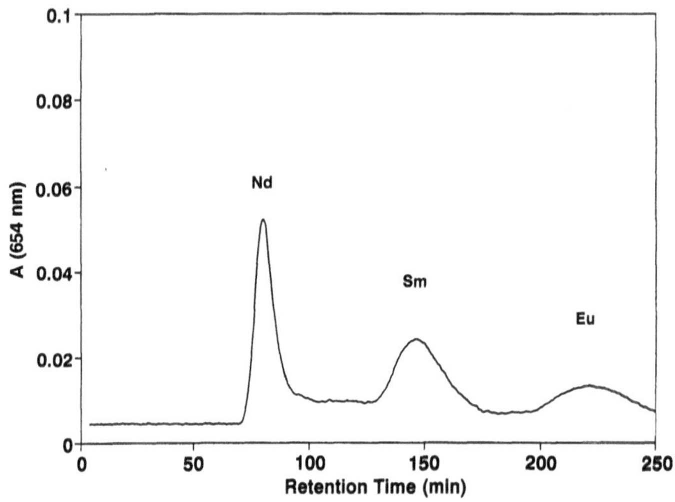


**Figure 3.** Separation of Nd, Sm and Eu with 0.1 M Cyanex 272 using the heptane-$H_2O$ phase pair at pH = 2.10 and $V_s/V_m$ = 0.25.

represent the distribution constants of the metal complex and the ligand respectively, between the organic and aqueous phases, $K_a$ the acid dissociation constant of the ligand in the aqueous phase, $K_d$ the dimerization constant of the ligand in the organic phase, and $\beta$ the stability constant of the complex in the aqueous phase.

The log $K_{ex}$ values for several tervalent lanthanides with these ligands are given in Table 1 which also lists values determined in micelles formed by a neutral surfactant Triton X-100 which will be subsequently discussed.

**Table 1.** Log $K_{ex}$ values in two phase systems and Triton X-100 micelles.

| METAL | CYANEX 272 | | HPMBP | |
|---|---|---|---|---|
| | Micelles | SX[a] | Micelles | SX[a] |
| $Pr^{3+}$ | -1.72 | -3.83 | -0.87 | -3.99 |
| $Eu^{3+}$ | -0.19 | -2.12 | -0.61 | -3.49 |
| $Tb^{3+}$ | 0.35 | -1.22 | -0.42 | -2.86 |
| $Ho^{3+}$ | 0.79 | -0.62 | -0.15 | -2.35 |
| $Yb^{3+}$ | 1.52 | 0.27 | -0.06 | -1.83 |

a. SX : solvent extraction. Cyanex 272 : Heptane-$H_2O$; HPMBP : Toluene-$H_2O$

$$K_{ex} = \frac{[M(HL_2)_3]_o [H^+]^3}{[M^{3+}][(HL)_2]_o^3} = \frac{K_{DC} K_a^6}{K_{DR}^6 K_d^3} \beta \tag{7}$$

$$K_{ex} = \frac{[ML_3]_o [H^+]^3}{[M^{3+}][HL]_o^3} = \frac{K_{DC}}{K_{DR}^3} K_a^3 \beta \tag{8}$$

It may be seen that the separation factors for Cyanex 272 are superior to those for HPMBP. For example, the separation factor for Pr and Eu with Cyanex 272 in the heptane-$H_2O$ phase pair is 1.7 and with HPMBP in the toluene-$H_2O$ phase pair is 0.5. This small difference is still sufficient to obtain the separation of Pr and Eu with HPMBP with a multistage technique like CPC. Complete separation of a mixture of heavy and light lanthanides is possible with both the ligands as indicated by their log $K_{ex}$ values. The differences in the extraction behavior of Cyanex 272 and HPMBP could be due to the differences in the various equilibrium constants that constitute $K_{ex}$, equations 7 and 8. The likely major contributing factor to this difference is β, the stability constant of the metal chelates in the aqueous phase. From the known values of the equilibrium constants[4,5], we may estimate that the β's for the Cyanex 272 complexes are several orders of magnitude larger than those for the HPMBP complexes. It is also evident from the log $K_{ex}$ values that the β increases form the light to the heavy lanthanides for both ligands, with a sharper increase in the case of Cyanex 272 than in the case of HPMBP.

A mixture of light and heavy lanthanides can be separated in a single experiment by using a pH gradient in the aqueous phase. An example of such a separation using 0.1 M Cyanex 272 in the heptane-$H_2O$ phase pair employing a continuous pH gradient is shown in Figure 4. The base line separation of all lanthanides except Sm and Gd were achieved in this experiment which serves to illustrate the value of CPC for the separation of tervalent lanthanides.

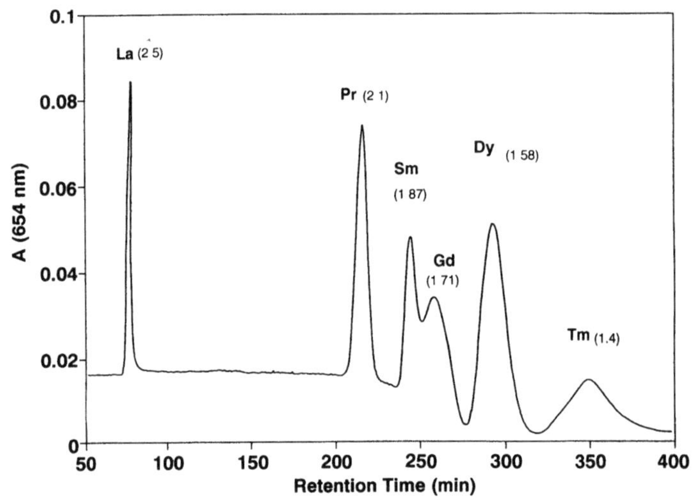


**Figure 4.** Separation of light and heavy lanthanides with 0.1 M Cyanex 272 using the heptane-$H_2O$ phase pair at $V_s/V_m$ = 0.25 and a continuous pH gradient in the aqueous mobile phase.

## CPC BAND WIDTHS AND CHEMICAL KINETICS

The band widths of the lanthanides as can be seen from Figures 2 and 3 are much broader than those encountered in conventional liquid chromatography for organic analytes. The CPC band widths for the lanthanides and other metals in separations involving complexation, are much larger than that for an organic analyte at the same D value. This can be readily seen in Figure 5 where the CPC band widths of 3-picoline and $PdCl_2(TOPO)_2$ in the heptane-$H_2O$ phase pair are compared at identical D values.[10]

The CPC band width in the case of 3-picoline is only due to diffusion and mass transfer as it simply extracts into and back-extracts from, the organic phase. In the case of the separation of metals by complexation, an additional factor namely the complex

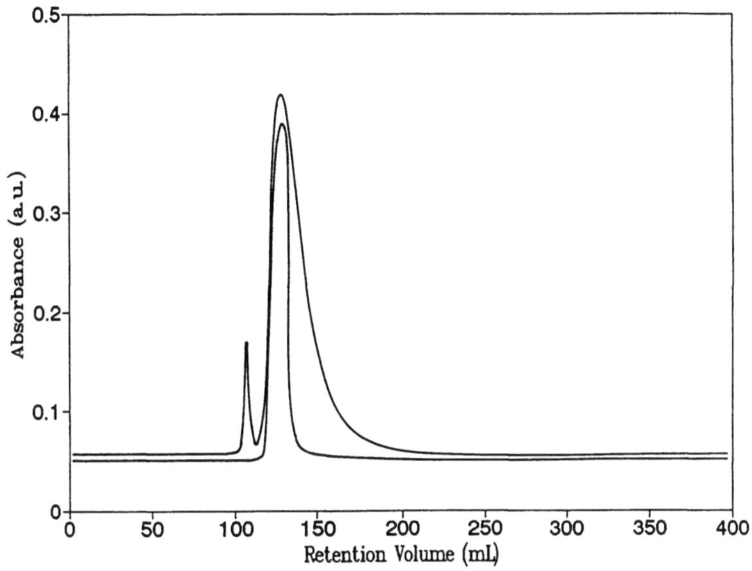


**Figure 5.** CPC chromatograms of 3-picoline and $PdCl_2(TOPO)_2$ in the heptane-$H_2O$ phase pair at identical D values and $V_s/V_m$ = 0.22.

formation and dissociation kinetics is also involved and slow kinetics will lead to broader band widths.[10] The CPC separations are usually conducted under a large excess of the ligand concentration over the metal and as such the metal complex formation reactions are rapid. This is indicated by the D values of the metals in the CPC experiments being equal to the equilibrium D values obtained by batch extraction under identical conditions. We may postulate that the additional band width in the case of metals is most likely due to the slow back-extraction kinetics. On the basis of this hypothesis, the $CETP_{obs}$ of the metal complexes may be expressed as a sum of the contributions from diffusion and mass transfer $CETP_{dif}$, and slow back-extraction kinetics $CETP_{ck}$,
equation 9.

$$CETP_{obs} = CETP_{dif} + CETP_{ck} \quad (9)$$

The $CETP_{dif}$ can be determined by using 3-picoline at identical D values of the metal complex. We have demonstrated that 3-picoline is a good model for determining $CETP_{dif}$ as the CETP of 3-picoline is similar to that for a large ion-pair $(NR_4)_2IrCl_6$ (R = heptyl).[11] We can determine $CETP_{ck}$ from the CPC chromatograms which according to our hypothesis should bear a direct correlation to the half-life of the back-extraction reaction under the conditions of the CPC experiments, equation 10.

$$CETP_{ck} \propto t_{1/2} \quad (10)$$

According to this relation, plotting the $CETP_{ck}$ from CPC measurements against the $t_{1/2}$'s determined by an independent kinetic experiment such as stopped-flow should result in a straight line. This relationship also indicates that the dependence of $CETP_{ck}$ on the concentrations of the species in the aqueous and organic phases should indicate the mechanism of the dissociation reaction and its rate limiting step, even without an independent kinetic study. In other words, CPC is not only a useful separation tool but also a very valuable method for discerning the mechanisms of metal complex formation and dissociation reactions. We have demonstrated the validity of this hypothesis for the Pd(II)-TOPO system[10], and we shall now examine the CPC chromatograms of the lanthanides to derive kinetic information on the back-extraction of the Cyanex 272 and HPMBP complexes.

## METALLOCHROMIC INDICATOR METHOD

The kinetics of the dissociation of the HPMBP and Cyanex 272 complexes of the tervalent lanthanides needs to be independently studied to obtain the $CETP_{ck}$ - $t_{1/2}$ correlation. These complexes do not have distinct uv-vis absorption spectra and their dissociation reactions are too fast to be monitored by spectrophotometry. We designed the "metallochromic indicator method" to study the dissociation reactions of tervalent lanthanides in micelles formed by neutral surfactants like Triton X-100.[12] The principle of the method is the rapid complexation by arsenazo III of the released tervalent lanthanide upon reaction of the lanthanide complex with proton. The formation of the lanthanide-arsenazo III complex is limited by the slow step in the dissociation reaction. As a result, the kinetics of the formation of the lanthanide-arsenazo III complex directly yields the rate constant for the dissociation of the metal complex provided the reaction of the lanthanide-arsenazo III complex with the excess ligand to reform the lanthanide-ligand complex is not significant. The metallochromic method is illustrated in equations 11 - 13 where $ML_3$ is the lanthanide complex and the subscript m represents the species in the micellar pseudophase.

$$ML_{3(m)} + 3H^+ \rightleftharpoons M^{3+} + 3HL \tag{11}$$

$$2M^+ + 2\,\text{arsenazo III} \rightleftharpoons (M\text{-arsenazo III})_2 \tag{12}$$

$$1/2\,(M\text{-arsenazo III})_2 + 3HL_m \rightleftharpoons ML_{3(m)} + 3H^+ + \text{arsenazo III} \tag{13}$$

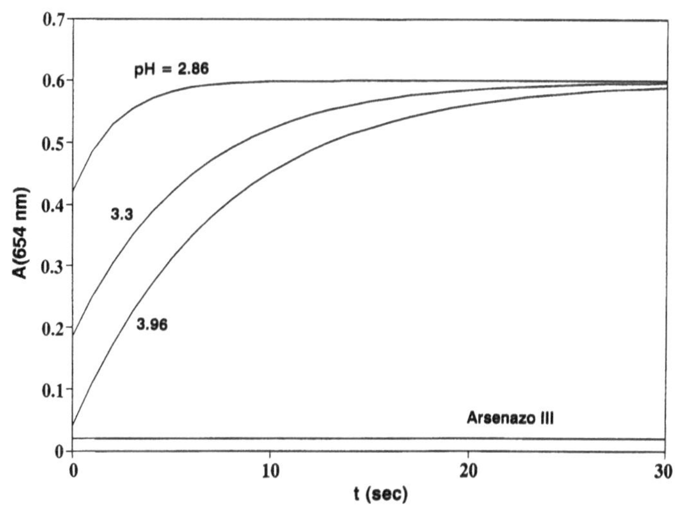


**Figure 6.** Absorbance at 654 nm vs. time plots for the Eu-Cyanex 272 complex formed from [Eu] = 2.2 x $10^{-5}$ M and [Cyanex 272] = 7.5 x $10^{-5}$ M at different initial pH values. The final pH after pH jump = 2.35 and the blank absorbance due to arsenazo III is also shown.

The lanthanide-arsenazo III complex is a dimer.[12] Typical absorbance vs. time changes for the Eu-Cyanex 272 complex is shown in Figure 6 at three different initial pH values with the final pH value being 2.35 in all the cases.

The final absorbances in all the cases are the same, but the initial absorbances are different. This is due to the different amounts of free metal present in the aqueous phase at the different pH values as dictated by the extraction equilibrium constant for the complex for extraction into the micellar pseudophase. The initial absorbance jump is instantaneous as the complexation of the free lanthanide by arsenazo III is rapid and this value yields the free concentration of the metal at the respective initial pH values and hence D value of the metal at that pH. This D value, defined in equation 14 can be used to calculate the extraction equilibrium constant (equations 7 and 8) for the various lanthanides for extraction into the micellar pseudophase by HPMBP and Caynex 272.

$$D = \frac{([M(III) - [M^{3+}]_{init})}{V_m [M^{3+}]_{init}} \qquad (14)$$

Here [M(III)] and $[M^{3+}]_{init}$ are the total and free concentrations of the lanthanide and $V_m$ the volume fraction of the micellar pseudophase (=$\phi$ (c-cmc); $\phi$ = molar volume of the micellar pseudophase; c = concentration of the surfactant; cmc = critical micelle concentration of the surfactant).

The log $K_{ex}$ values in Triton X-100 micelles for HPMBP and Cyanex 272 are shown in Table 1. The trend in the log $K_{ex}$ values parallels the two phase systems with Cyanex 272 exhibiting higher extractibilities and better separation factors than does HPMBP. These values in the micellar system are also two to three orders of magnitude larger than and the separation factors smaller than, the values in the two phase systems. It may also be seen that the log $K_{ex}$ values of the light lanthanides exhibit a much larger increase than the values for the heavy lanthanides, compared to the values in the two phase systems. The variation in the log $K_{ex}$ values for Cyanex 272 and HPMBP in the micelles parallels the variation in the two phase systems in the presence of an auxiliary ligand like trioctylphosphine oxide (TOPO) which forms a 1:1 adduct with these lanthanide complexes.[6,7] This indicates that Triton X-100 is most likely forming an adduct with these complexes providing the enhanced extraction in the micellar systems. A comparison of the log $K_{ex}$ values in the Triton X-100 micelles and the two phase systems indicates that Triton X-100 forms much stronger adducts than does TOPO. Further Triton X-100 like TOPO forms stronger adducts with the lighter lanthanides than with the heavier lanthanides which would explain the larger increase for the log $K_{ex}$ values of the lighter lanthanides in the micellar system. This analysis is important as it is necessary to understand the correlation between $CETP_{ck}$ and $t_{1/2}$ (vide infra).

## MECHANISM OF DISSOCIATION OF LANTHANIDE-HPMBP COMPLEXES

The mechanism of the dissociation of the HPMBP complexes of the lanthanides have been studied in the micelles formed by Triton X-100 by the metallochromic indicator method under pseudo first order conditions in the concentrations of proton and HPMBP.[9] The variation in the pseudo first order rate constant with the concentrations of the ligand, $H^+$ (directly proportional to $[H^+]$ and inversely proportional to [HPMBP]), and metal indicate the mechanism of dissociation of the $M^{3+}$-HPMBP complexes to be equations 15 - 17, with equation 17 being the rate limiting step.

$$ML_3 \underset{\text{fast}}{\overset{K_{-3}}{\rightleftharpoons}} ML_2^+ + L^- \tag{15}$$

$$ML_2^+ \underset{\text{fast}}{\overset{K_{-2}}{\rightleftharpoons}} ML^{2+} + L^- \tag{16}$$

$$ML^{2+} \underset{\text{slow}}{\overset{k_{-1}}{\rightarrow}} M^{3+} + L^- \tag{17}$$

The observed pseudo first order rate constant based on the above mechanism is equation 18 where HL represents HPMBP.

$$k_{obs} = \left(\frac{k_{-1}K_{-2}}{K_a}\right)\frac{[H^+]}{[HL]} \tag{18}$$

We can determine only the product $k_{-1}K_{-2}$ from the kinetic measurements assuming the $K_a$ value in the micellar pseudophase to be similar to that in the aqueous phase. These dissociation rate constants are listed in Table 2 along with the dissociation rate constants for the $M^{3+}$-Cyanex 272 complexes. The dissociation rate constant decreases from the light to the heavy lanthanides, that is the lighter lanthanides dissociate more readily than the heavier lanthanides. The rate constant for the dissociation of the $Yb^{3+}$ complex is about 20 times smaller than the dissociation of the $Pr^{3+}$ complex. This may be expected based on the order of the stability constants of these complexes which increases from the light to the heavy lanthanides.

We will subsequently discuss the mechanism of dissociation of the $M^{3+}$-Cyanex 272 complexes but will make here a brief comparison of the dissociation rate constants of the complexes of Cyanex 272 and HPMBP. The $M^{3+}$-Cyanex 272 complexes exhibit a much larger variation in the dissociation rate constants between the light and heavy lanthanides. There is a three orders of magnitude difference in the rate constants of dissociation of the $Pr^{3+}$ and $Yb^{3+}$ complexes of Cyanex 272 which is much larger than the variation for the HPMBP complexes of these metals. This may stem from the much larger differences in the stability constants of the light and heavy lanthanides in the case of Cyanex 272 compared to HPMBP. This, however does not translate into better CPC efficiencies for the separations of the light lanthanides with Cyanex 272 compared to HPMBP and a reverse situation for the heavy lanthanides. This is because the pseudo first order dissociation rate constant in the case of the Cyanex 272 complexes is only a function of the concentration of $H^+$ while in the case of HPMBP it is a function of both the concentrations of $H^+$ and HPMBP (vide supra). In addition, as will be discussed shortly the dissociation reactions of the $M^{3+}$-Cyanex 272 complexes is more complex than the $M^{3+}$-HPMBP complexes.

**Table 2.** The dissociation rate constants for the lanthanide complexes of Cyanex 272 and HPMBP in Triton X-100 micelles.

| METAL | CYANEX 272[a] | HPMBP[b] |
|---|---|---|
| $Pr^{3+}$ | 501 | 251 |
| $Eu^{3+}$ | 69.7 | 53.7 |
| $Tb^{3+}$ | 7.1 | 38.0 |
| $Ho^{3+}$ | 1.7 | 19.5 |
| $Yb^{3+}$ | 0.72 | 14.5 |

a. rate constants ($k_{3m}$) in $M^{-1}s^{-1}$. b. rate constants ($k_{-1}K_{-2}$) in $s^{-1}$.

It is evident from equation 18 that $t_{1/2}$ ($t_{1/2} = 0.693/k_{obs}$) for the dissociation of the $M^{3+}$-HPMBP complexes is directly proportional to [HPMBP] and inversely proportional to [$H^+$] which should also be the orders for the dependence of $CETP_{ck}$ on the concentrations of these species (see equations 9 and 10). The variation of the $CETP_{ck}$ values of the light and heavy lanthanides listed in Table 2 with the concentrations of $H^+$ and HPMBP clearly indicate this to be the case.

In the case of HPMBP the best CPC efficiencies are obtained when the separations are carried out at as low a concentration of the ligand as possible, taking into consideration the required capacity of the column for the analyte mixture on hand and the attainment of extraction equilibrium. The rate constants for the dissociation of the $M^{3+}$-HPMBP also indicate that the CPC efficiencies for the lighter lanthanides can be expected to be better than those for the heavier lanthanides which in fact is observed. The formation rate constants of the $M^{3+}$-HPMBP complexes also decreases from the light to the heavy lanthanides, requiring the use of larger concentrations of the ligand for the heavier lanthanides to attain extraction equilibrium. This in turn makes the base line separation of a mixture of heavy lanthanides difficult to obtain with HPMBP requiring

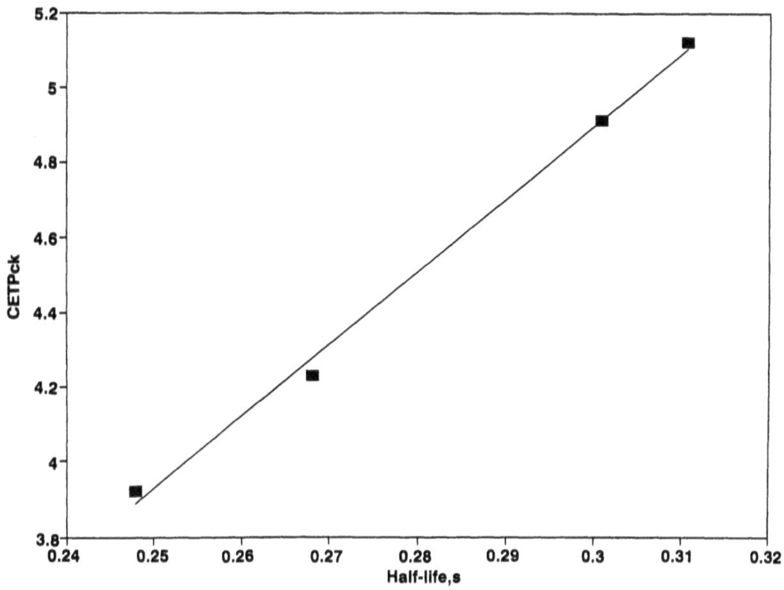


**Figure 7.** $CETP_{ck}$ vs. $t_{1/2}$ for the $Eu^{3+}$ complex of 1-phenyl-3-methyl-4-benzoyl-5-pyrazolone.

the use of gradient pH to achieve their complete separation. The linear correlation postulated in equation 10 is supported by the $CETP_{ck}$ vs. $t_{1/2}$ plot in Figure 7 for Eu, where the $t_{1/2}$ has been calculated using equation 18 under the conditions of the CPC experiment.

## MECHANISM OF DISSOCIATION OF LANTHANIDE-CYANEX 272 COMPLEXES

The dissociation kinetics in Triton X-100 micelles indicates the rate limiting step in the dissociation of the $M^{3+}$-Cyanex 272 complexes to be the reaction of the $M(HL_2)_3$ complex with $H^+$, equation 19.[12]

$$M(HL_2)_{3,m} + H^+ \underset{k_{3m}}{\overset{slow}{\rightarrow}} M(HL_2)_2^+ + (HL)_{2,m} \tag{19}$$

The rate constants for the dissociation of the complexes of the various lanthanides examined are listed in Table 2. As mentioned above, the dissociation rate constants decrease from the light to the heavy lanthanides which is reflected in the CPC chromatograms by the lower efficiencies (larger CETP values) for the heavy lanthanides compared to the light ones under identical conditions. The kinetic studies conducted as a function of the concentration of Cyanex 272 indicates that there are two kinds of $M^{3+}$-Cyanex 272 complexes extracted into the micellar pseudophase depending on the concentration of Cyanex 272, namely the 1:3 complex $M(HL_2)_3$, and the 1:2.5 complex $M(HL_2)_2L$. The rate limiting step in the dissociation of the $M(HL_2)L$ complex, equation 20 is similar to equation 19.

$$M(HL_2)_2L_m + H^+ \underset{k_{2.5m}}{\overset{slow}{\rightarrow}} M(HL_2)_2^+ \; HL_m \tag{20}$$

The equilibrium constant $K_{add}$ for the interconversion of the $M(HL_2)_2L$ and $M(HL_2)_3$ complexes, equation 21 can be calculated from the $K_{ex,3}$ (Table 3) and $K_{ex,2.5}$[12] values determined as discussed under the metallochromic indicator method above, equation 14. The $K_{add}$ as can be seen from equation 21 is $K_{ex,3}/K_{ex,2.5}$ and it has roughly the same value (~30) for the light and heavy lanthanides. The dissociation kinetics of the $M^{3+}$-Cyanex 272 complexes in Triton X-100 micelles is however complex due to the simultaneous dissociation of the two different complexes.

$$M(HL_2)_2L_m + 0.5(HL)_{2_m} \underset{fast}{\overset{K_{add}}{\rightleftharpoons}} M(HL_2)_{3_m} \tag{21}$$

The observed pseudo first order rate constant $k_{obs}$ is a composite of the rate constants $k_{3m}$ (Table 2) and $k_{2.5m}$[12], equation 22.

$$k_{obs} = \left\{ \frac{k_{3m} K_{add} [(HL)_2]_m^{0.5} + k_{2.5m}}{1 + K_{add} [(HL)_2]_m^{0.5}} \right\} [H^+] \tag{22}$$

The $M(HL_2)_2L$ complexes dissociate much more rapidly than the $M(HL_2)_3$ complexes as indicated by the $k_{2.5m}$ values. For example the $k_{2.5m}$ values for $Eu^{3+}$ and $Yb^{3+}$ complexes are 400 and 2.5 $M^{-1}s^{-1}$ respectively which are 4 - 6 times larger than the $k_{3m}$ values for these metals. This detailed.study of the kinetics of dissociation of the $M^{3+}$-Cyanex 272 complexes is essential to understand their behavior in the CPC separations.

## CORRELATION OF CPC INEFFICIENCIES WITH DISSOCIATION KINETICS FOR THE LANTHANIDE-CYANEX 272 COMPLEXES

The CPC experiments unlike the kinetic studies in the Triton X-100 micelles were conducted with high concentrations of Cyanex 272 (0.075 - 0.15 M) and very low concentrations of the metal ions (~ $10^{-5}$ M). It is reasonable to expect with such a large excess of the concentration of ligand over the concentration of the metal that $M(HL_2)_3$ will be exclusively formed in the heptane phase. This is supported by dependence of log D values determined from batch and CPC experiments on pH and the concentration of Cyanex 272. There is no indication of the formation of the $M(HL_2)_2L$ complex in these experiments. As such, we could expect the $CETP_{ck}$ of the lanthanides to bear a linear relationship with the concentration of $H^+$ ($t_{1/2}$ = 0.693/[$H^+$], equation 19), with the log $CETP_{ck}$ vs. pH plot having a slope of 1. Further, the $CETP_{ck}$ values should also be independent of the concentration of the ligand.

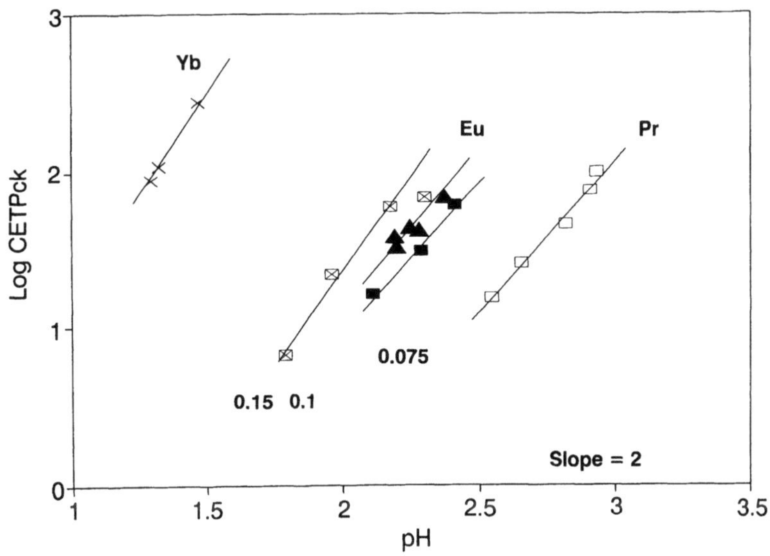


**Figure 8.** Log $CETP_{ck}$ vs. pH plots for Pr, Eu and Yb. The concentration of Cyanex 272 for Pr and Yb was 0.1 M and for Eu 0.075, 0.1 and 0.15 M concentrations of ligand were used.

The CPC experiments however reveal quite different dependencies, the log $CETP_{ck}$ vs. pH plots having a slope +2 and the log $CETP_{ck}$ vs. log $[(HL)_2]$ plots having a slope +0.5. The log $CETP_{ck}$ vs. pH plot is shown in Figure 8 for Pr, Eu and Yb all exhibiting a slope of +2.[13] It is also clear from this figure that the $CETP_{ck}$ values are not independent of the concentration of Cyanex 272 as illustrated by the $CETP_{ck}$ values of $Eu^{3+}$ at the various concentrations of Cyanex 272.

The clue to this apparent disagreement between the expected and observed dependencies of $CETP_{ck}$ on the concentrations of $H^+$ and Cyanex 272 lies in the kinetic studies conducted in Triton X-100 micelles. These studies revealed that two types of complexes $M(HL_2)_2L$ and $M(HL_2)_3$ are extracted into the micellar pseudophase and that the former complex has a higher rate constant for dissociation than the latter. The observed CPC behavior can be rationalized if the dissociation in the heptane-$H_2O$ system proceeds through the $M(HL_2)_2L$ complex formed rapidly from $M(HL_2)_3$ as the former complex dissociates much more rapidly than the latter. The dissociation of $M(HL_2)_2L$ by proton occurs through a rapid pre-equilibrium step where $M(HL_2)_2L$ is protonated and this protonated complex reacts with a second proton in a rate limiting step.

These reactions are shown in equations 23 - 25 where $K_c$ in equation 24 is the reciprocal of $K_{add}$ defined in equation 21 for the conversion of $M(HL_2)_2L$ to $M(HL_2)_3$.

$$M(HL_2)_3 \underset{\text{fast}}{\overset{K_C}{\rightleftharpoons}} M(HL_2)_2L + 0.5(HL)_2 \tag{23}$$

$$M(HL_2)_2L + H^+ \underset{\text{fast}}{\overset{K_p}{\rightleftharpoons}} M(HL_2)_2HL^+ \tag{24}$$

$$M(HL_2)_2HL^+ + H^+ \underset{\text{slow}}{\overset{k_{-1}}{\rightleftharpoons}} M(HL_2)HL^{2+} + (HL)_2 \tag{25}$$

The rate of dissociation of the $M^{3+}$-Cyanex 272 complex based on this mechanism is as written in equation 26 where the quantity within the braces is the pseudo first order rate constant $k_{obs}$. The $t_{1/2}$ and hence $CETP_{ck}$ clearly have the dependencies on the concentrations of $H^+$ and Cyanex 272 indicated in equation 27.

$$\text{Rate} = \left\{ k_{-1} K_C K_p \frac{[H^+]^2}{[(HL)_2]^{0.5}} \right\} [M(HL_2)_3] \tag{26}$$

$$t_{1/2} \propto \frac{[(HL)_2]^{0.5}}{[H^+]^2} \tag{27}$$

This mechanism explains the observed dependencies of log $CETP_{ck}$ on pH (slope +2) and log $[(HL)_2]$ (slope +0.5). The difference between micelles and the heptane-$H_2O$ system lies in the pre-equilibrium step, equation 24 where the complex $M(HL)_2L$ is protonated. The product $K_p[H^+]$ is much smaller than 1 in the two phase system while in the Triton X-100 micelles it is much larger than 1.

The dissociation reaction in the micelles occurs almost exclusively at the aqueous-micellar interface,[12] but in the two phase system could have interfacial and bulk aqueous components. The micellar pseudophase facilitates the protonation of $M(HL_2)_2L$ due to adduct formation between the surfactant molecules and the complex and in particular due to the oxyethylene chains of the surfactant molecules. If the dissociation reaction in the two phase system occurs predominantly in the bulk aqueous phase, then we may expect $K_p[H^+]$ to be much larger than 1 as in the case of the micellar pseudophase as the protonation of the $M(HL_2)_2L$ should proceed readily in $H_2O$. The kinetic results however indicate that $K_p[H^+]$ is much smaller than 1 in the two phase system leading to the conclusion that the dissociation reactions are predominantly interfacial in this medium as well.

The foregoing analysis not only reveals the subtle difference in the dissociation mechanism between micelles and two phase systems but also helps to determine the location of the reaction in the two phase system.

It is evident from Figure 9, that the $CETP_{ck}$ values bear a linear correlation to the $t_{1/2}$ values determined in micelles. A striking feature of this correlation is that this is not general, that is, all the points do not fall on a single straight line. In other words, the same CPC band width represents a different half-life for each lanthanide. This is partly due to the differences in the $K_{DC}$ values in the micellar and two phase systems as evident from equation 7. A more significant factor contributing to the different $t_{1/2}$ values for the lanthanides at the same $CETP_{ck}$ value is the subtle change in the mechanism of dissociation from the micellar to the two phase system.

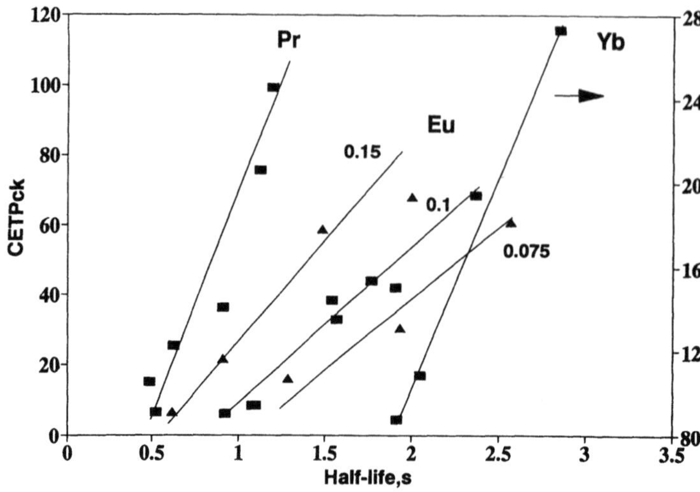


**Figure 9.** $CETP_{ck}$ vs. $t_{1/2}$ for the lanthanide-Cyanex 272 complexes. The ligand concentration in the case of Pr and Yb was 0.1 M and the Eu experiments were conducted at 0.075, 0.1 and 0.15 M.

## CONCLUSIONS

The studies with Cyanex 272 and HPMBP clearly show the value of CPC for achieving complete separations of the tervalent lanthanides despite the closeness in their extraction equilibrium constants. The CPC studies clearly show that slow kinetics of metal complex dissociation reactions limit the efficiencies of the CPC separations. This could be to some extent overcome by employing a pH gradient in the mobile phase. More significantly, the CPC band widths provide information on the kinetics of back-extraction of these complexes. A direct linear correlation exists between the CPC inefficiencies and the half-lives of the dissociation reactions measured independently in micelles. This correlation is not general due to the differences in the $K_{DC}$ values between the two phase and micellar systems, adduct formation between the lanthanide complexes and the surfactant molecules and the different degrees of participation of the bulk and interfacial reactions in the micellar and two phase systems (micellar system has exclusively interfacial kinetics and the two phase system has both bulk and interfacial kinetics). The correlation of the CPC inefficiencies with the half-lives of the dissociation reactions can also reveal subtle differences in the mechanism of the dissociation reaction in these two media and can further enable to determine the locus of the dissociation reactions in the two phase systems.

## REFERENCES


1. H. Freiser, Solvent extraction of tervalent lanthanides as chelates. A systematic investigation of extraction equilibria, *Solv. Extr. Ion Exch.* 6:1093 (1988).
2. S. Siekierski and I. Fidelis, Extraction Chromatography of Lanthanides, in: "Extraction Chromatography", T. Braun and G. Ghersini, eds., Elsevier Scientific Publishing Company, Amsterdam (1975).
3. D. W. Armstrong, Theory and use of centrifugal partition chromatography, *J. Liquid Chromatogr.* 11:2433 (1988).
4. Li Ke-an, S. Muralidharan and H. Freiser, Determination of the equilibrium constants of organophosphorus liquid-liquid extractants by inductively coupled plasma-atomic emission spectroscopy, *Solv. Extr. Ion Exch.* 3:895 (1985).
5. J. Stary and H. Freiser, "Equilibrium Constants of Liquid-Liquid Distribution Reactions. Part IV: Chelating Extractants", Pergamon Press, Oxford (1978).
6. Y. Komatsu and H. Freiser, Extraction separation of tervalent lanthanide metals with bis(2,4,4-trimethylpentyl)phsophinic acid, *Anal. Chim. Acta*, 227:397 (1989).
7. Y. Sasaki and H. Freiser, Mixed-ligand chelate extraction of lanthanides with 1-phenyl-3-methyl-4-acyl5-pyrazolones, *Inorg. Chem.* 22:2289 (1983).
8. R. Cai, S. Muralidharan and H. Freiser, Improved separation of closely related metal ions by centrifugal partition chromatography, *J. Liquid Chromatogr.* 13:3651 (1990).
9. G. Ma, H. Freiser and S. Muralidharan, unpublished results.
10. Y. Surakitbanharn, S. Muralidharan and H. Freiser, Centrifugal partition chromatography of palladium(II) and the influence of chemical kinetic factors on separation efficiency, *Anal. Chem.* 63:2642 (1991).
11. E. Ma, H. Freiser and S. Muralidharan, unpublished results.
12. K. Inaba, S. Muralidharan and H. Freiser, Simultaneous characterization of extraction equilibria and back-extraction kinetics: use of arsenazo III to characterize lanthanide-bis(2,4,4,-trimethylpentyl)phosphinic acid complexes in surfactant micelles, *Anal Chem.* 65:1510 (1993).
13. K. Inaba, H. Freiser and S. Muralidharan, Effect of kinetic factors on the efficiencies of centrifugal partition chromatographic separations of tervalent lanthanides with bis(2,4,4-trimethylpentyl)phosphinic acid as extractant, *Solv. Extrn. Res. Develop.* 1:13 (1994).

# REMOVAL OF PLUTONIUM AND AMERICIUM FROM HYDROCHLORIC ACID WASTE STREAMS USING EXTRACTION CHOMATOGRAPHY

Louis D. Schulte[1], John R. FitzPatrick[2], Richard R. Salazar[1], Bradley S. Schake[2] and Benjie T. Martinez[1]

[1]Aqueous Chloride Processing, Group NMT-3,
[2]Chemical Sciences and Technology, Group CST-3
Los Alamos National Laboratory
MS E511 P.O. Box 1663
Los Alamos, NM 87545

## ABSTRACT

Extraction chromatography is under development as a method to lower actinide activity levels in hydrochloric acid (HCl) effluent steams. Successful application of this technique for radioactive liquid waste treatment would provide a low activity feedstream for HCl recycle, reduce the loss of radioactivity to the environment in aqueous effluents, and lower the quantity and improve the form of solid waste generated.

The extraction of plutonium and americium from HCl solutions was examined for several commercial and laboratory-produced sorbed resin materials. Polymer beads were coated with n-octyl(phenyl)-N,N-diisobutylcarbamoylmethylphosphine oxide (CMPO) and either tributyl phosphate (TBP), or diamyl amylphosphonate (DAAP).

Distribution coefficients for Pu and Am were measured by contact studies in 1-10 M HCl, while varying REDOX conditions, actinide loading levels, and resin formulations. Flow experiments were run to evaluate actinide loading and elution under varied conditions. Significant differences in the actinide distribution coefficients in contact experiments, and in actinide retention in flow experiments were observed as a function of resin formulation.

## INTRODUCTION

Aqueous processing of Pu residues produces acidic effluent waste streams that require several treatment steps before the liquids may be discharged to the environment. Effluents from hydrochloric acid (HCl) processing streams are usually routed to controlled hydroxide precipitation for actinide waste polishing. This step coprecipitates many other metal hydroxides and salts with the actinides, producing an undesirable solid cake for storage. The liquid effluent from neutralization requires further treatment to reduce activity by a flocculation process that produces additional large volumes of transuranic (TRU) solid wastes.

The purpose of this work is to evaluate extraction chromatography techniques and materials as an alternative method to remove actinides from aqueous hydrochloric acid effluent streams. Octyl(phenyl)-N,N-diisobutylcarbamoylmethylphosphine oxide (CMPO) and related molecules have historically been examined for use in liquid-liquid extraction schemes for actinide removal from nitric acid media[1], with some efforts in hydrochloric acid[2-3]. Extraction chromatography using similar extractants has been used on an analytical scale to concentrate actinides for analysis.[4] However, testing of extraction chromatography techniques for larger-scale applications has received scant attention.[5-8] The potential advantages of applying these techniques for decontamination of radioactive aqueous effluents include generation of smaller quantities of solid residues in forms more suitable for storage, smaller quantities of TRU waste, more facile hydrochloric acid recycle, and efficient decontamination of aqueous effluents.

## EXPERIMENTAL

A Packard 2200CA scintillation counter was utilized to measure alpha activity in liquid samples. Aliquots of the actinide solutions were pipetted into 6 mL $H_2O$ and 14 mL Ultima Gold XR® scintillation cocktail. Experiments in HCl used either a purified Pu stock solution containing 15.3 mg/mL in 1.8 M HCl or a stock Am solution containing 0.50 mg/mL in 4 M HCl. Radiochemistry analysis of the Pu stock solution was within 6 % of the value obtained by scintillation counting. Duplicate scintillation samples were counted for all experiments.

Several resins from EIChroM Industries were tested in addition to resins prepared at Los Alamos. All resins reported in this study were based on octyl(phenyl)-N,N-diisobutylcarbamoylmethylphosphine oxide (CMPO) diluted in tributyl phosphate (TBP), diamyl amylphosphate (DAAP), or without a diluent. EIChroM resins included TRU-Spec® (13% CMPO, 27% TBP weight percent respectively), RE-Spec® (16% CMPO, 24% TBP), 20% CMPO/20% DAAP and 30% CMPO/10% DAAP. LANL prepared resins reported in this study include 40% CMPO on Amberlite® XAD-7. The LANL resin was prepared by methods similar to those reported in the literature.[4,7]

Contact experiments used Bio-Rad 10 mL polyethylene columns equipped with a 35 micron frit and an attached luer-lock stopcock. Columns were rotated at approximately 8 rpm. Liquid samples were removed by first draining a 1 mL portion through the frit to rinse the column tip and stopcock, followed by removing an aliquot for sampling. The typical experiment used a 5 mL solution volume, 0.5 g or 0.05 g of resin, and 0.25 M $NH_2OH$ or $NaNO_2$ to control REDOX conditions for plutonium experiments. Distribution coefficients (Kd values) discussed in this work are reported as actinide concentration per gram of resin divided by actinide concentration per mL of solution. A Microsoft EXCEL spreadsheet was utilized to calculate the Kd values and to correct for activity or volume changes due to reagent addition and/or sampling losses.

Flow experiments were performed in 1 cm ID Kontes Chromaflex® glass columns equipped with PTFE fittings and frits. A Cole-Palmer Masterflex® pump using PTFE & Norprene tubing (0.8 mm ID) was used pull solutions through the columns at controlled rates. The typical experiment used 2.5 g of resin (~4.7 mL column void volume). Flow rates were set at approximately 0.35 mL/min. (~13 minutes residence time). Actinide solutions for column loading were 25 mL (5 column void volumes) and sometimes contained 0.25 M REDOX reagent. Following the actinide loading cycle of approximately 5 column volumes, the column was washed with 5 column volumes of HCl at the same concentration as the loading cycle, followed by elution with 10 column volumes of a more dilute acid solution. Samples for analysis were collected in 5 mL increments (approximately 1 column volume) from column flow experiments.

## RESULTS & DISCUSSION

### Contact Experiments

Very good retention of Pu(IV) is observed for several of the resins containing the CMPO ligand, as evidenced by the large distribution coefficients (Kd's) seen in the contact studies (Figure 1). The final equilibrium Kd of approximately 10,000 for Pu(IV) does not vary dramatically at higher HCl concentrations (6-10 M) with this series of resins. The final equilibrium Kd at lower HCl concentrations (1-4 M) shows a trend of higher Kd for the resins that have more CMPO and less TBP or DAAP. Data for samples obtained at time intervals prior to equilibration show slower kinetics are observed for uptake of Pu(IV) on the resins that have more CMPO, and less TBP or DAAP as a diluent. The pure CMPO resin was very slow compared to the other resins, and it is not known if full equilibrium was reached for this resin even after 24 hours contact. Kinetics of Pu(IV) uptake are slowest when the HCl concentrations are high, the Pu loading level is high, and the ratio of solution to resin is large (not all data shown).[8]

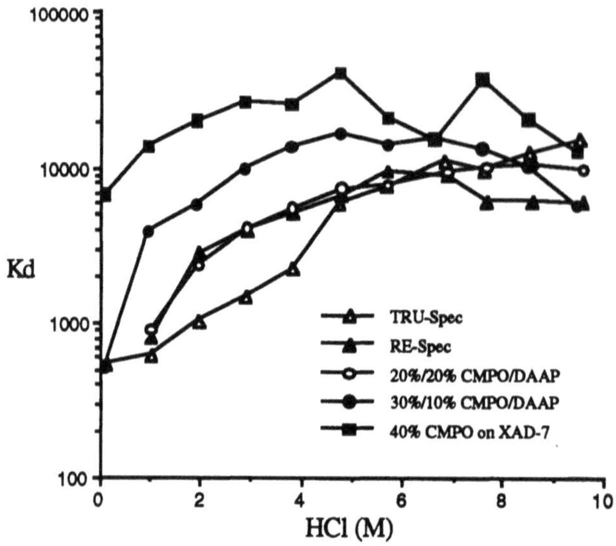


**Figure 1.** Pu(IV) Kd Comparison For Several Resins (24 Hour).

Distribution coefficients for Am(III) (Figure 2) are much lower than observed for Pu(IV) on these resins, and are significant only at higher HCl concentrations (4 M or above). For Am, the resin series again shows slower kinetics for the resins containing more CMPO, and less TBP or DAAP (not all data shown). Final equilibrium Kd values for Am(III) are significantly higher for the resins that have more CMPO and less TBP or DAAP. This is significantly different from the behavior observed for Pu(IV) at high HCl molarity where final Kd's are approximately 10,000 for all resins.

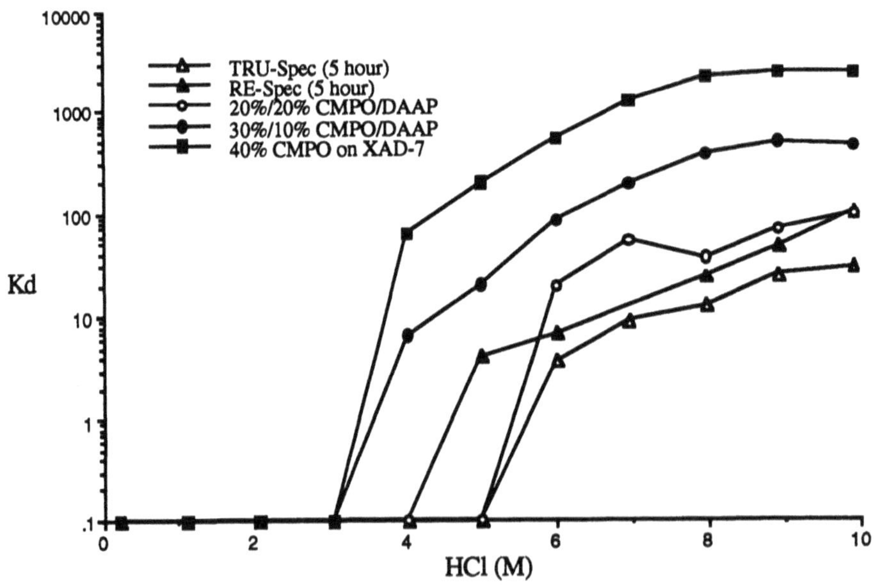


**Figure 2.** Am Kd Comparison For Several Resins (24 Hour).

Efficient use of these resins to remove actinides from process effluent streams will require a functional loading capacity significantly higher than that recommended for these resins in analytical concentration schemes (1-2 mg/ mL of resin or approximately 3-6 mg actinide/g of resin recommended for TRU-Spec). Figure 3 shows how the Pu(IV) Kd changed over time in a 6 M HCl contact study on TRU-Spec, as the Pu added to a set of experiments was varied over 2.5 orders of magnitude. The observed Kd's were nearly constant over the range of 3-30 mg Pu/g resin, and showed a predictable decrease as the resin loading capacity was exceeded at the higher loading levels. The reasons for the slightly lower Kd's at the lowest loadings in this study are not well understood. The maximum amount of Pu removed by TRU-Spec at 6 M HCl in this study was 66.3 mg/g resin.

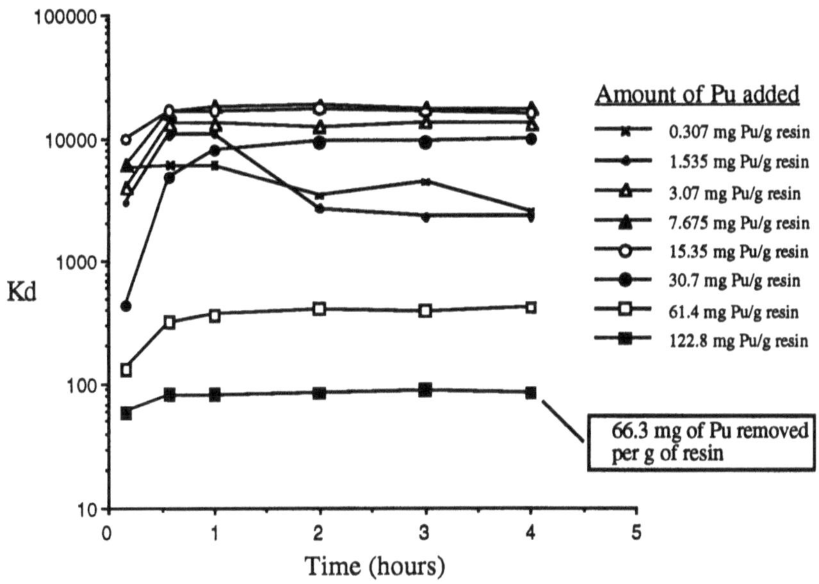


**Figure 3.** Pu(IV) Kd Vs Loading for TRU-Spec in 6 M HCl, 0.25 M $NaNO_2$.

Figure 4 shows a similar result for Pu(IV) loading experiment conducted for a series of resins tested at 7 M HCl and 12.24 mg Pu contacted with 50 mg resin. The resins followed a general trend of greater Pu removal for higher CMPO content. Analysis of this data at the 24 hour timepoint (Table 1), shows that the molar ratio of CMPO to Pu is approaching 1 for TRU-Spec, and is larger for the resins that contain a greater fraction of CMPO. The measured Pu(IV) loading capacity of the resins, 107 mg/g for the 30%/10% CMPO/DAAP resin, is approaching one-half the capacity of an anion exchange resin (typically ~1 mmole/g or ~240 mg actinide/g resin).

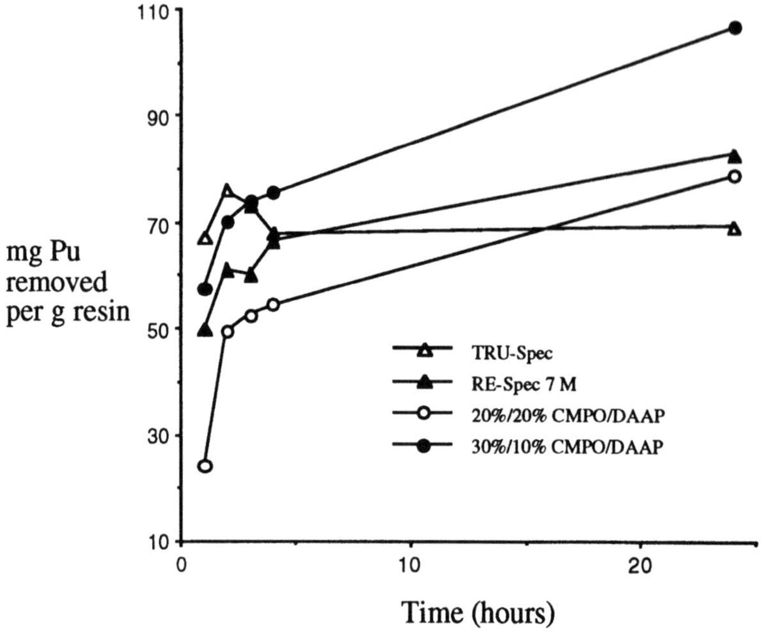


**Figure 4.** Pu(IV) removed from 7 M HCl for several resins over time.

**Table 1.** Analysis of maximum loading of Pu(IV) removed from a 7 M HCl solution containing 0.25 M $NaNO_2$, 12.24 mg of Pu and 50 mg of resin at 24 hours.

| | TRU-Spec | RE-Spec | 20%/20% CMPO/DAAP | 30%/10% CMPO/DAAP |
|---|---|---|---|---|
| % CMPO on resin (by weight) | 13 | 16 | 20 | 30 |
| Kd observed | 42.7 | 55.6 | 52.0 | 86.0 |
| mg Pu removed | 3.49 | 4.15 | 4.01 | 5.46 |
| mg Pu removed per g resin | 69.2 | 82.9 | 79.0 | 106.9 |
| Calc. mole ratio of CMPO/Pu | 1.10 | 1.13 | 1.49 | 1.65 |

In the loading experiments a relatively small fraction of the excess Pu in solution was removed by the resin. As the loading data was calculated by measuring the Pu that remained in solution, the experiment measured small changes in large numbers, leading to greater scatter and uncertainty than for other experiments in this report. The data for the pure CMPO resin showed low loading, but also unusual scatter, and was thus excluded from Table 1.

## Flow Experiments

Pu(IV) retention during the loading and wash cycles of flow experiments was generally very good for either TRU-Spec or the 30%/10% CMPO/DAAP resins at 6-8 M HCl and 0.25 M $NaNO_2$. Figure 5 shows the loading and elution profiles for 2.5 g of resin loaded with 0.77 mg of Pu dissolved in 25 mL of HCl (actinide loading cycle, column fractions 1-5), followed by 25 mL of HCl (wash cycle, column fractions 6-10), and 50 mL of elutrient (column fractions 11-20). Retention of Pu on either resin under the specified conditions in Figure 5 was over 99.9% for the loading and wash cycles. Figure 5 also shows comparative data for TRU-Spec resin eluted with 0.25 M hydroxylamine or ascorbic acid (AA) in 1 M HCl. Elution of Pu(IV) loaded resins with ascorbic acid as a reducing agent was more rapid and complete than elution with hydroxylamine in this and other studies. The elution of Pu(IV) from the 30%/10% CMPO/DAAP resin proved slower than elution from TRU-Spec resin for either hydroxylamine or ascorbic acid (only the ascorbic acid elution profile is shown).

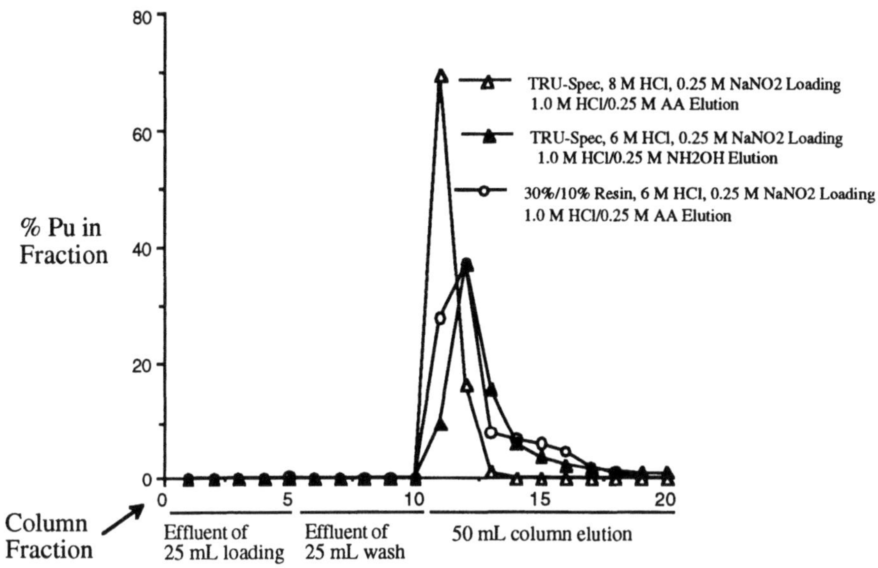


**Figure 5.** Pu (IV) Behavior on TRU-Spec and 30%/10% CMPO/DAAP Resins.

Pu(III) retention in 8 M HCl and 0.25 M $NH_2OH$ during the loading and wash cycles varied dramatically with the resin formulation as shown in Figure 6. The best Pu(III) retention was observed for the 30%/10% CMPO/DAAP resin. TRU-Spec showed a large breakthrough by the fourth column volume, while the 40% CMPO resin showed breakthrough on the very first column volume. Experiments on TRU-Spec resin at 6 M and 10 M HCl (not shown) showed relatively minor changes in Pu(III) breakthrough from the 8 M HCl result shown in Figure 6.

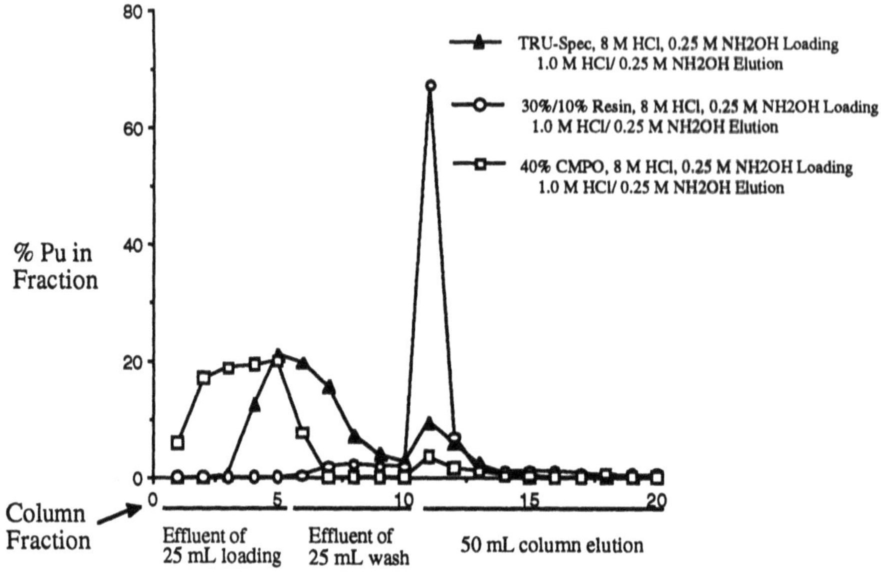


**Figure 6.** Pu(III) Behavior on Three Resins.

Am behavior in flow experiments on three different resins is shown in Figure 7. Am retention showed a trend similar to Pu(III) during the loading and wash cycles, with the best retention observed for the 30%/10% CMPO/DAAP resin. Retention of Am on the 30%/10% CMPO/DAAP resin was over 99.99% for the loading and wash cycles. TRU-Spec showed a large breakthrough of Am by the fourth column volume.

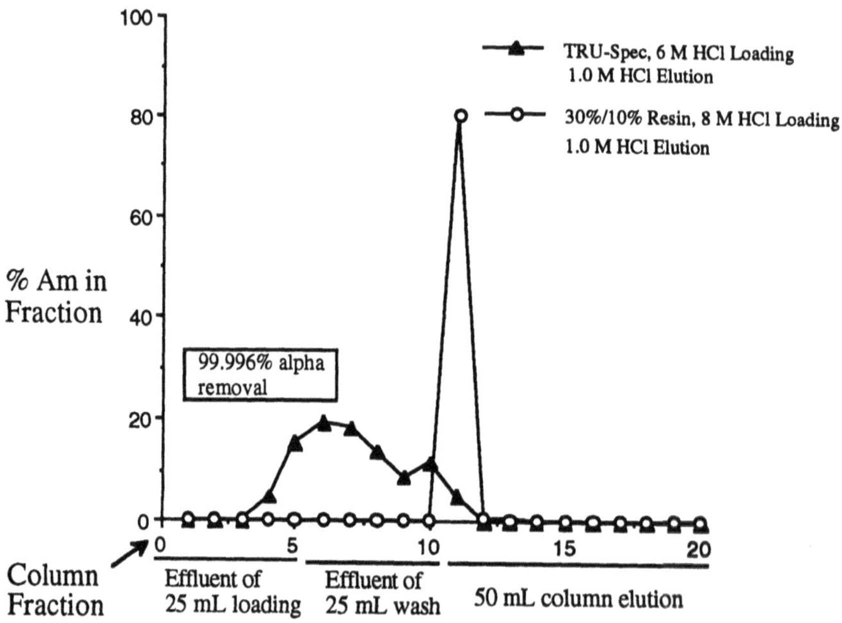


**Figure 7.** Am Behavior on TRU-Spec and 30%/10% CMPO/DAAP Resins.

The flow results for Pu(III) and Am(III) for the varied resins are quite similar. For flow experiments with either Pu(III) or Am(III), the superiority of 30%/10% CMPO/DAAP over TRU-Spec resin for actinide (III) retention are consistent with the large difference in Am Kd values observed in contact experiments (Figure 2) for these two resins. The poor Pu(III) retention of the 40% CMPO resin in Figure 6 cannot be explained by the Kd comparison made in Figure 2. The reason for the rapid Pu(III) breakthough observed for the 40% CMPO resin must be due to very slow kinetics, which were also apparent in the Am contact experiments. The effect is magnified at the moderately high loading of Pu in the flow experiments (0.77 mg of Pu on 2.5 g of resin).

## CONCLUSIONS

•Pu(IV) is effectively removed from HCl by any of several resins containing the CMPO ligand. Of the resins studied, TRU-Spec & RE-Spec had the best kinetics for Pu(IV) uptake, and most facile elution with a reducing agent. Ascorbic acid is superior to hydroxylamine for stripping Pu from these resins.

•Pu(III) and Am(III) are removed by CMPO resins only at higher HCl concentrations. The ratio of CMPO to diluent and the properties of the diluent play a large role in actinide(III) retention. More CMPO translates to higher actinide(III) Kd values.

•Kinetics of Pu(IV) uptake and elution varies more dramatically than that of Pu(III) or Am(III) with resin formulation. A larger ratio of CMPO to diluent appears to slow kinetics of uptake and elution.

•Contact loading studies with Pu(IV) and flow experiments indicate that the resins will function at loading levels much higher than recommended for analytical use. These observed actinide loadings make these resins an attractive option for HCl process stream decontamination.

•Of the series of resins tested, TRU-Spec appears to offer the best combination of properties for Pu(IV) decontamination from HCl. The 30%/10% CMPO/DAAP resin showed the best combination of properties for Pu(III) and/or Am(III) decontamination from HCl.

•Elution of Pu and Am from these resins is facile at 0.1 to 1.0 M HCl concentrations, if a reagent for reduction of Pu(IV) to Pu(III) is used. At these acid concentrations phosphonic acid degradation products of CMPO, observed to hinder actinide elution in more dilute acids, should provide little interference.

•Extraction chromatography may provide a method of concentration of actinides into smaller volumes of storable or treatable solid forms, decontamination of high acid stream effluents to activity levels that allow HCl recycle, and reduction in activity levels sent to wastewater treatment.

•The present formulation of 40% CMPO on XAD-7 showed slow kinetic uptake of actinides. This is undoubtedly due in part to the absence of TBP or DAAP to function as diluent and an aid to actinide transport from the aqueous phase to the organic phase of the bead. The larger resin bead size of the XAD-7 support material may also contribute to the slower kinetic uptake.

•The stability of these resins to large volumes of acids, radioactivity and repeated use in an effluent treatment environment are important factors that will need to be determined through additional study. This work is proceeding at full scale in our laboratories removing gram quantities of actinides from actual process effluents.

## ACKNOWLEDGMENTS

The authors would like to acknowledge Mike Gula of EIChrom® for helpful conversations and the gift of small samples of the 20:20 & 30:10 CMPO/DAAP resins. We also thank Karol L. Holmes and Sarah J. Rice for preparation of the CMPO coated XAD-7 resins. Acknowledgments to E. Phillip Horwitz of Argonne National Laboratory and Keith Fife, Mark Dinehart, Larry Avens, and Steve McKee of Los Alamos National Laboratory for helpful discussions.

## REFERENCES

1. Horwitz, E.P., Schulz, W.W.; in New Sep. Chem. Tech. Radioact. Waste Other Specific Appl., "The TRUEX Process: A Vital Tool for Disposal of U.S. Defense Nuclear Waste" and references therein. Editor(s): Cecille, L., Casarci, M., Pietrelli, L., Eds., Publisher: Elsevier, London, 1991, p 21-29.
2. Horwitz, E.P., Diamond, H., Martin, K.A.; "The Extraction of Selected Actinides in the (III), (IV) and (V) Oxidation States from Hydrochloric Acid by OFD(iB)CMPO: The TRUEX-Chloride Process". *Solvent Extr. Ion Exch.*, 5(3) (1987), 447-470.
3. Horwitz, E.P., Diamond, H., Martin, K.A.; "Extraction of Americium (III) by Octyl(phenyl)-N,N-diisobutylcarbamoylmethylphosphine Oxide". *Solvent Extr. Ion Exch.*, 5(3) (1987), 419-446.
4. Horwitz, E.P., Deitz, M.L., Nelson, D.M., LaRosa, J.J., Fairman, W.D.; "Concentration and Separation of Actinides from Urine Using a Supported Bifunctional Organophosphorus Extractant". *Analytical Chimica Acta.* 238 (1990) 263-271.
5. Yamada, W.I., Martella, L.L., Navratil, J.D.; "Americium Recovery and Purification Using a Combined Anion Exchange-Extraction Chromatography Process". *Journal of the Less Common Metals,* 86 (1982), 211-218.
6. Lumetta, G.J., Wester, D.W., Morrey, J.R., Wagner, M.J.; "Preliminary Evaluation of Chromatographic Techniques for the Separation of Radionuclides from High-Level Radioactive Waste". *Solvent Extr. Ion Exch.*, 11(4) (1993), 663-682.
7. Barney, G.S., Cowan, R.G.; "Separation of Actinide Ions from Radioactive Waste Solutions Using Extraction Chromatography". Westinghouse Hanford Report, WHC-SA-1520-FP, (1992).
8. Schulte, L.D., FitzPatrick, J.R., Salazar, R.R., Schake, B.S., Martinez, B.T.; "Cleanup of Hydrochloric Acid Waste Streams From Actinide Processes Using Extraction Chromatography". In Press, Separation Science and Technology, (1994).

# SORPTION BEHAVIOR OF URANIUM ONTO REILLEX™-HPQ ANION EXCHANGE RESIN FROM NITRIC AND HYDROCHLORIC ACID SOLUTIONS

Kent D. Abney,[1] Anthony B. Pinkerton,[1] Richard C. Staroski,[1]
Norman C. Schroeder,[1] Kenneth R. Ashley,[2] J. Michael Adams,[2]
and Jason R. Ball[2]

[1]Chemical Science and Technology Division (CST-12)
Los Alamos National Laboratory
Los Alamos, NM 87545

[2]Department of Chemistry
East Texas State University
Commerce, TX 75429

## ABSTRACT

Experimental distribution coefficients ($K_d'$) are reported for $UO_2^{2+}$ with the nitrate form of Reillex™-HPQ anion exchange resin as a function of nitric acid concentration. The values of $K_d'$ for $UO_2^{2+}$ at 1.00, 5.00, and 10.0 M $HNO_3$ are approximately 0.5, 10, and 9.0 mL/g dry resin, respectively; the maximum value is 13 mL/g at 7 M $HNO_3$. The plots are bell-shape curves; the interpretation is that $K_d'$ increases as $UO_2(NO_3)_3^-$ — the predominant sorbing species — is formed and sorbed onto the resin, decreasing as it is displaced by $NO_3^-$ ion under high nitric acid concentrations. Comparison data for Dowex™-1x8 and technetium are also presented. Experiments have shown that nitrate ion with sodium and aluminum as counter ions dramatically affects the sorption behavior of uranium. Sodium nitrate shifts the maximum sorption value from 7 to ~4 M $NO_3^-$. Aluminum nitrate increases the distribution coefficient exponentially through 8 M. Plots of elution volume (defined as the number of bed volumes needed to move the maximum concentration of the eluting uranyl-nitrate solution to the end of the column) as a function of nitric acid concentration between 1.00 and 10.0 M display a similar bell-shape curve for the $K_d'$-[$HNO_3$] data. Breakthrough volumes were determined for 0.0100 M $UO_2^{2+}$ in 2.0 to 10.0 M $HNO_3$. The 2.00 and 4.00 M $HNO_3$ solutions show nearly complete breakthrough at 1.5 and 2.5 bed volumes, respectively. The 6.00, 8.00 and 10.0 M solutions display similar behavior; 10% breakthrough occurs at approximately three

*Separations of f Elements*, Edited by K.L. Nash
and G.R. Choppin, Plenum Press, New York, 1995

bed volumes and 90% breakthrough at six bed volumes. Ordering of breakthrough volumes at 50% breakthrough is 8 M > 10 M ≈ 6 M >> 4 M > 2 M. The elution and breakthrough behavior can be explained by the same concepts as those governing $K_d'$ trends. Sorption behavior of U(IV) and $UO_2^{2+}$ chloride complexes were determined in hydrochloric acid media. The $K_d'$ values increase from ~1 mL/g at 1.00 M HCl to ~1100 g/mL at 7 M HCl and then remain constant to 9.5 M HCl.

## INTRODUCTION

During the past several years, Los Alamos National Laboratory has used Reillex™-HPQ to selectively separate plutonium from a variety of waste and scrap materials.[1–3] Reillex™-HPQ resin is a co-polymer of divinylbenzene and 4-vinylpyridine that has been subsequently methylated (≈75%) at the pyridine nitrogen to give pyridinium strong base anion exchange sites.[4] (This pyridinium functionally of Reillex™-HPQ resin is unique; other strong base anion exchange resins typically use quaternary alkyl amine resins.) Previous research using Reillex™-HPQ for plutonium revealed that at nitric acid concentrations typically used to elute metals from strong base anion exchange resins, plutonium sorbed the most strongly.[1] Because this resin proved stable under radiolytic and strong acid conditions, we initiated a study of other metal anions typically found in nuclear waste streams and recently reported the results of technetium separation studies using Reillex™-HPQ.[5]

Technetium as $TcO_4^-$ is one of the predominant radioactive components found in nuclear waste storage sites such as the Department of Energy's Hanford site, where uranium is the largest radioactive component by mass. Estimates indicate that within the single- and double-shell tanks, there is $1.44 \times 10^6$ kg of uranium.[6] Because of the massive amounts of uranium in the waste material as well as the presence of both technetium and uranium in the sludge at the bottom of the tanks, any technetium separations process used must also consider the presence of uranium. This chapter describes our efforts to determine the effect of uranium on the Reillex™-HPQ technetium separations process in highly acidic media.

## EXPERIMENTAL

### Reagents

The reagents used were analytical grade. All water used was deionized to 18 MΩ. The solutions of $HNO_3$ and HCl were prepared from concentrated acids and standardized against primary standard sodium hydroxide. All uranium oxides and metal were readily available at Los Alamos and were used without further purification. $UO_2(NO_3)_2$ was prepared by dissolving $U_3O_8$ in concentrated nitric acid and fuming to dryness. This process was repeated and followed by redissolving in 1.00 M $HNO_3$ to make a 0.100 M uranium stock solution. Concentrated $HNO_3$ and water were then added to an aliquot of the uranium stock solution to measure the distribution coefficient at the desired acid strength. Uranium(IV) chloride was prepared by dissolving uranium metal in 6.00 M HCl under refluxing conditions. After dissolution was complete, the solution was taken to dryness and redissolved in 5.00 M HCl to make a 0.100 M uranium solution. The uranium(VI) solution was prepared by dissolving $UO_3$ in 6.00 M HCl under refluxing conditions. This solution was taken to dryness after dissolution was complete and the solid was redissolved in 5.00 M HCl.

### Assay Techniques

We investigated several analytical methods before choosing direct current plasma atomic emission spectroscopy (DCP-AES). Typical measured uranium concentrations were in the 5–50 ppm range. We found that liquid scintillation counting was unsatisfactory both with and without alpha/beta discrimination unless the daughters were removed by column chromatography. UV-visible spectroscopy was accurate but did not give satisfactory precision. Inductively coupled plasma mass spectrometry (ICP-MS) did not give reproducible results without exhaustive attention to detail. Inductively coupled plasma atomic emission spectroscopy (ICP-AES) gave accurate and reproducible results but was not readily available for these studies. The DCP-AES was calibrated using a 1000 ppm uranium atomic absorption standard purchased from Aesar/Johnson Matthey, Inc. This standard was diluted to the working range of our experimental samples.

### Resin Treatment

We acquired 30–60 mesh size chloride form Reillex™-HPQ resin from Reilly Industries Inc.[4] The resin was washed with ≥20 bed volumes of water. All floating resin beads observed during the washing process were removed by decantation. The chloride form was used for measurements from HCl media. To ensure that the resin was in the chloride form, we treated the resin with 10 bed volumes of 10 M HCl. We used the nitrate form for the $HNO_3$ media and converted the resin to the nitrate form by passing at least 10 bed volumes of 1.0 M $HNO_3$ and 10 bed volumes of 0.10 M $HNO_3$ through the resin. Both treatments ensured protonation of the unmethylated pyridine sites. After conversion, the resin was washed with 20 bed volumes of water and then stored. Before use, the resin was dried at 60°C for 48 hours and kept in a desiccator over silica gel. (We discussed the details of this method of drying to achieve constant resin mass in an earlier publication.[5])

### Distribution Coefficient Determinations

We prepared each sample by accurately weighing 0.5 g of dried resin in a 30 mL plastic screwtop vial and adding 5.00 mL of the desired solution to it. Each of the 5.00 mL solutions was prepared by mixing the necessary amount of 15.64 ± 0.02 M $HNO_3$ with water and 1.00 mL of 0.100 M $UO_2(NO_3)_2$. The solution and resin were shaken for at least 1 hour at room temperature. We used a Bio-Rad Econo™ column fitted with a filter disk to separate the resin from the solution by gravity filtration. We analyzed the filtrates for uranium.

### Column Experiments

A 0.0600 M $UO_2^{2+}$ stock solution was prepared by dissolving $UO_2SO_4$ or $UO_2(NO_3)_2$ in 0.100 M nitric acid. We then prepared a 0.0200 M $UO_2^{2+}$ solution at the desired $HNO_3$ concentration by diluting standardized concentrated nitric acid with water and adding the appropriate amount of $UO_2^{2+}$ stock solution.

We used approximately 5.5 g of the Reillex™-HPQ resin in the nitrate form to prepare a 1.0 by 23 cm resin column for each elution experiment. We loaded 10 mL of 0.0200 M $UO_2^{2+}$ in the desired acid concentration onto the column that had been pretreated with the same nitric acid concentration. A small amount of the eluting acid was added to the top of the column to bring all of the $UO_2^{2+}$ solution into the column as a small plug of $UO_2^{2+}$. A Bio-Rad flow adapter connected to a Masterflex peristaltic pump was then placed on the column. Approximately 5 mL fractions of the eluent were collected at a flow rate of approximately 0.87 cm/min (0.66 mL/min). We analyzed the fractions at a wavelength of 414 nm on a Cary-15 UV-visible spectrophotometer.

The breakthrough experiments were performed in a similar way. Solutions of 0.0100 M $UO_2^{2+}$ in the desired acid concentration were prepared using solid $UO_2(NO_3)_2$ and the correct amount of nitric acid and then diluting this solution with water to the mark in a volumetric flask. The column setup, collection of fractions, and measurement of the absorbance of the fractions were all similar to those for the elution studies.

## RESULTS AND DISCUSSION

### Sorption Processes

Uranium sorption occurs by means of an exchange of nitrate anion with several possible uranyl nitrate complexes found in solution. Measurement of this process is complicated by these species' changes in concentration as a function of nitric acid and/or nitrate ion concentration. It has been observed that for $UO_2^{2+}$ sorption on Dowex™- 1x4 in nitric acid,[7] the sorption occurs only at high acid conditions, whereas the sorption of $UO_2^{2+}$ on Reillex™-HPQ in sulfuric acid media occurs only at low pH.[8]

In concentrated nitric acid solutions, the $UO_2^{2+}$ ion will complex with varying numbers of $NO_3^-$ ligands to give species of the general formula $UO_2(NO_3)_n^{-(2-n)}$. However, stability constants have not been reported for species with n > 4. We have measured the distribution coefficients, $K_d'$, for $UO_2^{2+}$ on Reillex™-HPQ as a function of nitric acid concentration. The results imply that species of n ≥ 3 are formed (*vide infra*). Equation (1) describes this sorption process.

$$R^+NO_3^- + UO_2(NO_3)_3^- \rightleftharpoons R^+UO_2(NO_3)_3^- + NO_3^- \qquad (1)$$

Equation 1 represents the exchange reaction for $UO_2(NO_3)_3^-$ with one resin site. These results are consistent with the recent results of Viers and Marsh.[9] The definition of the distribution coefficient for $UO_2(NO_3)_3^-$ can be described by Equation (2). Equation (3) defines the experimental distribution coefficient, $K_d'$. The $(U(VI))_{total}$ is the total uranium(VI) in $V_{sol}$ mL of solution before contact with the resin, and $(U(VI))_{sol}$ is the analytical concentration of uranium(VI) in $V_{sol}$ mL of solution after contact with the resin. The units associated with $K_d'$ are mL/g.

$$K_d = \frac{\left(R^+UO_2(NO_3)_3^-\right)}{\left[UO_2(NO_3)_3^-\right]} \qquad (2)$$

$$K_d' = \frac{\dfrac{(U(VI))_{total} - (U(VI))_{sol} \text{ X } V_{sol}}{\text{g of dry resin}}}{\dfrac{(U(VI))_{sol} \text{ X } V_{sol}}{V_{sol}, \text{ mL}}} \qquad (3)$$

The use of $K_d'$ rather than $K_d$ shows that the experimentally determined distribution coefficient is a composite of all sorbing species.

**Approach to Equilibrium Measurements**

One variable considered in the measurement of batch distribution coefficients was the time required to establish equilibrium for the uptake of uranium. As is shown in Table 1, equilibrium conditions are established within 15 minutes. These results are similar to those found for technetium, for which uptake on Reillex™-HPQ reaches equilibrium within 5 minutes.[5] Conversely, plutonium uptake on Reillex™-HPQ requires at least 2 hours before equilibrium is nearly established; in the case of Dowex™-1x4, equilibrium is still not established after 8 hours.[2]

**Table 1.** Distribution coefficient as a function of time for $UO_2^{2+}$ in 8 M $HNO_3$ at 23°C.

| Contact time (min) | $K_d'$(mL/g) |
|---|---|
| 5 | 7.42 |
| 10 | 9.64 |
| 15 | 11.17 |
| 30 | 12.25 |
| 45 | 12.70 |
| 60 | 13.16 |
| 120 | 11.47 |
| 1140 | 12.57 |

## Distribution Coefficient Measurements

The $K_d'$ *vs* [$HNO_3$] graph (Figure 1) is bell shaped and shows a maximum at approximately 7 M $HNO_3$. Also indicated on this graph are the analogous values for Dowex™-1x4 that were reported previously.[10] Although Dowex™- 1x8 and Reillex™-HPQ show similar behavior, from our technetium results we had anticipated that Reillex™-HPQ would have larger $K_d'$ values. The fixed pore size in Reillex™-HPQ or its more rigid structure may limit the number of sites that can interact/sorb $UO_2(NO_3)_3^-$, as is described in Equation 1.

These data provide a contrast to the technetium data in Figure 2. Technetium exhibits large $K_d'$ values at low nitric acid concentrations, whereas $UO_2^{2+}$ does not. Technetium typically is eluted under strong acid conditions. For nuclear processing flow sheets, the use of strong oxidizing acids in conjunction with organic ion exchangers is commonly considered a large drawback. The pyridine functionality decreases the likelihood of explosion hazards associated with this ion exchange resin.[2] Based upon the results in Figure 2, it is evident that technetium and uranium separations are best effected at low acid concentrations. Process flow sheets used for the selective separation of uranium would sorb uranium at 7–8 M $HNO_3$ and would elute using dilute $HNO_3$.

## Salt Effects upon Sorption

The effect of nitrate-containing salt concentration upon the sorption of uranium is presented in Figures 3 and 4. In Figure 3, the values of $K_d'$ in 1.00 M $HNO_3$ with added sodium nitrate are compared to the values in nitric acid alone as the source of nitrate ion. The addition of sodium nitrate initially enhances formation of $UO_2^{2+}$-nitrate anionic complexes. Hence, the peak maximum shifts to lower total nitrate ion concentration. However, further addition of nitrate in the form of $NaNO_3$ allows the "free nitrate concentration" to effectively compete for the resin exchange sites and thus lowers the $K_d'$ values.

Figure 4 shows that the value of $K_d'$ in 1.00 M $HNO_3$ increases—and does not decrease with the addition of aluminum nitrate. This can be explained if one assumes that aluminum

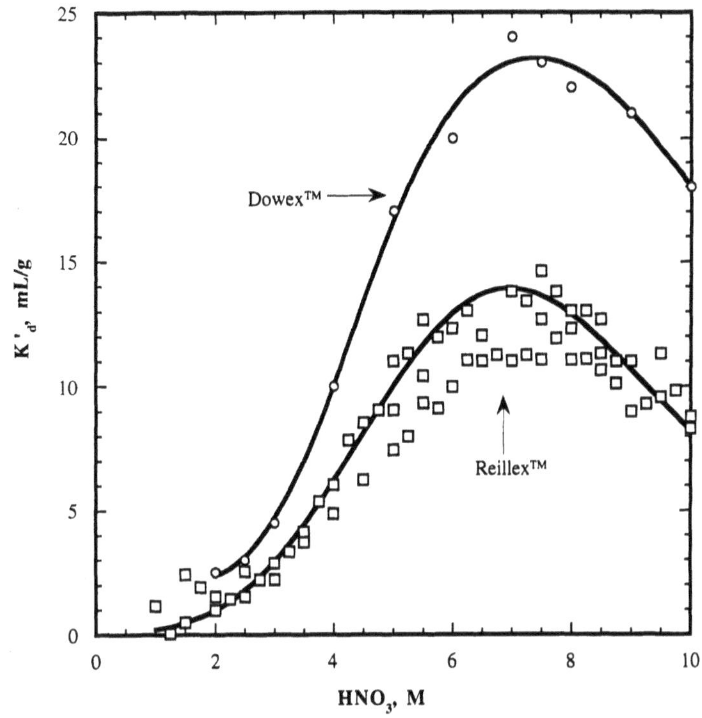


**Figure 1.** Experimental distribution coefficients ($K_d'$) *vs* nitric acid concentration for $UO_2^{2+}$ with Dowex™-1x8[10] and Reillex™-HPQ. The curves are only for illustration.

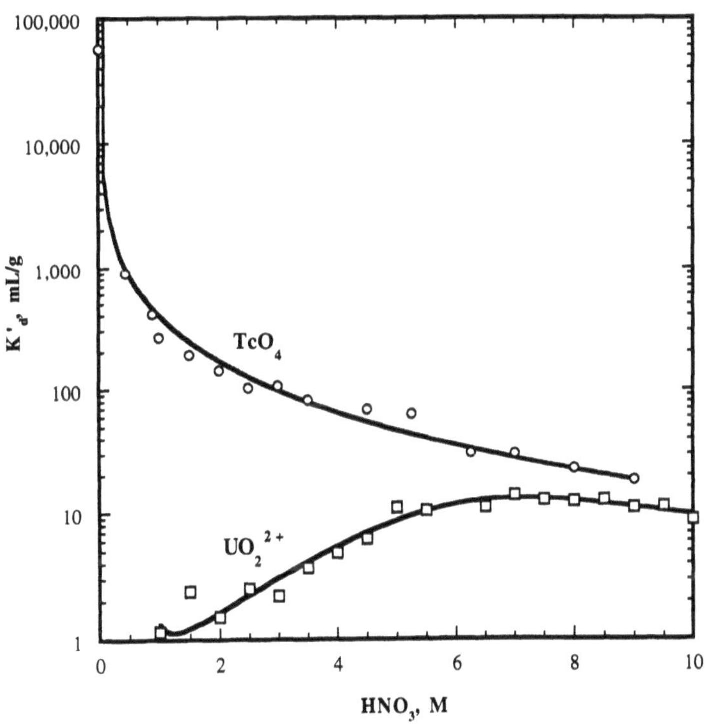


**Figure 2.** Comparison plots of $K_d'$ values for $UO_2^{2+}$ and $TcO_4^-$ *vs* nitric acid concentration with Reillex™-HPQ. The curves are for illustration only.

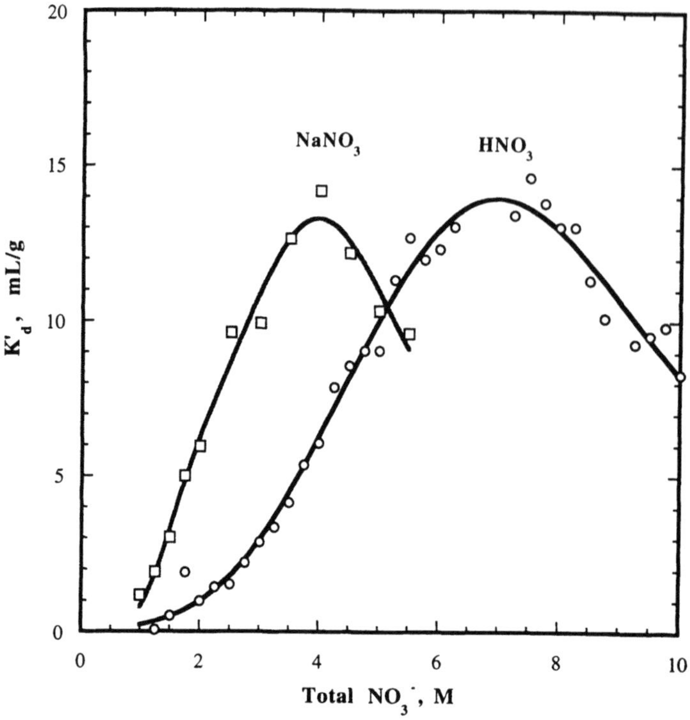


**Figure 3.** Comparison plots of $K_d'$ on Reillex™-HPQ in the presence of nitric acid and a mixture of 1.00 M nitric acid with added sodium nitrate. The curves are for illustration only.

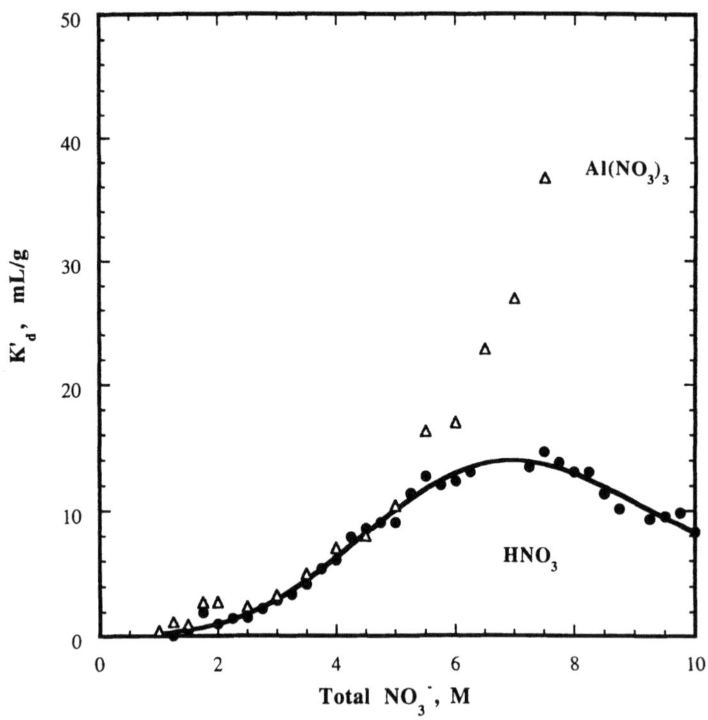


**Figure 4.** Comparison plots of the values $K_d'$ for $UO_2^{2+}$ sorption on Reillex™-HPQ in the presence of nitric acid and 1.00 M nitric acid with added aluminum nitrate. The curves are for illustration only.

nitrate complexes form at the higher nitrate concentrations, resulting in a lower effective "free nitrate concentration" with which to displace the sorbed uranyl nitrate anionic species. As a consequence, the distribution coefficient values continue to increase. As noted above, the values are greater than those observed in nitric acid only. Thus, not only is the "free nitrate" concentration decreased, but higher anionic complexes of uranium nitrate are formed.

## Mathematical Modeling

An examination of the plots of $K_d'$ values *vs* [$HNO_3$] reveals that the values of $K_d'$ increase with $HNO_3$ concentration to a maximum at ≈7 M and then begin to decrease. This behavior can be qualitatively described by the following reasoning. The ion $UO_2^{2+}$ is a cation and therefore cannot be sorbed onto the Reillex™ resin. As the $NO_3^-$ concentration increases, anions form of the general formula $UO_2(NO_3)_n^{(2-n)}$, where n ≥ 3. These are the sorbed species. However, as the $NO_3^-$ concentration continues to increase, there is also the displacement of the sorbed $UO_2(NO_3)_n^{(2-n)}$ ions by $NO_3^-$ ions. These two effects are counter to each other and result in a maximum of $K_d'$ at ≈7 M $HNO_3$.

This behavior can also be quantitatively described. Reaction (4) is the general reaction for the formation of the $UO_2(NO_3)_n^{(2-n)}$ species. Equation (5) is the definition of the overall stability constant ($\beta_n$). In this equation and the ones following, the activity and other nonideal effects are neglected. It is well understood that this raises some question about the validity of the treatment. However, in our opinion, the uncertainty associated with any nonideal treatment justifies the phenomenological approach presented here.

$$UO_2^{2+} + nNO_3^- \rightleftharpoons UO_2(NO_3)_n^{(2-n)} \qquad (4)$$

$$\beta_n = \frac{\left[UO_2(NO_3)_n^{(2-n)}\right]}{\left[UO_2^{2+}\right]\left[NO_3^-\right]^n} \qquad (5)$$

Equation (1) represents the resin equilibrium that is assumed to be dominant and Equation (6) gives the equilibrium constant for this reaction. Equation (7) gives the concentration of the dominant uranium(VI) species in terms of the total uranium(VI) present in solution.

$$K' = \frac{\left(R^+UO_2(NO_3)_3^-\right)\left[NO_3^-\right]}{\left(R^+NO_3^-\right)\left[UO_2(NO_3)_3^-\right]} \qquad (6)$$

$$\frac{\left[UO_2(NO_3)_3^-\right]}{(U(VI))_{sol}} = \frac{\beta_3\left[NO_3^-\right]^3}{1+\beta_1\left[NO_3^-\right]+\beta_2\left[NO_3^-\right]^2+\beta_3\left[NO_3^-\right]^3+\beta_4\left[NO_3^-\right]^4} \qquad (7)$$

The definition of the distribution coefficient for $UO_2(NO_3)_3^-$ can be described by Equation (2). Equation (3) defines the experimental distribution coefficient, $K_d'$. Combining the previous equations results in Equation (8). The experimentally determined $K_d'$ can be related to the independent variable, $NO_3^-$. It is assumed that $[NO_3^-] = [HNO_3]$.

$$K_d' = \frac{K'\left(R^+NO_3^-\right)\beta_3\left[NO_3^-\right]^2}{1+\beta_1\left[NO_3^-\right]+\beta_2\left[NO_3^-\right]^2+\beta_3\left[NO_3^-\right]^3+\beta_4\left[NO_3^-\right]^4} \quad (8)$$

The solid line in Figure 5 is the nonlinear least squares fit of the data to Equation (8).[11] The fit of the data yielded the following values and standard deviations for the parameters: $K'(R^+NO_3^-)\beta_3 = 0.45 \pm 0.055$; $\beta_1 = 3.90 \times 10^{-4} \pm 6.75$; $\beta_2 = 5.11 \times 10^{-4} \pm 18.2$; $\beta_3 = 3.22 \times 10^{-4} \pm 14.3$; and $\beta_4 = 3.11 \times 10^{-4} \pm 0.868$. This apparently poor result is the best fit of the several variations of Equation (8) that we tried. This calculation includes the assumption that $UO_2(NO_3)_4^{2-}$ was the major species and that both $UO_2(NO_3)_3^-$ and $UO_2(NO_3)_4^{2-}$ were important. Values of 0.27, $5.88 \times 10^{-3}$, and $1.32 \times 10^{-4}$ have been reported for $\beta_1$, $\beta_2$, and $\beta_3$.[12] Although the agreement among the $\beta_n$ values is not good, the values are not inconsistent.

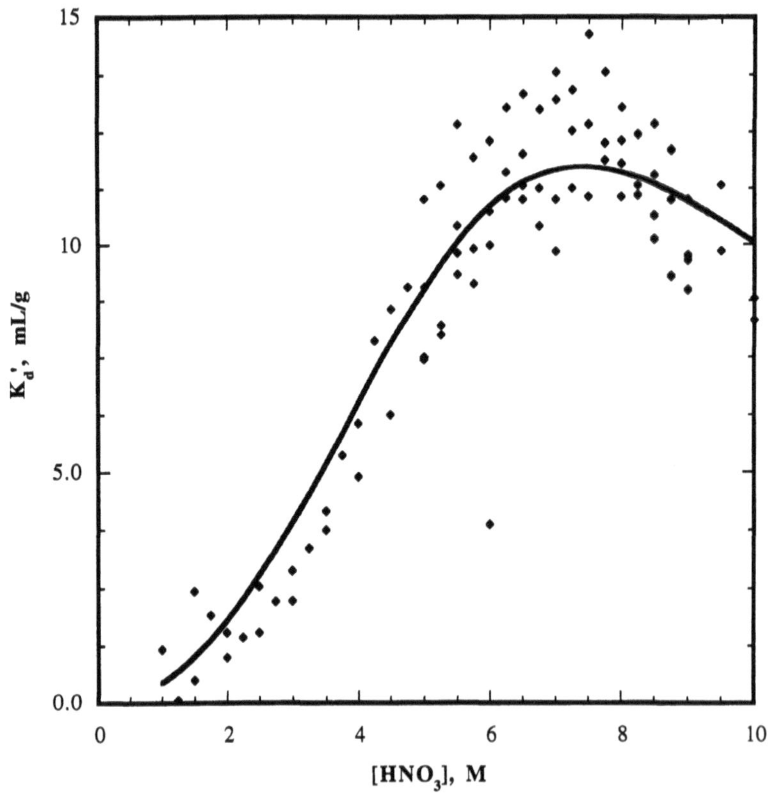


**Figure 5.** The relationship between $K_d'$ and nitric acid concentration for the sorption of $UO_2^{2+}$ onto Reillex™-HPQ. The line is the nonlinear least squares fit of the data to the model discussed in the text.

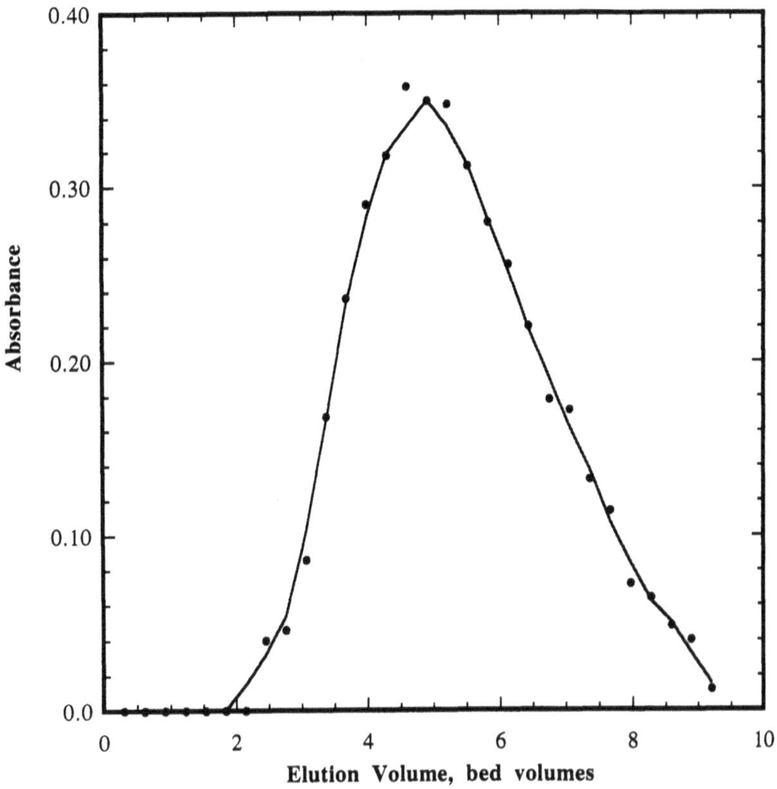


**Figure 6.** The elution profile for uranium(VI) in 8.0 M $HNO_3$.

## Elution Behavior

We performed elution studies by loading onto the column 0.20 mmol of $UO_2^{2+}$ at the desired nitric acid concentration. The Reillex™-HPQ resin column contained nitric acid in the same concentration. The loading was 1.1 to 3.3% of the total exchange capacity of the column (5.5 g resin x 3.34 meq/g dry resin). The $UO_2(NO_3)_n^{(2-n)}$ species were then eluted with this same concentration of nitric acid. The [$HNO_3$] was varied from 1.00 to 10.0 M for these experiments. Figure 6 displays the elution profile for 8.0 M $HNO_3$. The elution volume, in bed volumes (BV), is taken as the maximum of these curves.

We expect that two opposing factors determine the elution volume for the uranium(VI) species. As the nitric acid concentration is increased, formation of species with increasing number of nitrate ligands will be favored, which results in increased elution volume. In contrast, elution is easier because of the increased nitrate anion—the displacing anion—concentration. These two effects should give rise to a bell-shape curve for the elution volume *vs* [$HNO_3$] plot, similar to the observed plot of $K_d'$ *vs* [$HNO_3$], and Figure 7 shows exactly this!

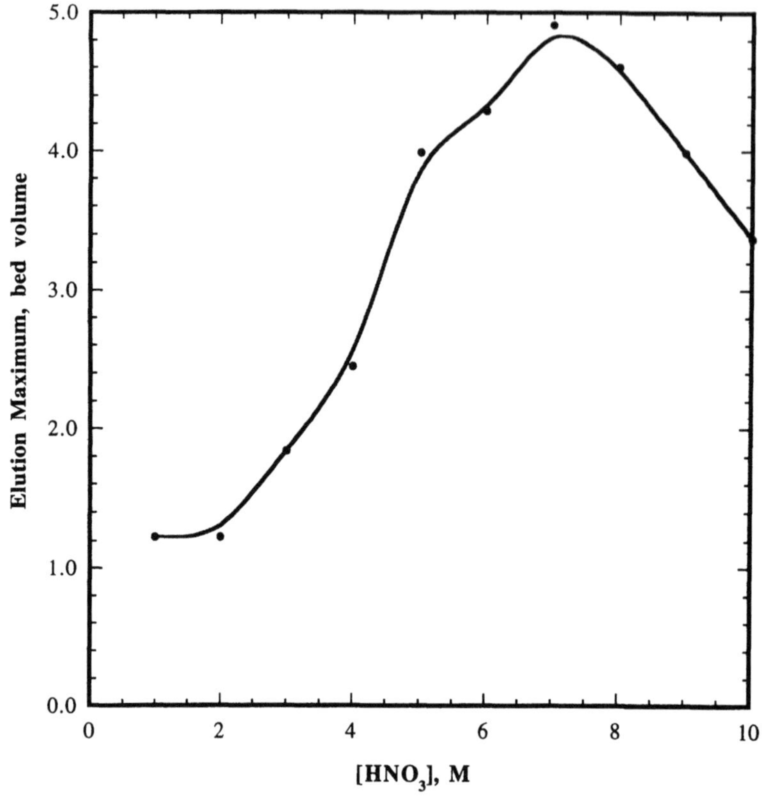


**Figure 7.** The elution volume measured in bed volumes as a function of the nitric acid concentration.

## Breakthrough Behavior

The percent breakthrough is measured by ratioing the $UO_2^{2+}$ absorbance at 414 nm of the effluent to that of the feed solution. Figure 8 is a composite plot for the percent breakthrough *vs* elution volume for the various nitric acid concentrations we studied. The ordering of the breakthrough volumes at 50% breakthrough are 8 M > 10 M ≈ 6 M >>4 M >2 M. This order is expected from the $K_d'$ measurements and the elution behavior with the $UO_2^{2+}$ Reillex™-HPQ system.

## Sorption from Chloride Media

We investigated the sorption of $UO_2Cl_n^{(2-n)}$ (n ≥ 3) and $UCl_m^{(4-m)}$ (m ≥ 5) species in HCl on Reillex™-HPQ with the results depicted in Figure 9. As anticipated from previous work using Dowex™-1x8, the values of $K_d'$ are much greater than those for nitric acid:[12] this results from the formation of higher charged anionic complexes and the smaller size of the chloro complexes compared to the nitrato complexes. The magnitude of the $K_d'$ values for Reillex™-HPQ and Dowex™-1x8 are between 1,000 and 2,000 mL/g for both uranium(VI) and uranium(IV). Thus, an HCl medium would be useful for uranium separations if the corrosive nature of HCl toward steel could be overcome.

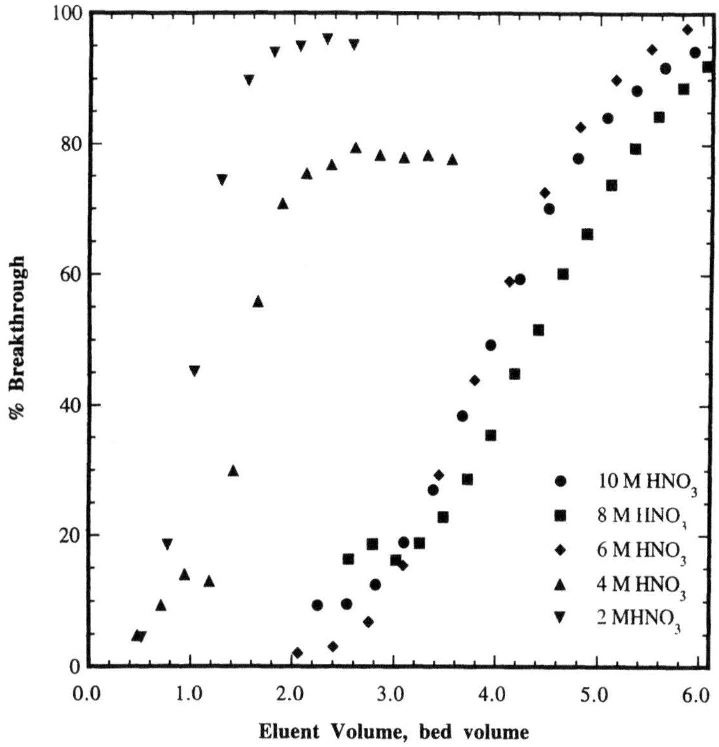


**Figure 8.** The percent breakthrough as a function of eluent volume for various nitric acid concentrations.

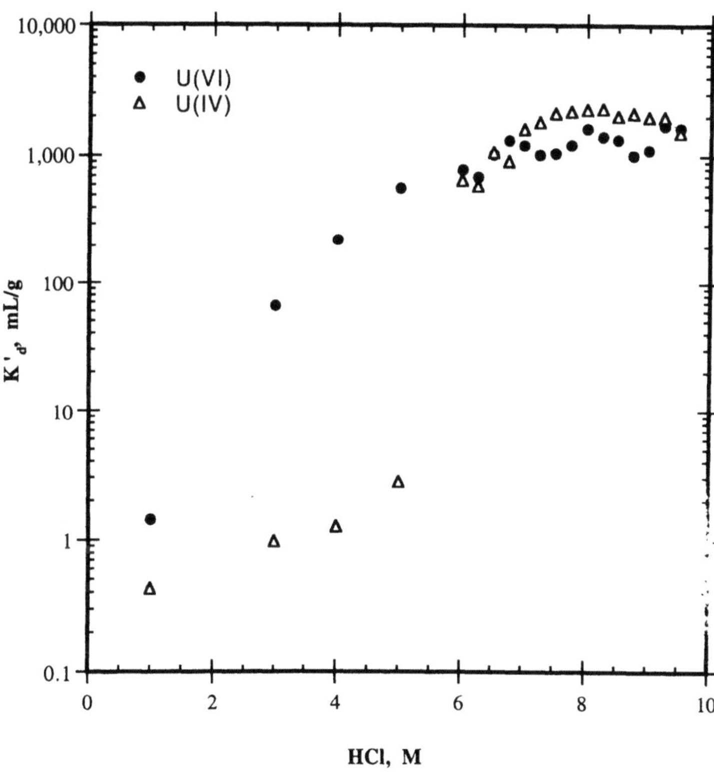


**Figure 9.** The values of $K_d'$ as a function of [HCl] for $UO_2^{2+}$ and uranium(IV) in HCl.

## CONCLUSIONS

The sorption behavior of uranium onto Reillex™-HPQ closely resembles that observed for Dowex™-1x8. The maximum uptake occurs at approximately 7 M $HNO_3$. The plots are bell shaped curves that are interpreted to imply that $K_d'$ increases as $UO_2(NO_3)_3^-$, the predominant sorbing species, is formed, sorbed onto the resin, and then decreases as it is displaced by $NO_3^-$ ion under high nitric acid concentrations. Elution of the uranium is best achieved at low acid concentrations where, at 1 M $HNO_3$, the elution is nearly complete after one column volume. Breakthrough experiments have shown that a greater number of bed volumes is required to achieve breakthrough using 8 M $HNO_3$ than when using other acid strengths. This result is consistent with increased uptake of uranium at this acid concentration, as was observed for the batch distribution coefficient measurements. The uranium sorption behavior is also distinctly different than that of technetium. Sorption of technetium occurs most strongly at dilute acid conditions. Thus, for high-level nuclear waste such as tank supernate or slightly acidified sludge leachate, technetium separation from uranium should occur. The presence of significant quantities of nitrate in the sodium and aluminum forms may be a complicating factor in this process. We have shown that nitrate significantly affects the uptake of uranium and could complicate the separations procedure. Sodium nitrate shifts the maximum sorption value from 7 to ~4 M $NO_3^-$. Aluminum nitrate increases the distribution coefficient exponentially through 8 M. Finally, we have shown that a 2 orders of magnitude difference in uptake is observed between the nitrate and chloride sorption data. Thus, for any uranium separations process flow sheet, the use of HCl should be considered.

## ACKNOWLEDGMENTS

This work was funded by the Efficient Separations and Processing Integrated Program, ESP-IP, Office of Environmental Management, EM-54, of the Department of Energy Office of Technology Development.

## REFERENCES

1. S. F. March, Solvent Extc. Ion Exch., **7**, 889 (1989).
2. S. F. March, "The Effects of Ionizing Radiation on Reillex™-HPQ, a New Macroporous Polyvinylpyridine Resin and on Four Conventional Polystyrene Anion Exchange Resins," Los Alamos National Laboratory report LA-11912 (1990).
3. S. F. March, "The Effects of In Situ Alpha-Particle Irradiation on Six Strong-Base Anion Exchange Resins," Los Alamos National Laboratory report LA-12055 (1991).
4. Reilly Industries, Inc., 1510 Market Square Center, 151 N. Delaware St., Indianapolis, IN 46204.
5. K. R. Ashley, J. R. Ball, A. B. Pinkerton, K. D. Abney, and N. C. Schroeder, Solvent Extc. Ion Exch., **12**, 239 (1994).
6. J. R. Morrey and J. L. Swanson, "A Primer on Hanford Defense Tank Wastes and Prospects for Advanced Chemical Separations" Pacific Northwest Laboratory report (1991).

7. D. B. James, "Equilibrium Anion Exchange of Uranium(VI) on Dowex 1x4 form Nitric Acid Solutions," Los Alamos Scientific Laboratory report LA-3540 (1966).
8. M. Chanda and G.L.Rempel, Reactive Polymers, **18**, 141 (1992).
9. D.K. Veirs, C.A. Smith, B.D. Zwick, S.F. Marsh, and S.D. Conradson, " Characterization of the Nitrate Complexes of Pu(IV) Using Absorption Spectroscopy, $^{15}$N NMR, and EXAFS," Los Alamos National Laboratory document LA-UR-93-3281 (1993).
10. J. Krtil, J. Mencl, and A. Moravec, Radiochem. Radioanal. Lett., **21**, 115 (1975).
11. The least squares fit of the data to the various equations was done using the program KaleidaGraph™ for the Macintosh® computer. The uncertainties reported with the numbers are the estimated standard deviation of that number.
12. I. G. Grenthe, J. Fuger, R. J.M. Konings, R. J. Lemire, A. B. Muller, C. N. Cregu, and H. Wanner, "Chemical Thermodynamics," North-Holland, New York (1992).

# ISSUES IN THE SCALE-UP OF AQUEOUS BIPHASIC EXTRACTION OF URANIUM FROM CONTAMINATED SOILS*

D. J. Chaiko,[1] Deepak,[1] J. Gartelmann,[1] Y. Vojta,[1] E. Thuillet,[2] and C. J. Mertz[1]

[1]Argonne National Laboratory
Chemical Technology Division
9700 South Cass Avenue
Argonne, Illinois 60439-4835

[2]Institut National des
Sciences et Techniques Nucléaires
CEA, Saclay
Gif-Sur-Yvette 91190

## INTRODUCTION

We are evaluating the use of aqueous biphasic extraction (ABE) for solid and liquid waste treatment and the recovery of uranium from contaminated soil.[1-5] The extraction process involves the selective partitioning of solutes or fine particulates between two immiscible aqueous phases. The biphasic systems that we have been working with consist of immiscible polyethylene glycol (PEG) and salt solutions. Each phase contains at least 70 to 80 wt.% water. Some inorganic salts that promote biphase formation with PEG solutions include the sodium/potassium salts of sulfate, carbonate, phosphate, and hydroxide.[6]

The selective partitioning of particulates in ABE systems is based on physicochemical interactions between the particle surface and the liquid phases, rather than bulk phase properties like density. Consequently, particle size should be small enough so that particle settling due to gravity is slow compared to the rate of liquid/liquid phase separation. In general, we have found that an upper size limit of about 50 $\mu$m can be accommodated. Bench-scale tests have shown that, in general, ultrafine metal oxide particulates (such as $PuO_2$, $UO_2$, $CeO_2$, $Fe_2O_3$, $TiO_2$, and $Al_2O_3$) partition preferentially into the salt-rich phase in PEG/salt systems.[2-4]

The only exceptions to this partitioning behavior that we have found so far are crystalline silicates (e.g., quartz and cristobalite), layered silicates (e.g., clays such as kaolinite and montmorillonite), and glass. Clays partition into the PEG-rich phase with partition coefficients of about 100, regardless of pH (we looked at pH's ranging from 2 to12). On the other hand, the partition behavior of crystalline silicates and glass is dependent on the degree of surface ionization. At pH's near the point-of-zero charge (PZC), these materials partition into the PEG-rich phase, while partitioning into the salt-rich phase occurs when the pH is either above or below the PZC.[3]

*Work supported by the U.S. Department of Energy, Office of Technology Development, under Contract No. W-31-109-ENG-38.

In addition to pH, phase transfer agents (PTA's) can be employed to alter the partitioning behavior of ultrafine particulates.[7] For example, water soluble complexants, such as Arsenazo III, can be used to transfer $PuO_2$ or $UO_2$ particulates from the salt-rich phase into the PEG-rich phase in the PEG/$Na_2SO_4$ system. With the use of a PTA and proper control of pH, it is possible to selectively recover $PuO_2$ from $PuO_2$/silica and $PuO_2$/graphite mixtures.[2-3]

In this chapter, we describe our efforts to scale up this separation technology for the removal of uranium from contaminated soil. We summarize results from countercurrent extraction tests in a pilot-scale Karr column. We discuss the key issues for successful operation of the extraction column with feed solutions containing high concentrations of suspended soil particles. Also discussed are issues in solid/liquid separation of column effluents and secondary waste treatment.

## BACKGROUND

The former Feed Materials Production Center, now called the Fernald Environmental Management Project (FEMP), at Fernald, Ohio, produced uranium metal for use in defense reactors until production was halted in 1989. During its operation, the soil at the site became contaminated with uranium from a variety of sources, including deposition of airborne uranium particulates and spills of solvents and process effluents.[8] There are an estimated 2,000,000 $yd^3$ ($1.6 \times 10^6$ $m^3$) of uranium-contaminated soil that needs to be treated. Soil clean-up levels have not yet been set.

The major constituents of the soil at the FEMP site are quartz, calcite, dolomite, and clay.[8] Because of the high levels (>80 wt.%) of silt (<53 to 2 $\mu$m) and clay (<2 $\mu$m), this soil is not directly amenable to treatment by conventional soil washing techniques. Buck et al.[9-11] have identified a number of uranium(VI)-bearing phases in the soil. These include uranium silicates, uranium phosphates (e.g., autonites), uranium oxide ($UO_3$), and uranium associated with calcium fluorite. In addition, Buck et al. also identified a number of uranium(IV)-bearing phases, such as uranium silicide, uranium oxide ($UO_2$), and uranium phosphite [$U(PO_3)_4$]. While the uranyl species are easily leached with aqueous carbonate solutions, some of the uranium phases, like uranium phosphite, are extremely refractory and do not readily dissolve in carbonate solutions even in the presence of an oxidizing agent. The more refractory uranium phases are present as discrete particles with sizes in the micron and submicron range.[11]

Test-tube-scale studies performed in our laboratory indicated that aqueous biphasic extraction showed potential for the removal of both soluble and refractory forms of uranium from Fernald soils.[4-5] We examined a number of PEG/salt systems, which included sodium and potassium carbonate, sodium hexametaphosphate [$(NaPO_3)_6$], and sodium sulfate.[4] In addition, two different PEG molecular weights (1500 and 3400) were used. The most suitable system, in terms of phase separation characteristics and partitioning selectivity, was PEG/$Na_2CO_3$ at temperatures above 35°C. Partitioning of soil with an initial uranium concentration of 545 mg/kg produced a uranium concentrate in the salt-rich phase that contained 1-2 wt.% of the soil feed and a uranium concentration of 904 mg/kg soil. The clean soil partitioned into the PEG-rich phase and had a uranium concentration of 138 mg/kg soil. Because of the tendency for the cleaned soil to become recontaminated by uranium particles adhering to the test tube walls, it was felt that continuous, countercurrent extraction would produce a significantly lower residual uranium concentration in the cleaned soil.

In all of the test-tube scale partitioning tests that we conducted, significant amounts of uranium dissolved into the salt phase. In some of the tests, over 70% of the uranium in the soil dissolved into the salt phase during the 1 to 2 min it took for the extraction.

This is undoubtedly because a significant portion of the uranium in the soil is in the form of soluble uranyl species.[12] Uranium dissolution has the unfortunate effect of reducing the mass of uranium recovered as solids in the salt rich phase and increasing the burden on any secondary processes for aqueous waste treatment. In general, the ABE process is ideally suited to the removal of highly refractory particulate contaminants that are in the micron size range and that are well liberated from the soil matrix.

## EXPERIMENTAL

### Materials

The reagents used to generate the biphase systems were polyethylene glycol (PEG-1500), with an average molecular weight of 1500, and sodium carbonate. They were obtained from Aldrich and Fisher Scientific, respectively. Both chemicals were reagent grade. Nonionic, cationic, and anionic polymeric flocculants were provided by Nalco Chemical Co. (Naperville, IL.) and Cytec Industries (West Paterson, NJ). Also, $^3$H-labeled polyethylene glycol (MW 4000) and $^{14}$C-labeled sodium carbonate were purchased from NEN Research Products (Boston, MA).

The uranium-contaminated soil sample was provided by FEMP (Fernald, Ohio). The soil sample used in our tests was collected from near the waste incinerator at the FEMP site.

### Methods

We prepared the soil for the extraction studies by dispersing it in deionized water containing sodium pyrophosphate as a dispersant at a concentration of 200 mg/kg of slurry. The slurry was screened at 200 mesh, and the <200 mesh fraction was filtered by a centrifuge and air dried. Analysis of this soil fraction by delayed neutron counting (Activation Laboratories Ltd., Ontario, Canada) indicated a uranium concentration of 471 mg/kg of soil.

Partition coefficients were calculated as the ratio of the species concentration in the PEG-rich phase to that in the salt phase.

The adsorption of PEG onto soil and kaolinite clay was measured by equilibrating a known amount of soil (particle size <100 mesh), or kaolinite (<325 mesh), with aqueous PEG solutions that were previously spiked with $^3$H-labeled PEG-4000. The specific activity of the radiotracer was 2.0 mCi/g. The suspensions were equilibrated for 20 min in a thermostated water bath with intermittent mixing. The samples were then centrifuged for 15 min, and aliquots of the supernatant were removed for counting in a Tri Carb scintillation counter, model 2500CA. The PEG adsorption densities were calculated from the reduction in radiotracer concentration in the PEG/$Na_2CO_3$ solution after the addition of soil.

## THE KARR COLUMN AND ITS OPERATION

A 3.89-m-tall Karr column with an internal diameter (ID) of 2.54 cm was purchased from the Otto H. York Company (now available from Glitsch Technology Corp., Parsippany, NJ) and installed in our laboratory. The column body is made of borosilicate glass. The top and bottom portions of the column consist of expanded 5.08-cm ID glass sections, respectively. The column is jacketed, and in our studies was maintained at 40°C by circulating water from a constant temperature bath. A schematic diagram of the column is shown in Fig. 1.

The extraction zone is agitated by reciprocating vertical strokes of the perforated plates which are mounted on a rod. Our column is equipped with a total of 101 stainless steel plates spaced approximately 5 cm apart from each other. The maximum reciprocation rate that can be accommodated with the liquid/liquid systems that we are testing is about 1.33 Hz. The column is set up with a fixed stroke length of 1.9 cm.

We operate the column with the less-viscous, salt-rich phase as the continuous phase. Typically, the average diameter of the dispersed-phase droplets is about 1 mm. We have found that, with care, the column can be operated, without flooding, with a liquid/liquid density difference as low as 3-5%. The design of the column, together with its hydrodynamic characteristics, permits rapid transport of fine solid particulates through the column without solids settling within the column. We had estimated the mean residence time of dispersed phase droplets within the extraction zone to be approximately 2 min.

During extraction, vibration of the dispersed-phase droplets containing suspended soil particles can be seen. In addition, the dispersed-phase droplets continuously undergo a cycle of size reduction and agglomeration. We have found that rapid mass transfer between the two phases can be achieved even with a high degree of solids loading (e.g., >20 wt.%) in the dispersed phase.

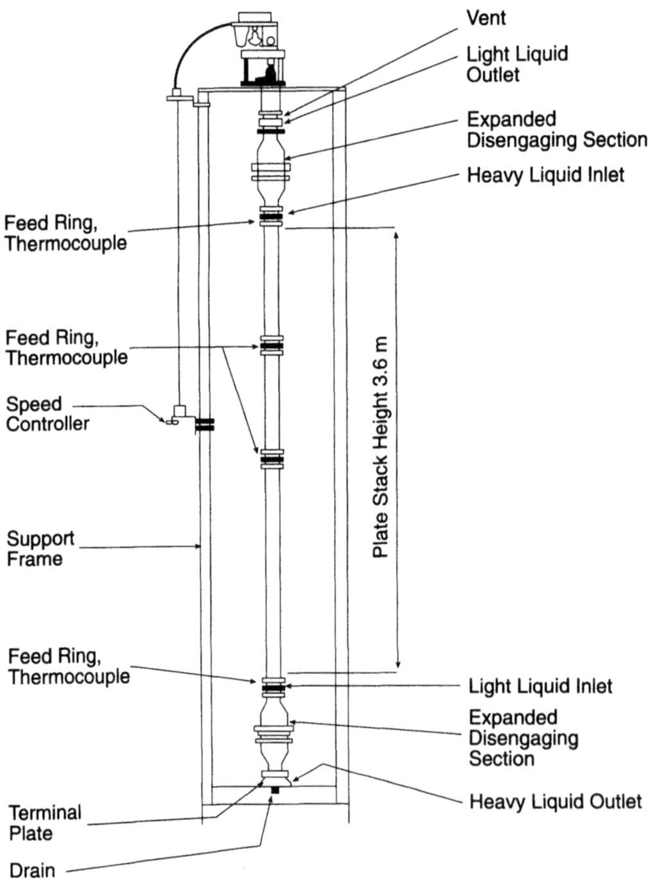


**Figure 1.** Schematic diagram of pilot-scale Karr column.

Efficient mass transfer within the extraction zone of the column is a function of many physical and physicochemical variables. These include the liquid flow rates, degree of agitation in the column, difference in density of the two liquid phases, viscosity, interfacial tension, temperature, and the phase separation time. The effects of most of these variables are interconnected and are discussed next.

**The Phase Diagram and Tie Lines**

Unlike conventional oil/water extraction systems, the two immiscible phases in an aqueous biphasic system share a common diluent--water. Therefore, the potential for process upset due to excessive mass transfer of water must be carefully considered. To prevent the net mass transfer of water between liquid phases, we prepared the PEG-rich and salt-rich phases at compositions defined by the tie lines from the appropriate PEG/salt/water phase diagram.

The phase diagram for the PEG-1500/$Na_2CO_3$ system at 40°C with three tie lines (A, B, and C) is shown in Fig. 2. The binodal curve (connected triangles in Fig. 2) defines the phase boundary, with the two-phase regime located to the right of the binodal, and the single-phase regime on the left. The upper and lower intersection points of the tie line with the binodal curve define the compositions of the PEG-rich and salt-rich phases, respectively.

For convenience, we determined the tie lines by using a radiotracer technique with $^{14}C$-labeled sodium carbonate and $^{3}H$-labeled PEG-4000. The radiotracers enable us to rapidly and easily measure the partition coefficients of carbonate and PEG. The compositions of the two immiscible aqueous phases are then determined from the measurement of liquid/liquid phase ratio and the appropriate mass balance equations. The procedure is described in detail elsewhere.[2] Ideally, the tie lines should be determined from chemical analysis of the equilibrated phases. To verify the validity of our approach, we contacted equal volumes of the two phases defined by tie-line A in Fig. 2 and measured the change in the liquid/liquid phase ratio. The volume increase or decrease of the two phases, as a percentage of the total volume of the system, was only 2-3%. This small error in the tie line is probably due to slight differences in the partition coefficients of PEG-1500 and the $^{3}H$-labeled PEG-4000.

A change in temperature will affect the position of the binodal and, hence, performance of the extraction column. An increase in temperature causes the binodal to shift to the left, as shown in Fig 3. Increasing temperature increases interfacial tension and reduces the viscosity of the liquid phases. Both effects improve phase separation rates. Therefore, to enhance the rate of mass throughput in the Karr column, we operate at 40°C instead of room temperature.

Whether tie-line A, B, or C is chosen, the viscosity of the salt-rich phase is little changed. However, the viscosity of the PEG-rich phase increases measurably as one moves further away from the critical point (i.e., the point where the length of the tie line is zero). Of course, increasing viscosity would adversely affect mass transport within that phase. Equally important, however, is that increasing viscosity affects the physical entrainment of the salt-rich phase in the PEG-rich stream exiting the column. The liquid-phase viscosity, together with the density difference and interfacial tension, determines the rate at which the PEG droplets arriving at the phase-inversion zone coalesce and form the continuous PEG phase.

The logarithm of the interfacial tension between the PEG and salt phases is proportional to the length of the tie lines.[13] Therefore, the farther away from the critical point, the larger the interfacial tension. For efficient mass transfer, a large number of droplet dispersion/coalescence cycles must take place within the column. This is achieved by the proper balance of column agitation rate with liquid phase flow rates. The optimum

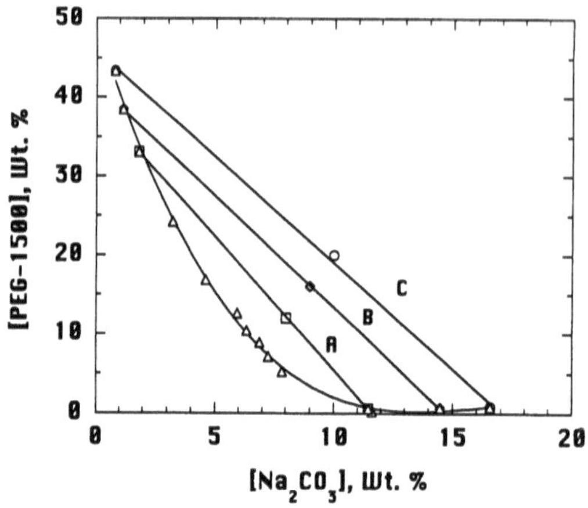


**Figure 2.** Phase diagram with tie lines for the PEG-1500/$Na_2CO_3$/water system at 40°C.

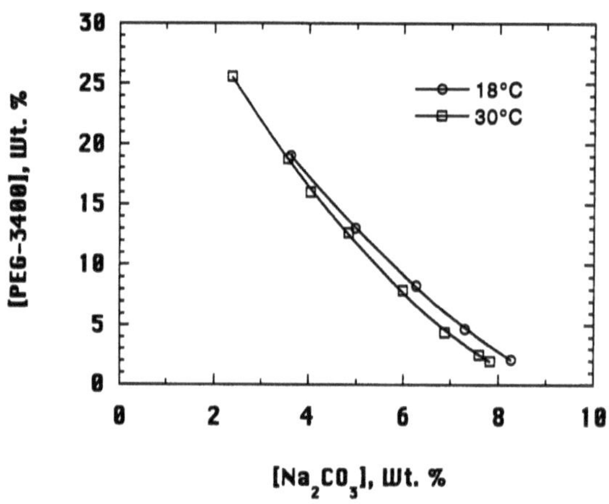


**Figure 3.** Effect of temperature on biphase formation.

balance point will, in turn, depend on the interfacial tension and the densities of the two phases.

The rate of phase separation, which is dependent on liquid phase viscosities, interfacial tension, liquid phase densities, etc., is an important variable that directly impacts column throughput. This parameter[14-15] has been characterized by a dimensionless dispersion number ($N_{Di}$), defined as follows:

$$N_{Di} = \frac{1}{t_{Di}} \sqrt{\frac{\Delta z}{g}} \tag{1}$$

where $t_{Di}$ is the time for phase separation under gravitational field g, and $\Delta z$ is the height of the two-phase system. The higher that $N_{Di}$ is, the higher the process throughput one can expect.

Faster phase separation implies a lower chance for physical entrainment of salt in the PEG out-flow stream. To characterize our extraction system, we determined the dispersion number for the biphase formed by tie-line A in Fig 2. Approximately equal volumes (~ 35 mL) of PEG and salt phases were vortexed together for one min and then allowed to separate in a graduated 100-mL cylinder. At 40°C, complete phase separation occurred within 75 s, yielding a dispersion number of 15.5 x $10^{-4}$.

As expected, the dispersion number decreases with addition of soil to the biphase system. For example, when 1 wt.% (expressed as a percentage of total PEG phase) of Fernald soil was added to the PEG phase, the phase separation time of the biphase system increased to 120 s, giving an $N_{Di}$ value of 9.7 x $10^{-4}$. With or without solids, however, the dispersion numbers are comparable with more common oil/water extraction systems which usually exhibit $N_{Di}$ values of 1-15 x $10^{-4}$.

Based on viscosity, interfacial tension, and phase separation time, the selection of a tie line for process operation can be difficult. We have found that tie-lines A and B, shown in Fig. 2, permit operation of the column at reasonable liquid flow rates without flooding. The determining factor in deciding which of the two tie lines to use is the density difference between the PEG and salt phases. The compositions of PEG and salt phases, along with the corresponding densities of the phases for tie-lines A, B, and C, are listed in Table 1.

Clearly, a longer tie line provides a greater difference in the density of the two phases. The manufacturer of the Karr column recommends a liquid/liquid density difference of at least 5%. Therefore, manufacturer guidelines would dictate that tie-line A is approximately the shortest tie line permissible for countercurrent extraction with the PEG-1500/$Na_2CO_3$ system. However, when we apply this liquid/liquid extraction technique for particulate separation, new possibilities arise, as we will describe below.

## CONTAMINATED SOIL TREATMENT IN THE KARR COLUMN

During the aqueous biphasic extraction process, the clean soil partitions to the PEG-rich phase, while the uranium contaminants partition selectively to the salt-rich phase. Therefore, to minimize the amount of mass transfer required for soil decontamination, the soil feed is dispersed in the PEG phase. The effective density ($d_e$) of the PEG/soil slurry is given by:

$$d_e = \frac{100\ d_s\ d_p}{(d_p - d_s)\ W + 100\ d_s} \tag{2}$$

**Table 1.** Composition and density of phases on PEG/$Na_2CO_3$ tie lines.

| | PEG-Rich Phase | | | Salt-Rich Phase | | |
|---|---|---|---|---|---|---|
| Tie Line | [PEG-1500], wt.% | [$Na_2CO_3$], wt.% | Density, kg/$m^3$ | [PEG-1500], wt.% | [$Na_2CO_3$], wt.% | Density, kg/$m^3$ |
| A | 33.11 | 1.82 | 1080 | 0.62 | 11.45 | 1120 |
| B | 38.46 | 1.15 | 1080 | 0.71 | 14.48 | 1150 |
| C | 43.33 | 0.08 | 1090 | 0.80 | 16.56 | 1180 |

where W is the amount of suspended soil (wt.%) in the PEG phase, and $d_s$ and $d_p$ are the densities of soil and PEG phases, respectively.

An average density for the Fernald solid is assumed to be 2500 kg/m$^3$. Hence, at the critical values of W = 6.3, 10.7, and 13.5 wt.% for tie-lines A, B, and C, respectively, the density of the PEG/soil slurry is equal to that of the salt phase. This means that, if the column is to be operated with the PEG phase as the less-dense phase, the soil loading in the PEG phase must be less than the appropriate critical value of W. This, however, severely restricts the solids throughput of the column.

We propose an alternative method of operation that greatly increases the throughput of the column. By loading the soil in the PEG phase at a much greater amount than the critical value of W, the PEG phase is effectively made heavier than the salt phase. In this approach, the salt-rich phase is introduced to the column from the lower inlet (see Fig. 1), while the PEG-rich phase, containing the contaminated soil, is fed from the upper inlet. As the PEG phase travels down the column, the uranium is extracted into the salt phase, and the density of the PEG phase is gradually reduced. However, because the contaminated soil contains only 100-1000 mg uranium/kg soil,[8] we expect only a small fraction of the soil mass to transfer into the salt phase. We carried out experiments to determine the dependence of column operation on the solids concentration in the PEG-rich phase. The details of those column runs are discussed below.

**Experiment A: 1 wt.% Soil in PEG Phase**

A 1 wt.% soil loading in a PEG phase with density of 1080 kg/m$^3$ gives a PEG/soil slurry density of 1086 kg/m$^3$. To maximize the liquid-phase density difference, while not inordinately increasing the viscosity of the PEG phase, we selected tie-line B for this experiment. This provides a salt-phase density of 1150 kg/m$^3$, and a liquid-phase density difference of approximately 5.9%. Figure 4 shows a flow diagram for this experiment.

During startup, the column was filled with the salt phase, and then a salt flow rate of 65 mL/min was established. The lighter PEG phase, without soil, was introduced into the column at a low rate of 22.5 mL/min. The salt phase was gradually drained from the underflow port until the dispersed PEG phase hold up in the column was approximately 20% by volume. Then, by increasing the agitation speed to 0.65 Hz, a liquid/liquid interface was established about 10 cm below the overflow port at the top of the column.

The soil feed was prepared by dispersing 60 g of soil in a portion of the PEG phase at a solids concentration of 10 wt.%. The slurry was then added to the PEG feed tank, giving a final solids density of 1 wt.%. This slurry was fed to the column through the less-dense feed inlet at the bottom of the column. During extraction, the salt phase flows countercurrent to the PEG phase, and the uranium particles are stripped from the PEG phase droplets into the salt phase. At the same time, the clean soil remains dispersed in the PEG phase and eventually exits the column at the light-phase overflow (see Fig. 4). During the run, we continuously filtered the column effluents with in-line centrifuges. In principle, if more solids are to be treated, we can return the PEG phase from the centrifuge effluent back into the feed tank for continuous recycle. In this experiment, the salt-phase underflow, after centrifuge filtration, was continuously recycled back to the column.

The feed slurry was pumped into the column for approximately 230 min. After all of the slurry had been fed into the column, additional PEG phase (without any soil) was pumped to the column for 40 min to flush the residual solids from the column. The soil was separated from the PEG overflow and the salt underflow in the centrifugal filters, and was then washed and dried for chemical analysis. The liquid phases were filtered at 0.2 $\mu$m and analyzed for dissolved uranium.

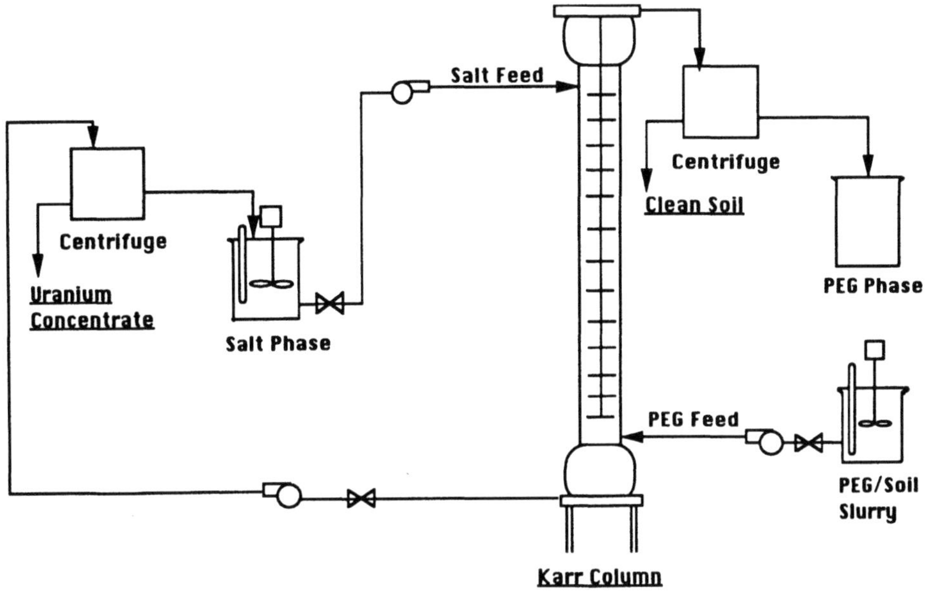


**Figure 4.** Flow diagram for 1 wt.% soil dispersed in the PEG phase.

## Experiment B: 20 wt.% Soil in PEG Phase

At sufficiently high loading of soil in the PEG-rich phase, the salt phase then becomes the lighter phase. In this experiment, the liquid-phase compositions were defined by tie-line A, which offers the lowest salt-phase density (see Table 1) and, hence, permits density inversion at a lower soil loading. With 20 wt.% soil in PEG, the effective density of this phase becomes 1218 $kg/m^3$. The salt-phase density is 1120 $kg/m^3$, giving a density difference of 8.75%.

In this experiment (see Fig. 5), the salt phase was pumped into the light phase inlet and exited at the column overflow port at a flow rate of 65 mL/min. A total of 288 g of Fernald soil was dispersed in 1150 g of PEG phase. The 20 wt.% PEG/soil slurry was pumped into the column at a flow rate of 13 mL/min, and the agitation rate was set at 1 Hz. The soil traveled down the column, while remaining dispersed in the PEG-phase droplets, and formed an interface at the bottom of the column. After the interface was established at the desired position, the PEG phase was then pumped from the column at a rate required to maintain a constant interface height.

## RESULTS AND DISCUSSION

To take full advantage of the high selectivity that aqueous biphasic extraction offers, one must ensure that particle movement within the extraction column is influenced by surface chemistry and not gravitational settling. This necessity places an upper limit on the particle size that can be treated. The relationship between the way the column is operated and particle size limit is discussed in more detail below.

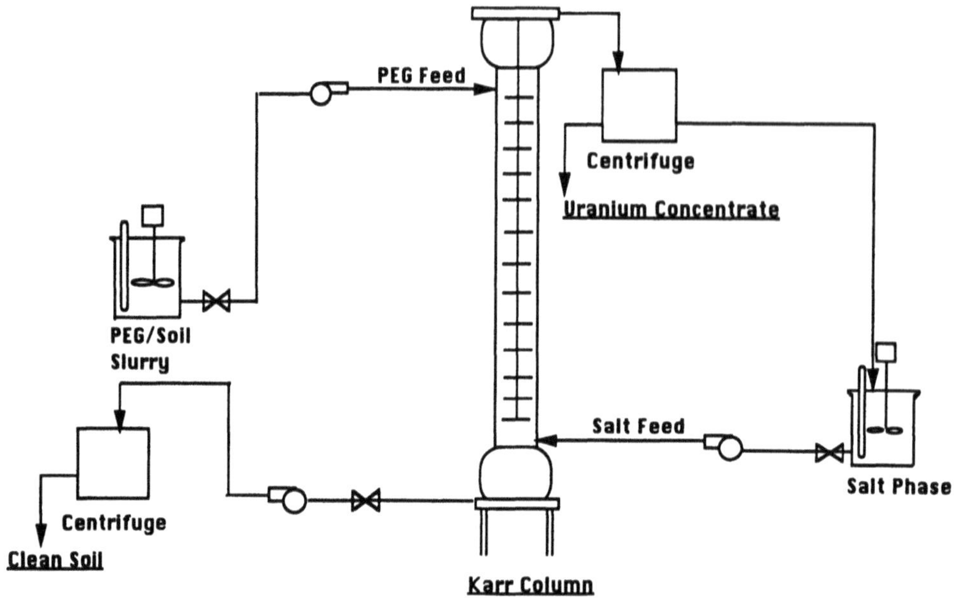


**Figure 5.** Flow diagram for 20 wt.% solids dispersed in the PEG phase.

In experiment A, 1 wt.% soil was suspended in the PEG-rich phase, and the clean soil was carried upward in the extraction column against the direction of gravity. Therefore, this method of column operation is more suited to treating fine particle sizes (e.g., <200 mesh). However, this limitation is relaxed in the alternative method of operating the column whereby the PEG phase becomes the more-dense phase. This is because the clean soil moves downward through the column in the direction of gravity. We have since treated Fernald soil with a top particle size of 100 mesh (147 $\mu$m) with no apparent problems. The partitioning behavior of the particulate uranium, which is only a few microns in diameter, is unaffected by the large size of the soil particles.

In both of the column extraction experiments, we determined the dissolved uranium content in the salt and PEG phases exiting the column. The results, presented in Table 2, indicate an increase in dissolved uranium concentration with time. The salt phase in experiment A was continuously recycled, and as expected, the uranium concentration in that phase increased progressively. The uranium concentration in the PEG phase, however, remained either below or just at the detection limit of the delayed neutron counting technique used to measure the uranium concentration.

In experiment A, 60 g of soil was introduced to the column in approximately 5.5 L of PEG phase. Additionally, approximately 7 L of salt phase was used. In experiment B, 288 g of soil was introduced with only 1.065 L of the PEG phase. The total salt-phase volume, including column holdup, was approximately 7.5 L. Because of the greater amount of soil, the total amount of uranium dissolved and the uranium salt-phase concentration were higher in experiment B than in A. However, there is no indication that the salt phases became saturated with dissolved uranium in either experiment. In experiment B, the salt-phase recycle was begun at 52 min into the run. The continual buildup of dissolved uranium in the salt phase during the course of the run indicates that the column never reached steady state.

An attractive feature of this extraction system is that the mass of solids recovered in the salt phase is only 1-2 wt.% of the initial feed (see mass balance data in Table 3). Complete recovery of the solids in either the uranium concentrate or clean soil streams was not possible because a portion of the soil (e.g., humic acids) dissolved in the salt and PEG phases, and a portion was recovered separately during column shutdown. However, in both column runs, the clean soil recovered with the PEG phase was lower in uranium concentration than the solids recovered from the salt phase. The uranium concentration in the uranium concentrate was lower than that in the soil feed due to dissolution of uranium into the liquid phases. In effect, the soil is decontaminated during column extraction by two processes: (a) dissolution of soluble uranyl species into the salt and PEG phases and (b) partitioning of refractory uranium particulates into the salt phase. This combination of decontamination mechanisms permits us to reach lower residual uranium concentrations in the clean soil than attainable through carbonate leaching alone. For example, when we leached this soil with 20 wt.% $Na_2CO_3$ at 40°C for 2 h using a liquid/solid weight ratio of 8.8, we reduced the uranium concentration in the soil to only 146 mg/kg.

The experimental results in Table 3 clearly indicate that high solids loading in the PEG phase has little effect on the degree of uranium extraction from the soil. In addition, the high solids loading in experiment B provided an order of magnitude increase in the soil throughput in the column. Our more recent column tests indicate that it is possible to increase the soil loading in the PEG phase to 30-35 wt.% and the feed flow rate to at least 24 mL/min. This gives a solids processing capability of approximately 660 g/h with the 2.54-cm ID column. A full-scale column of 1.5-m ID would be capable of handling 2400 kg/h.

**Table 2.** Dissolution of uranium during column extraction.

| Experiment A | | | Experiment B | | |
|---|---|---|---|---|---|
| Time, min | [Uranium], mg/kg Salt Phase | PEG Phase | Time, min | [Uranium], mg/kg Salt Phase | PEG Phase |
| 0 | 0 | 0 | 0 | 0 | ---[1] |
| 25 | 1.0 | <0.1 | 15 | 5.1 | |
| 50 | 1.3 | <0.1 | 30 | 9.1 | |
| 70 | 1.4 | <0.1 | 52[2] | 13.0 | |
| 105 | 1.8 | <0.1 | 85 | 17.3 | |
| 135 | 2.1 | <0.1 | | | |
| 205 | 2.6 | <0.1 | | | |
| 230 | 3.1 | 0.1 | | | |

[1]The composite PEG phase contained 3.7 mg uranium/kg.
[2]Salt-phase recycle back into the column began at this point.

**Table 3.** Mass balance on solids from column extraction.

| | Experiment A Mass, g | [U], mg/kg | Experiment B Mass, g | [U], mg/kg |
|---|---|---|---|---|
| Soil Feed | 60 | 471 | 288 | 471 |
| Uranium Concentrate | 1 | 105.9 | 5 | 218 |
| Clean Soil | 49 | 80.7 | 175.5 | 77.7 |
| Column Residual[1] | 0 | --- | 77.5 | 88 |
| Soil Loss | 10 | --- | 30 | --- |

[1]Refers to soil remaining in column after shutdown.

Maximizing extraction efficiency requires minimization of physical entrainment of one phase in the other phase exiting the column (referred to as "other-phase carryover"). This limits the effective partition coefficient, and if the other-phase carryover is excessive, additional unit operations would be required to separate the liquid phases before they could be recycled back to their respective process streams. As shown in Table 4, the entrainment of the PEG-rich phase in the salt-rich phase was below detection limits in both experiments. On the other hand, the amount of salt-rich phase entrained in the PEG phase exiting the column varied from 1 to 7.5 vol.%, with an average of 3.9 vol.% for the case where the PEG feed contained 1 wt.% solids. For the case where the PEG feed contained 20 wt.% solids, the salt-phase entrainment in the PEG-phase underflow ranged from 0 to 5.9 vol.%, with an average of 3.2 vol.%. Interestingly, increased solids concentration does not seem to affect salt-phase entrainment significantly.

With high solids loading, we noticed that the entrained salt phase contained high concentrations of humates and uranium (83.1 mg/kg). Normally, the PEG phase has a much darker color than the salt phase due to the partitioning behavior of dissolved humic acids. Earlier measurements showed that the partition coefficient of humates is greater than one in a variety of PEG/salt biphase systems.[2] In this case, however, the entrained salt phase was considerably darker than the PEG phase. One possible explanation for this apparent reversal in partitioning is that saturation of the PEG phase by the high concentration of dissolved humic acids caused their partition coefficient to drop below one.

## Silica Partitioning

In test-tube-scale experiments with model systems, we measured the partition coefficients for various silicate species, including silicic acid, quartz, and clay. The results of these measurements (see Table 5) contradict the partitioning behavior of the crystalline silicates (e.g., quartz with a system pH of 8.5) in the soil during the column extractions.

**Table 4.** Solvent entrainment during column operation.

| Experiment | Salt Entrainment in PEG Phase, vol. % | PEG Entrainment in Salt Phase, vol. % |
|---|---|---|
| A | 3.9±3.1[1] | 0 |
| B | 3.2±1.9[2] | 0 |

[1]Reported variation represents the standard deviation of 5 measurements.
[2]Reported variation represents the standard deviation of 9 measurements.

**Table 5.** Partitioning of silica species in PEG-3400/$Na_2SO_4$ system at 25°C.

| Species | System pH | Partition Coefficient |
|---|---|---|
| $SiO_2$ | 8.5 | <1.0 |
| | 2-3 | ~100 |
| Kaolinite | 2-12 | ~100 |
| Montmorillonite | 2-12 | ~100 |
| $Si(OH)_4$ | 3 | 0.64 |
| | 6 | 0.38 |

We believe that the quartz partitioning to the PEG-rich phase during the column extraction is due to surface activation of these soil particles by absorbed humic acids. This is consistent with our observation that, when the soil is preleached with $Na_2CO_3$ (i.e., conditions that cause the dissolution of humic matter), the crystalline silicates, like quartz, partition selectively to the salt phase. Under these conditions,[4] approximately 11 wt.% of the soil was recovered in the salt phase with a uranium concentration of only 15 mg/kg. The amount of quartz recovered in the salt phase roughly corresponds to the amount of quartz in the soil. The identity of the mineral phases recovered in the salt phase was verified by electron microscopy.[4]

The soil organics (i.e., humic and fulvic acids) are believed to have an aromatic ring structure of di- and tri-hydroxyl phenols that are linked by -O-, -NH-, -N-, and -S- bonds.[16] These acids also contain significant numbers of free OH groups. These materials, with molecular weights ranging from a few hundred to more than 300,000, have a large number of reactive functional groups. It is therefore not surprising that the humic acids adsorb onto mineral surfaces, turning a hydrophilic mineral surface into a hydrophobic one.[17] If quartz particles are rendered hydrophobic by the adsorption of humic acids, one can expect them to partition into the PEG phase. In any case, the partitioning of quartz and clay into the PEG phase is an obvious advantage since this increases the amount of material recovered in the clean soil stream.

**Solid/Liquid Separation**

Because of the high clay content of the Fernald soil, solid/liquid separation is expected to dominate capital equipment and processing costs. To minimize the size of filtration equipment, we are exploring the use of flocculants for enhancing the rate of solid/liquid separation. Because most of the soil is recovered from the PEG-rich phase, our initial studies were focused on this process stream. In particular, whether the high PEG concentration in that effluent stream would interfere with flocculant adsorption onto the soil particles. We tested a number of polymeric flocculants that were recommended by Nalco and Cytec. A partial list is given in Table 6.

The test procedure involved adding the flocculant to the PEG phase containing 15 wt.% soil. After gently mixing the slurry in a 100 mL graduated cylinder and equilibrating in a 40°C water bath, the settling rate of the soil was measured. In the absence of any flocculant, the settling rate of the soil was only 0.026 m/h (see Table 6). The settling rate was little improved by the Cytec flocculants; however, with the Nalco 7877 flocculant we observed the highest settling rate, which was 2.04 m/h. With a solids concentration of 20-30 wt.%, the PEG-phase effluent could be treated directly by vacuum filtration without the need for preconcentration in a thickener. We are now in the process of measuring filtration rates using the best performing flocculant as a filter aid.

**Table 6.** Settling rate from PEG-rich phase with 50 mg flocculant/kg slurry. The solids concentration was 15 wt.%.

| Flocculant | Polymer Type | Settling Rate, m/h |
|---|---|---|
| None | | 0.026 |
| Magnifloc 1596C | Cationic Polyacrylamide | 0.069 |
| Magnifloc 496C | Cationic Polyacrylamide | 0.091 |
| Magnifloc 1555C | Cationic Polyacrylamide | 0.249 |
| Nalco 7871 | Non-ionic, Latex | 0.207 |
| Nalco 9857 | Cationic, Latex | 0.124 |
| Nalco 7877 | Anionic, Latex | 2.04 |

**Table 7.** Adsorption of PEG onto soil and kaolinite at 40°C.

| Dispersed Solids | Adsorption Density, mg PEG/g Solid |
|---|---|
| Fernald Soil[1] | 8.4 - 14.2 |
| Kaolinite[2] | 1.84 - 2.03 |

[1]Equilibration time was 20 min.
[2]Equilibration time was 1 h.

We expect the solids content in the filtered cake to be typically 60-70 wt.%. Therefore, the PEG phase entrained in the pores of the solid cake represents 30-40% of the filtrate mass and must be recovered to maintain economic viability. A water wash would be used to recover dissolved uranium and PEG from the void volume in the filter cake. The PEG phase can then be recycled after adjustment to the tie-line composition.

In addition to physical entrainment of PEG phase within the filter cake, another possible mechanism of PEG loss is through physical and chemical adsorption of PEG onto soil particles. The results of adsorption measurements with $^3$H-labeled PEG-4000, however, suggest that this mode of polymer loss is likely to be fairly insignificant. With 20 and 30 wt.% solids dispersed in the PEG phase (composition given by tie-line A in Fig. 2), we found a maximum of 8.4-14 mg PEG adsorbed per gram of soil at 40°C (see Table 7). Also shown in Table 7 are adsorption data for kaolinite.

## Secondary Waste Treatment

In one sense, the presence of organic matter in the soil is beneficial in that humic acid coatings cause silica to partition to the PEG phase. However, the humic acids are also a source of concern in that their build-up in process streams could lead to possible process upset. For this reason, we have begun to examine methods of removing soil organics from process effluents. It is well known[16] that humic acids are readily precipitated from solution at pH $<3$. However, if we adopt this approach, it would result in the addition of a large amount of mass to the process streams and would alter the composition of the biphase system. It should be noted that below a pH of 9-10, a biphase is not formed in the PEG/carbonate system. In addition, fulvic acids remain soluble even at low pH.

The conventional method for removing humic acids from drinking water involves coagulation/flocculation by alum, followed by sedimentation. As an alternative, for increased efficiency of humic acid removal, Zhang et al.[18] flocculated the colloidal humic matter with polyaluminum chloride and a polyelectrolyte and then used membrane ultrafiltration to separate the humic matter. They could extract as much as 92% of the humic matter measured by the total organic content, at a permeation flux of 250 L $h^{-1}m^{-2}$ with a $10^6$ molecular weight cut-off membrane. However, their treatment is at a pH of 7, and humic acid content in drinking water supplies is much lower than that in our column effluents. Furthermore, ultrafiltration requires regeneration of the membrane, which would require washing it with, for example, NaOH and a large amount of water.

Our preliminary studies indicate that it is possible to precipitate most of the humic matter from the salt phase with $Ca^{++}$ addition. This is consistent with the results of Orlova et al.,[19] who demonstrated that the optically dense humic fraction precipitates with $CaSO_4$ addition. We have also found that the optically dense humic-acid fraction can be selectively precipitated from the salt phase with the addition of small amounts (e.g., 20 vol.%) of methanol. We are now examining these options further for both humate and uranium removal from the PEG and salt-phase effluents.

## CONCLUSIONS

Preliminary test results using a Karr column for countercurrent extraction indicate that aqueous biphasic extraction is able to remove both soluble uranyl species and refractory uranium particulates from soils with high clay content. The uranium concentration in the soil was reduced from 471 mg/kg to approximately 78 mg/kg. Examination of the clean soil by electron microscopy indicates that some of the uranium particulates still remain in the cleaned soil. We believe that adsorbed humic acids are acting as phase transfer agents and are responsible for the partitioning of uranium particulates into the PEG phase. This surface activation mechanism also explains our ability to recover crystalline silicates, like quartz, in the PEG phase. This leads to soil recoveries of 97-98% from the clean soil fraction. We are now examining ways to selectively alter the surface chemistry of the uranium phases for improved recovery of uranium particulates in the salt phase.

High column capacity was attained by suspending high concentrations of soil in the PEG phase, effectively making the PEG phase the more-dense phase. Variation of solids concentration from 1 to 20 wt.% had no effect on column performance. Solids throughput reached 190 g/h in a 2.54-cm-ID Karr column. We have since increased the throughput further, above 660 g/h, by using 30-35 wt.% solids in the PEG phase and increasing the flow rate. Our results with the 2.54-cm column indicate that the solids throughput of a full-scale column (1.5-m diameter) would exceed approximately 2400 kg/h. We are now conducting an economic analysis of the process in conjunction with the preparation of a more detailed flowsheet design.

## ACKNOWLEDGMENTS

The authors wish to thank R.A. Leonard in the Chemical Technology Division for helpful discussions on contactor design and operation.

## REFERENCES

1. D.J. Chaiko, L. Reichley-Yinger, E.R. Orth, E.H. Van Deventer, G.F. Vandegrift, M. Krumpolc, J.E. Helt, R.D. Coleman, S.N. Kakar, T.S., Tasi, K. Horken, W. Killian, and N.F. Sather, Development of a process for treating red water by organic/inorganic separation and biodegradation, in: "Proceedings 14th Annual Army Environmental R & D Symposium," USATHAMA Report CETHA-TE-TR-900055, p. 303 (1989).

2. D.J. Chaiko, R. Mensah-Biney, C.J. Mertz, and A.N. Rollins, "Actinide Recovery Using Aqueous Biphasic Extraction: Initial Development Studies," ANL-92/36, Argonne National Laboratory, Argonne, IL (1992).

3. D.J. Chaiko, R. Mensah-Biney, and C.J. Mertz, and A.N. Rollins, Beneficiation of Pu residues by ultrafine grinding and aqueous biphasic extraction, *Sep. Sci. and Technol.* 28:765 (1993).

4. D.J. Chaiko, R. Mensah-Biney, and E. Van Deventer, Soil decontamination by aqueous biphasic extraction, in: "Removal of Uranium from Uranium - Contaminated Soils, Phase 1: Bench Scale Testing," ORNL-6762, Oak Ridge National Laboratory, Oak Ridge, TN (1993).

5. D.J. Chaiko, R. Mensah-Biney, and D. Gupta, Soil decontâmination using aqueous biphasic separation, in: "Proc. Waste Management '94," Vol. 3, Laser Options, Tucson, AZ, p. 2257 (1994).

6. K.P. Ananthapadmanabhan and E.D. Goddard, Aqueous biphase formation in polyethylene oxide-inorganic salt systems, *Langmuir* 3:25 (1987).

7. K.P. Ananthapadmanabhan and E.C. Goddard, "Process for the separation of solid particulate matter," U.S. Patent 47225358 (1987).

8. S.Y. Lee, and J.D. Marsh, Jr., Characterization of uranium contaminated soils, in: "DOE Fernald Environmental Management Project Site: Results of Phase I Characterization," ORNL/TM - 11980, Oak Ridge National Laboratory, Oak Ridge, TN (1992).

9. E.C. Buck, N.L. Dietz, J.K. Bates, and J.C. Cunnane, Analytical electron microscopy examination of uranium contamination at the DOE Fernald operation site, in: "Proc. Waste Management '93," Vol. 1, Laser Options, Tucson, AZ, pp. 797-801 (1993).

10. E.C. Buck, N.R. Brown, N.L. Dietz, and J.C. Cunnane, Supporting soil remediation at Fernald by electron beam methods, in: "Proc. Waste Management '94," vol. 3 Laser Options, Tucson, AZ, p. 2131 (1994).

11. E.C. Buck, N.R. Brown, and N.L. Dietz, Distribution of Uranium Bearing Phases in Soils from Fernald, in: *Mat. Res. Soc. Symp. Proc.* 333:437 (1994).

12. P.G. Allen, J.M. Berg, J. Chisholm-Brause, S.D. Conradson, R.J. Donohoe, D.E. Morris, J.A. Musgrave, and C.D. Tait, Determining uranium speciation in contaminated soils by molecular spectroscopic methods: Examples from the uranium in soils integrated demonstration, in: "Proc. Waste Management '94," Vol. 3, Laser Options, Tucson, AZ, p. 2063 (1994).

13. P.A. Albertsson, "Partitioning of Cell Particles and Macromolecules," 3rd ed., Wiley-Interscience, New York (1986).

14. R.A. Leonard, G.J. Bernstein, R.H. Pelto, and A.A. Ziegler, Liquid-liquid dispersion in turbulent Couette flow, *AIChE J.* 27:495 (1981).

15. R.A. Leonard, R.W. Bane, M.J. Steindler, R.A. Wigeland, and A.A. Ziegler, Operation with 3 liquid phases in a staged liquid-liquid contactor, *Sep. Sci. Technol.* 18:1563 (1983).

16. H.L. Bohn, B.L. McNeal, and G.A. O'Connor, "Soil Chemistry," Wiley Interscience, New York (1985).

17. M. Abe, Y. Kaneko, W. Agui, and K. Ogino, Removal of humic substances dissolved in water with carbonaceous adsorbents, *Sci. of the Total Envir.* 117/118:551 (1992).

18. X.C. Zhang, M. Masumoto, H. Sunahara, and A. Akazawa, Removal of humic acids by the process of flocculation-ultrafiltration, *Env. Technol.* 11:1007 (1990).

19. N.Y. Orlova, L.G. Bakina, and T.A. Plotnikova, Interaction of humic acids with calcium and implications for the liming of soils, *Eurasian Soil Sci.* 24:12 (1992).

# PLUTONIUM AND AMERICIUM SEPARATION USING ORGANOPHOSPHORUS EXTRACTANT ABSORBED ONTO FERROMAGNETIC PARTICLES

L. Nuñez and G. F. Vandegrift

Chemical Technology Division
Argonne National Laboratory
9700 South Cass Avenue
Argonne, Illinois 60439

ABSTRACT

Polymeric coated ferromagnetic particles with an absorbed layer of octyl(phenyl)-N,N-diisobutylcarbamoylmethylphosphine oxide (CMPO) diluted by tributyl phosphate (TBP) are being evaluated for application in the separation and the recovery of low concentrations of americium and plutonium from nuclear waste solutions. Due to their chemical nature, these extractants selectively complex americium and plutonium contaminants onto the particles, which can be recovered from the solution using a magnet. Physical and chemical characterization of the extractant-absorbed particles were performed by gamma and liquid scintillation counting, transmission electron microscopic (TEM) micrograph, and energy disperse spectroscopy (EDS). Plutonium and americium separations have been performed at various $HNO_3$ concentrations. Parameters such as absorbed layer thickness and separation abilities were studied to determine the limitation and capacity of the process.

The status of the chemistry and application of the process to DOE remediation efforts for actinide decontamination will be discussed.

## INTRODUCTION

Nuclear fuel reprocessing and nuclear weapons production have generated high-level waste (HLW) containing low concentrations of long-lived radioisotopes, most importantly the transuranic (TRU) elements. Separating out the TRU elements can alleviate the burden of disposing of large volumes of TRU wastes. The TRUEX solvent extraction process has

*Separations of f Elements*, Edited by K.L. Nash
and G.R. Choppin, Plenum Press, New York, 1995

been demonstrated to separate TRU elements from radioactive liquid waste. This process uses organophosphorous compounds such as tributyl phosphate (TBP) and octyl(phenyl)-N,N-diisobutylcarbamoylmethylphosphine oxide (CMPO).[1-4] This low concentration of extracted TRU elements can be processed into a final waste form that is based on a glass matrix and deposited into a deep geological repository, while the bulk of the nuclear waste can be disposed of in a near-surface, (more economical) low-level waste (LLW) facility. Thus, TRU separation processes are necessary and can achieve high cost savings. One waste site at Hanford is the residence of millions of gallons of tank waste from the plutonium finishing plant (PFP).

Magnetically-assisted chemical separation (MACS) processes use magnetic particles prepared with an extractant-specific, organic complexant-containing solvent. The organic complexes are selective extractants for feed waste components like Cs, Sr, or TRU elements. The particles are prepared by coating iron or another ferromagnetic material with either an organic polymer or ion-exchange resin. The ion-exchange resin is attached to the particle by an adhesive or by direct bonding. Organic solvents can be adsorbed onto the polymeric surface by contacting the particles with the solvent in a volatile diluent that is subsequently removed by evaporation. These coatings selectively separate the contaminants onto the particles due to their chemical nature. Once loaded, the particles can be recovered from the tank using a magnet.

Once particle recovery is achieved, the contaminants can either (1) be left on the loaded particles and added to the glass feed slurry or (2) be stripped into a small volume of solution to regenerate the extracting particles. For example, the selective extractant coating can be washed from the particles with hexane or alcohol, which acts to dilute and thus diminish the extraction power of CMPO and TBP for TRU complexation. It is then stripped of activity by dilute acid. The organic solution can then be evaporated to regenerate the particles. The greatest benefits of the technology are the simple separation of radionuclides from various waste streams using a compact, cost-effective process that does not produce additional large waste streams. This process is applicable to many DOE liquid TRU wastes.

The actual process development studies in this paper are: (1) particle morphology and effects on sorption of organophosphorous extractants, (2) optimizing separation capabilities of the magnetic particles, (3) temperature effects on extraction, (4) stripping of particles for regeneration, and (5) acid degradation effects on particles.

## PROCESS DESCRIPTION

The MACS process is still in the developmental stage. Figure 1 shows a diagram of the MACS concept. In this concept the dissolved sludge or aqueous stream is added to a treatment tank, which can also be used for the stripping stage. Magnetic particles coated

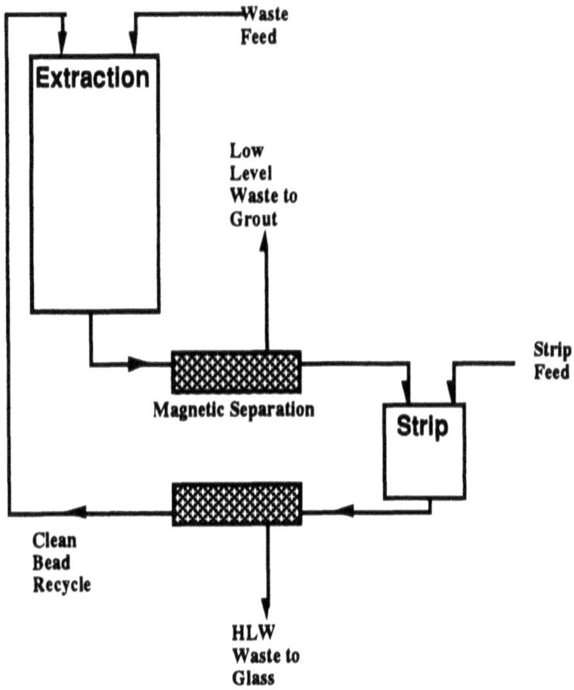


**Figure 1.** Concept Diagram of MACS Process

with extractants are added to the treatment tank to separate the TRU and/or other contaminants from the dissolved sludge or liquid waste. To maintain particle suspension, the tank contents can be mixed by mechanical stirring, magnetic fluidization, or other methods. After extraction, the particles are magnetically separated by one of three methods: (1) placing a magnetic field around the treatment tank, (2) pumping the solution through a magnetic filter (e.g., commercially available magnetic units), or (3) introducing a magnet into the tank. The waste solution, now decontaminated, is withdrawn and can be sent to grout. The TRU-loaded particles are either sent to a vitrification process or left in the tank for stripping with a smaller volume of liquid (compared to the original feed volume). The TRU contaminants are separated from the magnetic particles and sent to a vitrification process (e.g., glass melter) while the particles are recycled for further extraction.

The MACS process has the advantages of (1) less complex equipment, compared with solvent extraction and ion exchange and (2) simplicity that will facilitate scale-up. One of the Hanford tank wastes that the MACS process can successfully treat is the acid-dissolved waste from their neutralized PFP. The Hanford Site currently has many tanks filled with PFP wastes, each tank containing approximately one million gallons.[5]

A variety of applications of superparamagnetic (ferromagnetic) particles has increased their use for detection and monitoring in assay developments and analysis. Immunological analysis, chemiluminescense,[6] and immunomagnetic separation of food bacteria[7] and genes[8] are areas where superparamagnetic particles have found wide applications. Magnetic particles are available from several commercial suppliers. These particles come in a variety of sizes and magnetic properties with different functionalized and nonfunctionalized polymeric coatings. On the basis of our own experience, the literature, and discussions with the suppliers' technical experts, we have chosen the materials best suited for our application.

## EXPERIMENTAL

### Reagents

All the reagents used to prepare the simulant waste were ACS reagent grade. The reagent-grade TBP was obtained from Aldrich Chemical Company used without further purification, but the CMPO was recrystallized for purity.[9] All radioactive tracers were obtained from the Chemical Technology Division at ANL. The superparamagnetic or magnetic particles were manufactured from either CORTEX, BioMag, or Bang companies, and pretreated with filtered deionized water to eliminate any preservatives.

### Preparation of PFP Simulant

Table 1 shows the composition and ion concentrations of the PFP simulant waste prepared.

**Table 1.** PFP Simulant Waste

| Component | Molarity |
|---|---|
| $H^+$ | 1.50E+00 |
| $Be^{2+}$ | 7.00E-05 |
| $Na^+$ | 4.00E-02 |
| $Mg^{2+}$ | 6.00E-02 |
| $Al^{3+}$ | 4.30E-01 |
| $K^+$ | 3.00E-03 |
| $Ca^{2+}$ | 6.00E-02 |
| $Cr^{3+}$ | 4.00E-04 |
| $Mn^{2+}$ | 3.00E-03 |
| $Fe^{3+}$ | 3.00E-02 |
| $Ni^{2+}$ | 8.00E-02 |
| $Cu^{2+}$ | 3.00E-04 |
| $Zn^{2+}$ | 6.00E-04 |
| $Pb^{2+}$ | 5.00E-04 |
| $F^-$ | 9.00E-02 |
| $SO_4^{2-}$ | 1.00E-02 |
| $NO_3^-$ | 3.05E+00 |

## Microscopy and Electron Dispersive Spectroscopy (EDS)

The MACS particle samples were prepared for electron microscopy: dried, embedded in epoxy, and sliced into 500- to 1000-Å-thick sections with an ultramicrotome. Samples were then examined in a JEOL 2000FXII Transmission Electron Microscope (TEM) at 200 kV. Samples that were suitable for electron dispersive spectroscopic (EDS) analyses and experiments were performed in tandem using a NORAN ultrathin window system in the horizontal port of the JEOL 2000. The morphology studies included resin effects on the particles and particle morphology.

## Coating Procedure

All particles in this study exhibit superparamagnetic properties because of magnetite incorporated in their structure. The particles range from 0.1 to 25 μm and exhibit a wide variety of shapes. The commercially purchased particles have a polymeric coatings. The polymeric coating that demonstrated the most promise for CMPO/TBP adsorption was charcoal in a cross-linked N, N-methylene bis-acrylamide. The particles were washed with 0.02-μm-filtered deionized water several times to eliminate preservatives and additives, dried, and weighed. Then, 0.02-μm-filtered hexane or ethanol was used to disperse the particles and act as a volatile solvent for the coating. A dissolved CMPO/TBP mixture (> 0.6 M CMPO) was added to the particles, which were dispersed by an ultrasonic bath, dried in a rotoevaporator at 80-100°C, and heat-treated in an oven at 100-150°C until a constant weight was obtained.

## Extraction Process

Particle suspensions of ~1 wt % or less were contacted with nitric acid or simulant solutions containing $^{241}$Am or $^{238}$Pu. This low particle-mass-to-solution ratio was chosen to allow this process to compete with more traditional separation techniques, such as solvent extraction and ion exchange. The suspensions were placed for a few minutes in a sonic cleaning bath to enhance the dispersion. Then they were moved to a temperature bath at 25°C for thermal equilibration, and vortexed to further disperse the solid suspension. These two steps were repeated twice. Then the suspensions were centrifuged and placed in a rare-earth permanent magnetic rack to induce physical separation of the particles from the solution. An aliquot of the solution was taken for alpha-beta counting with a Packard 2200CA TRI-CARB liquid scintillation analyzer ($^{238}$Pu) or gamma counting ($^{241}$Am) on a Packard (Minaxi γ Autogamma 5000 Series) 3-in. (7.62-cm) NaI crystal detector. Mass balance calculations were used to determine partition coefficients ($K_d$). All nitric acid solutions were titrated using the Metrohm 670 Autotitrator with a glass electrode containing 3 M KCl.

### Stripping Measurements

The radionuclide-loaded magnetic particles were dried in preparation for stripping experiments. Fresh stripping agent was added and contacted, following a procedure similar to that used in the extraction experiments. The volume of stripping agent was no more than 1/100 that of the aqueous phase volume used for extraction experiments.

### Nitric Acid Dissolution

Even though the magnetite is often coated by a polymer, the high solubility of magnetite in nitric acid required us to study a series of experiments to determine the dissolution from the coated particles. Six prepared particle samples were weighed, and each mixed with 5 mL of a nitric acid solution (0.02 - 8 M). These samples were mixed following the extraction procedure and allowed to settle for 336 hours (2 weeks). They were magnetically separated, and 1-mL samples of the aqueous aliquots were analyzed for iron using atomic emission spectroscopy (AES). An aqueous phase sample without any particles present was analyzed as the background.

## RESULTS AND DISCUSSION

### Morphology Studies and Magnetite Distribution Experimentation

The capacity of the particles to absorb the organophosphorous extractants and maintain physical integrity under chemical (acid) and physical (gamma irradiation)[10] conditions was examined using TEM. The combination of the TEM results, $K_d$ values, gamma irradiation, and magnetic measurements was used to select a suitable particle type for MACS applications. The $K_d$ measurements are determined by Eq (1).

$$K_d = \left(\frac{C_i - C_f}{C_f}\right) x \frac{V}{M} \quad (1)$$

where
V = contact volume of TRU solution in mL
m = mass of particles in grams
$C_i$ = initial stock solution counts before contact
$C_f$ = final solution counts after contact

Commercially available coated superparamagnetic particles were chosen that were judged as best for the MACS process. An important criteria for selecting a particle is a heavy radionuclide loading ability. This requirement was not an original application for which the particles were developed. For a biochemical assay applications, maximizing absorption is unnecessary. Loading capacity is obviously an important concern for the task of recovering a particular component of the nuclear waste stream.

Another important consideration for applications of the MACS particles is the absorption or binding strengths exhibited between (1) the magnetite core and the polymeric coating and (2) the polymeric coating and the absorbed organophosphorous extractants. Loss of interfacial integrity in either case under HLW tank conditions is undesirable. The chemical durability of the magnetite core is also required for separation. The pretreated waste streams can be either a harsh acid or base environment, and dissolution of the magnetite would behave in a similar manner to physical separation of the magnetite and polymer. In this report, we intend to show the durability of the particles in an acidic environment using the $K_d$ values and information obtained by TEM and other experiments.

**Resin Effects and Inner-Outer Polymeric Sphere Magnetite Distribution**

In the first stage of this work, TEM sample preparation techniques were examined to determine whether samples could be produced without chemical interaction with the embedding resin used to mount them for ultramicrotomy and to provide a baseline view of samples before complexation or exposure to radiation. Five samples were selected (Table 2) four of which were examined by TEM. Sample 3, the charcoal coated sample, was not used for microtomy because particles were quite large (up to 25 μm).

**Table 2.** Particle Type Selected for TEM Studies

| Sample | Description | Size, μm | Manufacturer |
|---|---|---|---|
| 1 | DVB[a]/acrylic acid (58%) bound to magnetite core (35%) | 0.68 | Bang |
| 2 | Magnetite particles bound to polyacrolein core | 1-10 | Cortex |
| 3 | Charcoal-coated magnetite particles | 1-25 | Cortex |
| 4 | Magnetite particles embedded in/DVB[b] | 12 | Bang |
| 5 | Acrylamide-based magnetite particles with charcoal-dextran coating | 12 | BioMag |

[a]DVB = divinylbenzene.
[b]S = styrene.

Each sample (1, 2, 4, and 5) was embedded in Medcast, Araldite, and LR White embedding resins. All particles sectioned well, and there was no significant difference between particles embedded in the different epoxy resins. However, magnetite crystallites were observed to be bound to the surface of particles in sample 2 and appeared to loosen from LR White epoxy prior to curing. The samples in TEM micrographs in Figure 2 were embedded in Medcast or Araldite.

Sample 1, divinylbenzene-magnetite particles, is pictured in Figure 2a. The magnetite cores are rather large (0.2 μm) and irregularly shaped. The polymer appears to be randomly bound to the surface of the core. The divinylbenzene polymer demonstrated low sorption of the organophosphorous extractants, with $K_d \leq 100$ for Am in 2 M nitric acid.

Sample 2, polyacrolein-magnetite particles, is pictured in Figure 2b. The magnetite particles are extremely small (<1 nm) where the magnetite is attached to the outer surface of the polymer rather than being located in the core. Two deleterious effects are likely. First, the magnetite particles obstruct access of the organophosphorous solution to the absorbent polymer core. The effective surface area of the particle is dramatically reduced and absorption is limited by how quickly the solution can percolate or diffuse through the narrow spaces between the magnetite particles. Secondly, an exposure to radiation that is likely to embrittle the polymer may result in loss of the very small magnetite particles attached to the surface. The nonporous polyacrolein polymer demonstrated insignificant sorption of the organophosphorous extractants, with $K_d \leq 10$ for Am in 2 M nitric acid.

Sample 4, styrene/divinylbenzene-magnetite particles, is pictured in Figure 2c. The particles are uniform in shape, with magnetite particles embedded as clumps within the absorptive polymer. Most of the surface is available for absorption. The styrene/divinylbenzene polymer showed low sorption of the organophosphorous extractants, with $K_d \leq 100$ for Am in 2 M nitric acid.

Sample 5, acrylamide-based magnetite particles, is shown in Figure 2d. The particles here are randomly shaped, and the distribution of magnetite particles is also irregular. Some particles appear to be a mass of magnetite, while others have very little magnetite attached. It is possible that a significant number of particles would not be magnetically separable because few magnetite particles are present. The charcoal/dextran polymer showed low sorption of the organophosphorous extractants, with $K_d \leq 100$ for Am in 2 M nitric acid, but the polymer combination of charcoal in a cross-linked N, N-methylene bis-acrylamide gave $K_d \geq 4000$ for similar tracer solution. This high partition coefficient was attributed to the chemical affinity of the polymer and the large surface area of the particles.

## Nitric Acid Dependence

The nitric acid dependency of the MACS process is important for applying of this process to waste streams having a wide range of acidities. Dissolution of solid or sludge waste is usually achieved by acid hydrolysis. The chemical stability of the TRU under acidic condition is well-documented.[11-12]

With the aide of TEM, the physical features of the particles were determined allowing for the selection of the charcoal in a cross-linked N, N-methylene bis-acrylamide particle type for the MACS process.

The partition coefficients for various coatings were monitored using $^{241}$Am(III) and $^{238}$Pu(IV) tracers. Samples were tested in 0.001, 0.01, 0.1, 2, 5 and 8 M nitric acid with CMPO concentrations of 1.0 M. The results are presented in Figure 3. In the region below 1 M $HNO_3$ the $D_{Am}$ and $D_{Pu}$ are close in values, this is not due to a change in the Pu oxidation states, instead attributed to experimental uncertainty in $K_d$ values below 1 M $HNO_3$.

Figure 3 shows that the two curves have similar shapes. The $K_d$ values for Am and Pu increase slowly with $HNO_3$ concentration, reaching peaks at 2 M and 5 M $HNO_3$, respectively. These results suggest that the process can be applied in a batch mode for a wide range of high nitric acid concentrations without a large loss in extraction capabilities.

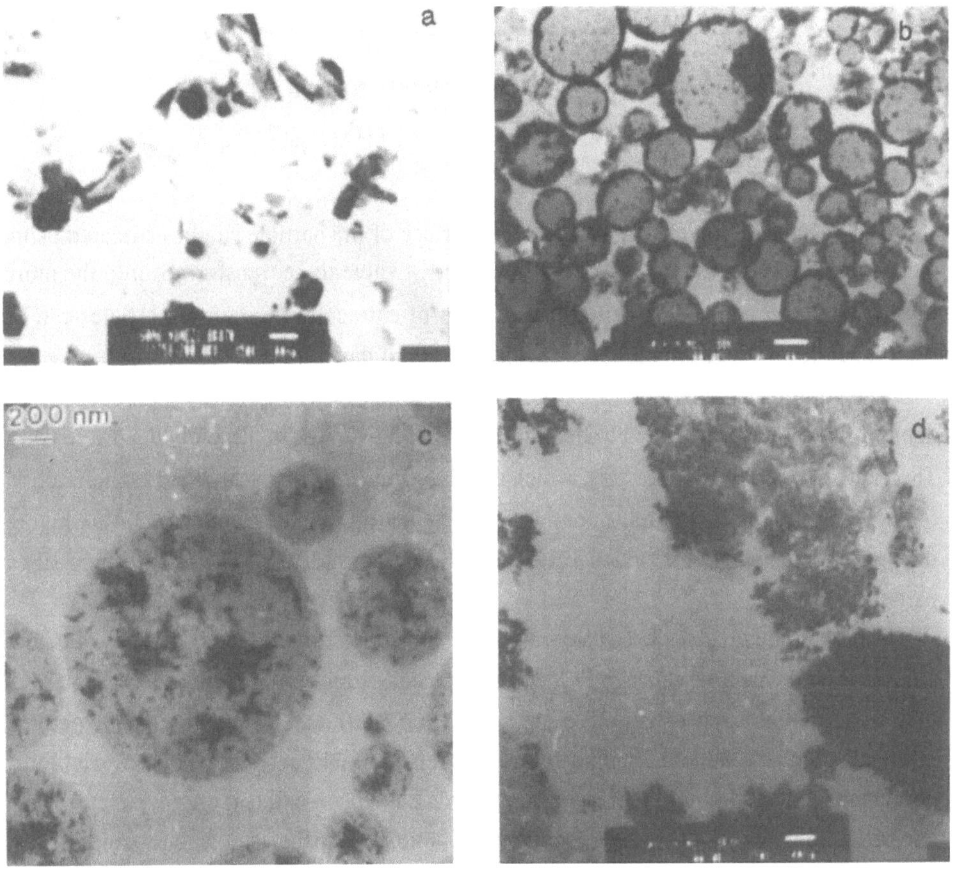


**Figure 2a. TEM Micrographs.** Sample 1, which is comprised of large, irregular-shaped magnetite particles (0.2 μm) with polymer randomly bonded to surface core. **2b.** Brightfield Image of sample 2 (polyacrolein) embedded in Araldite. Magnetite particles are small (<10 nm) and attached to the outer surface of the particles. **2c.** Mag-encapsulated particles. The magnetite particles are small (~10 nm) and are embedded as agglomerates within the polymer. **2d.** Dextran-coated particles are shown here. Polymer particles are randomly shaped, with a non-uniform polymer coating. In most cases, it does not appear that the magnetite and polymer are tightly bound.

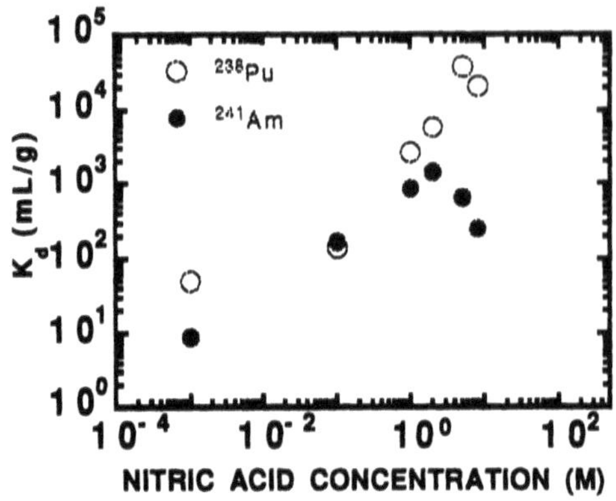


**Figure 3.** $K_d$ vs. Nitric Acid Concentration for $^{241}Am$ and $^{238}Pu$ for 1.0 M CMPO/TBP Coatings on Magnetic Particles at 25°C

The volume of solvent absorbed on the surface of the particle can be estimated using the solvent coverage fraction, which allows the $K_d$ value to be transformed into the more traditional distribution ratio. The ratio of gram-of-extractant to gram-of-particles can be determined from the coating process and the density of the particles.

The volume occupied by the extractant and particles was previously reported.[13] The density of particles is assumed to be in the range of 2.5-2.7 g/cm$^3$ (mainly due to $Fe_3O_4$). The value of $f$ (conversion factor), which can vary from 25 to 60, is defined as the fraction of the gram of particles per mL of extractant. The Am and Pu distribution ratios are higher than traditional solvent extraction values and those found for CMPO absorbed onto silica particles.[14]

Although the precise chemical equilibrium is not determined, the $K_d$ values obtained are much larger than D values obtained in a solvent extraction process. Increases in $K_d$ may be a result of a synergistic effect between CMPO/TBP and the polymer on the particle in the absence of a diluent, (such as normal paraffinic hydrocarbons (NPH) or trichloroethylene (TCE). Although CMPO/TBP systems have been extensively investigated, the current studies without a standard diluent are unique. Further studies are necessary to determine the precise driving mechanism that aids in the extraction. The MACS process shows promise in using more efficiently the standard organophosphorous extractants such as CMPO and TBP for TRU separation applications.

## Nitric Acid Dissolution

Magnetite is soluble in nitric acid, but the polymeric coating can partially protect it. Dissolution of the magnetite would inhibit particle recovery and hinder application of the MACS process. A series of experiments were performed to determine the extent of dissolution of the magnetite from the particles as a function of nitric acid concentration.

The hydrolysis of magnetite increased with acid concentration, as evidenced by ICP analysis of dissolved iron. The dissolution occurs linearly (Figure 4) throughout the nitric acid range, including the expected MACS process ranges of 2-4 M $HNO_3$. The presence of exposed magnetite, which allows for its dissolution, is evident in TEM micrographs, where a significant amount of magnetite particles are in the outer sphere of the polymeric material for irregularly shaped particles. The information provided by these experiments is limited to dissolved iron, not magnetite remaining on the particles.

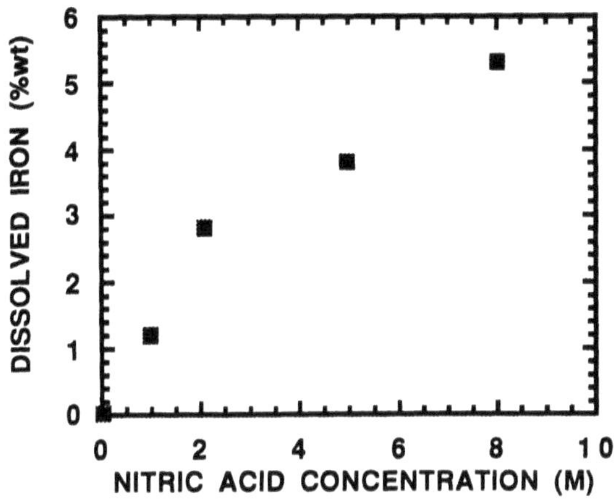


**Figure 4.** Dissolved Iron in Supernatant vs. Nitric Acid Concentration

Figure 4 shows a steady increase in the amount of dissolved iron in the supernatant with increasing nitric acid concentration. In shorter contact times (minutes), the effect is expected to be smaller, but it remains a major concern for process recycling.

## Temperature Dependence

Due to the possible application of the MACS process to wastes that have been stored in tanks for decades, the true operation temperature may vary between 20-50°C. The MACS process must evaluate the temperature dependence within such a temperature range.

The extraction of americium was monitored as a function of temperature for 1 M CMPO/TBP coating. Three temperatures were used for the experiments (10°, 25°, and 50°C). Figure 5 illustrates the general trend.

All three $K_d$-vs.-nitric acid curves (10°, 25°, and 50°C) have the same general shape. In many solvent extraction systems,[15] the distribution ratios of the metals decrease with increasing temperature, especially between 0-60°C. The MACS particles show negligible temperature dependency between 10-50°C for the extraction of americium.

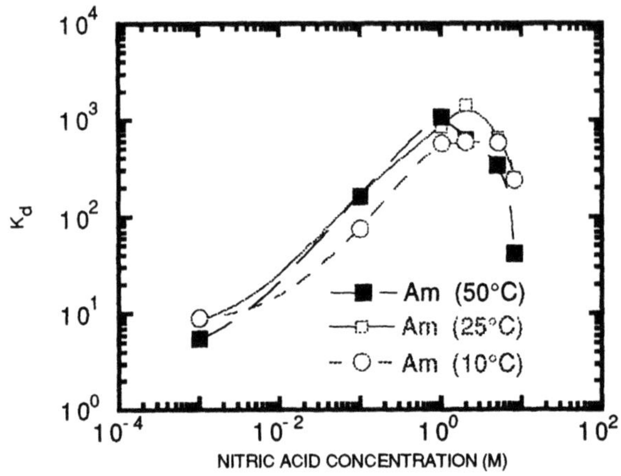


**Figure 5.** The $K_d$ Values for 1 M CMPO vs. Nitric Acid for 10°, 25°, and 50°C

## PFP Simulant Waste Stream

The $K_d$ values for Pu(IV) in PFP simulant were on the order of 20,000. However, Am showed very low values of $K_d$, 100. Figure 6 shows the results for various saturated particles of CMPO/TBP. The $K_d$ values show little dependence on CMPO concentration for both Am and Pu (Figure 6). The solution chemistry of Pu and Am favors complexation with fluoride. These fluoride complexes have very low extractability with CMPO; however, results from this study show that Pu is highly extractable even in the presence of fluoride. The decrease in the $K_d$ for Am is probably due to some effects other than fluoride competition for the actinide. High iron concentration present in the simulant solution may interfere with Am extraction. The Am separation was then predicted for the addition of oxalic acid solution using the Generic TRUEX Model (GTM) to complex the free iron.[1-4] This model allows the determination of TRUEX solvent extraction chemistry based on a thermodynamically correct mechanism.

Using the PFP waste stream components and the GTM speciation with oxalic acid addition, the concentration of free iron was calculated to be 0.001 M for a 0.28 M oxalic acid solution. This addition resulted in an increase of the $K_d$ at room temperature from 100 to 3500 for 1.5 M CMPO particles.

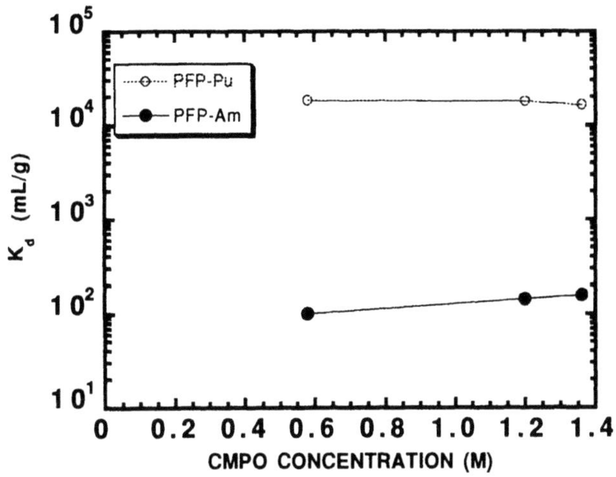


**Figure 6.** Am and Pu Separation Using PFP Simulant Solution at 25°C with 1.5 M CMPO/TBP Coated Particles

## Stripping Measurements

Because the $K_d$ values obtained using magnetic particles are quite large, an efficient recovery stage is required. The usual stripping process for nitrate waste streams involves using dilute nitric acid for Am and an aqueous phase complexant or reducing agent for Pu. However, because the MACS process results in large $K_d$ values (partly due to the high concentration of CMPO), dilute nitric acid would be inefficient for stripping Am. A three-step stripping process is necessary. In the first step, complexants (CMPO and TBP) and the TRU elements are stripped from the particles using a hydrocarbon or an alcohol. In the second step, the TRU are stripped from the organic phase, which is now a dilute CMPO solution, using an appropriate aqueous phase. Finally, the volatile hydrocarbon or alcohol (e.g., octanol) is evaporated leaving the CMPO/TBP residual on the particle surface.

Table 3 gives the stripping $K_d$ values for plutonium. Some of the $K_d$ values not reported gave negative values and were attributed to an error in the mass balance assumptions (e.g., carry over of phases, conservation of particle mass). The values were derived from the mass balance equation:

$$g_p[\mathrm{cpm}/\mathrm{g}]^{\mathrm{original}} = g_p[\mathrm{cpm}/\mathrm{g}]^{\mathrm{final}} + (\mathrm{mL})[\mathrm{cpm}/\mathrm{mL}]^{\mathrm{final}} \quad (2)$$

assuming that there is negligible loss in particle mass before and after stripping.

The results are promising, but work is needed to develop a stripping step that can obtain a stripping stage $K_d$ value of 1/1000 for an efficient recovery process. This value is calculated as follows:

$$K_d^{strip} = \frac{\left[\left(\frac{g_p}{V}\right)K_d^{ext}C_{final}^{ext} - C_{final}^{strip}\right]}{\left(\frac{g_p}{V}\right)C_{final}^{strip}} \tag{3}$$

where, $g_p$ is the mass of the particles in grams, $V$ is the volume of waste solution in mL, and $C_{final}^{strip}$ and $C_{final}^{ext}$ are the final concentrations after stripping and extraction, respectively, and $K_d^{ext}$ is the extraction partition coefficient.

**Table 3.** Stripping $K_d$ for $^{238}Pu$ in PFP Simulant

| Sample | Stripping Solvent | $K_d$ Loading | $K_d$ Stripping | O/A ratio | Volume Reduction |
|---|---|---|---|---|---|
| 1.2 M CMPO | ethanol | 1850 | 0.73 | 0.01 | 400 |
| 1.0 M CMPO | butyl alcohol | 3323 | 4.75 | 0.008 | 100 |
| 1.0 M CMPO | ethanol | 1553 | 0.65 | 0.008 | 100 |

We were able to easily reduce the volume by a factor of 400 from the original feed solution. Since the volume of solvent on the particles is small, having organic phase/aqueous phase ratios (O/A) of 0.01 or smaller suggest that higher volume reductions are attainable. For a one batch operation, the stripping stage recovers greater than 99% of the Pu and up to 92% of the Am, for an overall recovery of greater than 95% of TRU.

## SUMMARY

The MACS process is a promising efficient separation technology that uses organophosphorous extractants absorbed on superparamagnetic particles. The large $K_d$ values obtained for plutonium and americium illustrate the efficiency of separation using the CMPO/TBP on superparamagnetic particles, which surpasses traditional liquid/liquid extraction processes for similar waste streams. The studies of plutonium and americium extraction as a function of nitric acid concentration suggested that the extraction CMPO/TBP species are different from those modeled by the GTM for the TRUEX process. They indicate that further studies are needed to model the data and determine a precise mechanism. Stripping the particles using alcohols instead of a stripping agent and hydrocarbons has been successful with large volume reduction. The most efficient stripping process for such high TRU $K_d$ values is a three-stage process. Activities are underway to optimize these stages. Magnetite dissolution in acid seems to be a limitation on the recyclable batch process. We are investigating the possibility of designing particles with a smaller fraction of exposed magnetite to minimize the solid/liquid interface. On the other hand, for applications of MACS on basic solutions containing TRU, exposure of the magnetite can lead to an added sorption mechanism.[15] At this point, we have succeeded in stripping at least 95% of the TRU from the particles in batch testing.

## ACKNOWLEDGMENTS

The authors would like to thank C. Bradley for the TEM micrographs and S. Aase and M. Kaminski for solution preparation and $K_d$ determination. This work is supported by the U. S. Department of Energy under Efficient Separation Processes Integrated Program (ESPIP), Contract W-31-109-Eng-38.

## REFERENCES

1. G. F. Vandegrift, R. A. Leonard, M. J. Steindler, E. P. Horwitz, L. T. Basile, H. Diamond, D. G. Kalina, and L. Kaplan, "Transuranic Decontamination of Nitric Acid Solution by the TRUEX Solvent Extraction Process-Preliminary Development Studies," Argonne National Laboratory Report, ANL-84-45(1984)

2. L. E. Trevorrow and G. F. Vandegrift, "Alternatives for Disposal of Raffinate from the TRUEX Process," Argonne National Laboratory report, ANL-89-27 (1989).

3. D. B. Chamberlain and R. A. Leonard, "TRUEX Hot Demonstration: Final Report," Argonne National Laboratory Report, ANL-89-37 (1989).

4. G. F. Vandegrift, D. B. Chamberlain, C. Conner, J. M. Copple, J. A. Dow, L. Everson, J. C. Hutter, and R. A. Leonard, "Development and Demonstration of the TRUEX Solvent Extraction Process," Argonne National Laboratory Report, ANL-84-45 (1993).

5. T. V. Stabler and A. L. Siegel "Chemiluminescence Immuoassy of Aldosterone in Serum," *Clin. Chem.* 37, Vol. 11, 1987 (1991).

6. E. L. Skjerve, L. M. Rorvik, and O. Olsvik "Detection of Listeria Monocyntogenes in Foods by Immunomagnetic Separation," *Appl. Environ. Microbiol.* 56, 3478 (1990).

7. A. Lund, Y. Wasteson, and O. Olsvik "Immunomagnetic Separation and DNA Hybridization for Detection of Enterotoxigenic Escherichia coli in a Piglet Model," *Journal of Clin. Microbiol.* 29, Vol. 10, 2259 (1991).

8. D. J. Chaiko, P. K. Tse, and G. F. Vandegrift, "Modeling of the Aqueous and Organic Phase Speciation for Solvent Extraction Systems," Mineral and Mat. Soc. pp 246 (1988).

9. J. M. Cleveland, The Chemistry of Plutonium, American Nuclear Society, La Grange Park, IL (1979).

10. L. Nuñez and G. F. Vandegrift, Argonne National Laboratory, unpublished information, (1994).

11. L. R. Morss in The Chemistry of the Actinide Elements, 2nd Ed., eds., J. J. Katz, G. T. Seaborg, and L. R. Morss, Chapman and Hall, New York, Chapter 17, Vol. II (1986).

12. L. Nuñez, B. A. Buchholz, and G. F. Vandegrift, submitted to: Proceedings of the Eighth Symposium on Separation Science and Technology, Gatlinburg, TN (1993).

13. L. Pietrelli, A. Salluzzo, and F. Troiani, New Separation Chemistry Techniques for Radioactive Waste and Other Applications, "Actinide Removal by Means of Octyl(phenyl)-N,N-diisobutylcarbamoylmethylphosphine Oxide (CMPO) Sorbed on Silica," Elsevier Sci. Publishing Co, London (1991).

14. G. M. Ritcey and A. W. Ashbrook, Solvent Extraction- Principles and Applications to Process Metallurgy: Part I, Elsevier Sci. Publishing Co, Amsterdam (1984).

15. R. L. Klochen, "Actinide Removal from Aqueous Solution with Activated Magnetite," Rocky Flats Plant Report, RFP-4100 (1987).

# MAGNETIC SEPARATION FOR ENVIRONMENTAL REMEDIATION

Ann R. Schake, Larry R. Avens, Dennis D. Padilla, David A. Romero, and Laura A. Worl
Advanced Technology Group, NMT-6, MS E510
Los Alamos National Laboratory, Los Alamos, NM 87545

F. Coyne Prenger and Dallas D. Hill
Advanced Engineering Technology Group, ESA-EPT, MS J576,
Los Alamos National Laboratory, Los Alamos, NM 87545

Thomas L. Tolt
Lockheed Environmental Systems and Technologies Company
Las Vegas, Nevada 89119

Magnetic separation is a physical separation process that segregates materials in a mixture on the basis of magnetic susceptibility. Because all actinides and their compounds and fission products are paramagnetic, and most host materials such as water, graphite, soil, and sand are diamagnetic, magnetic separation methods can be used to extract the actinides from these hosts, concentrating the toxic materials into a low volume waste stream. The technology relies only on physical properties, and therefore separations can be achieved while producing little or no secondary waste.

We are currently developing the magnetic roll or drum-type separator and the high gradient magnetic separator (HGMS) for applications to soil decontamination, liquid waste treatment, underground storage tank waste treatment and chemical processing residue concentration. The magnetic roll separator is used to separate dry powder mixtures and is most effective on particles ranging from 90 to 850 microns in size. HGMS involves passing a slurry through a magnetized volume and is used to separate solids from other solids, liquids or gases. It is most effective on small particles (90 microns and smaller), and is complementary to the roll separator. Our results with nonradioactive surrogates and radioactive materials show that over 90% extraction of the paramagnetic components can be achieved.

## INTRODUCTION

Magnetic separation is a physical separation process that segregates materials on the basis of magnetic susceptibility. High Gradient Magnetic Separation (HGMS) is a form of magnetic separation used to separate solids from other solids, liquids or gases. HGMS uses large magnetic field gradients to separate ferromagnetic and paramagnetic particles from diamagnetic host materials. The technology relies only on physical properties, and therefore separations can be achieved while producing a minimum of secondary waste.

Actinide and fission product wastes within the DOE weapons complex pose challenging problems for environmental remediation. Because the majority of actinide

*Separationf of f Elements,* Edited by K.L. Nash
and G.R. Choppin, Plenum Press, New York, 1995

complexes and many fission products are paramagnetic, while most host materials are diamagnetic, HGMS can be used to concentrate the contaminants into a low volume waste stream. We are currently developing HGMS for applications to soil decontamination, liquid waste treatment, underground storage tank waste treatment, and actinide chemical processing residue concentration.[1] In 1992 Los Alamos National Laboratory entered into a Cooperative Research and Development Agreement (CRADA) with Lockheed Environmental Systems & Technologies Company to develop High Gradient Magnetic Separation technology for soil decontamination.

Application of HGMS usually involves passing a slurry of the contaminated mixture through a magnetized volume. A diagram of the method is shown in Figure 1. Field gradients are produced in the magnetized volume by a ferromagnetic matrix material, such as steel wool, expanded metal, iron shot, or nickel foam. The matrix fibers become trapping sites for ferromagnetic and paramagnetic particles in the host material. The particles with a positive susceptibility (paramagnetic) are attracted toward an increasing magnetic field gradient and can be extracted from particles with a negative susceptibility (diamagnetic), which react in the opposite direction, moving away from the areas of high field gradients. Table 1 lists several contaminants and host materials and indicates the magnetic susceptibility of each. The extracted paramagnetic contaminants are flushed from the matrix fibers when the magnetic field is reduced to zero or when the matrix canister is removed from the magnetic field. The actinide-containing concentrate can be processed for disposal. Because only a fraction of the original bulk requires disposal, significant cost savings can be achieved.

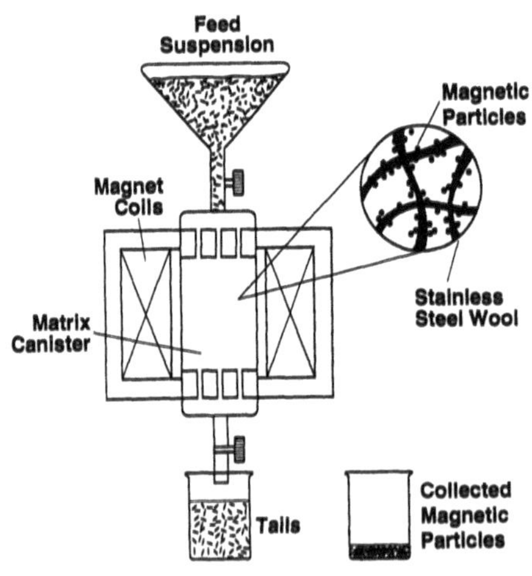


**Figure 1.** Schematic diagram of High Gradient Magnetic Separation.

**Table 1.** Volume Magnetic Susceptibility of Selected Compounds

| Compound/Element | Susceptibility x $10^6$ (SI) |
|---|---|
| FeO | 7178 |
| $Fe_2O_3$ | 1479 |
| $UO_2$ | 1204 |
| $Cr_2O_3$ | 844 |
| Am | 707 |
| Pu | 636 |
| U | 411 |
| $PuO_2$ | 384 |
| Th | 84 |
| $UO_3$ | 41 |
| $ThO_2$ | -7.5 |
| $ZrO_2$ | -8.0 |
| $H_2O$ | -9.0 |
| MgO | -11.4 |
| $CaCO_3$ | -13.0 |
| $SiO_2$ | -13.6 |
| Graphite | -14.0 |
| $Al_2O_3$ | -18.0 |

## EXPERIMENTAL

The production of both high magnetic fields (>4 tesla) and large field gradients using superconducting magnet technology now makes it possible to separate small (>0.1μm) paramagnetic particles. Our current HGMS equipment includes one conventional-coil separator and three superconducting high gradient magnetic separators. The conventional coil (one inch bore diameter) and one three inch bore superconducting magnet are installed on a vent hood in the plutonium facility at Los Alamos. This treatability lab is used to handle low-level radioactive samples such as contaminated soils and uranium and plutonium spiked experiments. A second three inch bore superconducting magnet is currently installed at the Los Alamos waste treatment facility where it is being used for experiments with applications of HGMS to waste water treatment. This magnet will soon be moved and interfaced with a glove box in the plutonium facility to investigate applications of HGMS to plutonium chemical processing residues. Finally, a six-inch bore superconducting magnet is being used in a non-radioactive area for non-radioactive and surrogate tests and for prototype development for soil decontamination applications through the CRADA.

The conventional coil magnet is a Pacific Electric Motor Co. crossfield electromagnet with a maximum field intensity of 2.0 tesla. All three superconducting magnets were built by Cryomagnetics, Inc. with maximum field strengths of 8 tesla. The field is axial with respect to the flow through the matrix.

In a typical HGMS experiment, a 10 weight % aqueous slurry of the contaminated mixture is prepared. A surfactant is added, the pH adjusted, and the slurry is sonicated; each of these steps is to aid in particle liberation. A sample of the feed slurry is taken and the remainder of the feed is pumped through the magnetic matrix using a peristaltic pump at superficial velocities of 0.1-1.0 cm/s. After each pass the canister housing the matrix is physically removed from the area of the magnetic field and flushed in the reverse direction at four times the forward velocity.

Feed samples, effluents and backflushes are collected and analyzed for each HGMS experiment. Samples are analyzed for uranium or plutonium concentration using alpha spectroscopy (Controls for Environmental Pollution, Santa Fe, NM), gamma spectroscopy (LESAT, Las Vegas, NV) and Instrumental Neutron Activation Analysis, INAA (Activation Laboratories, LTD, Ancaster, Ontario, Canada). HGMS experiments are conducted using both the conventional and superconducting magnets, with magnetic fields orientations both parallel and perpendicular to the slurry flow.

## HIGH GRADIENT MAGNETIC SEPARATION ANALYTICAL MODEL

More than ten variables independently affect the HGMS process (*vida infra.*); an analytical model for the process was needed to select appropriate bench-scale experiments and to effectively analyze the resulting data. In addition, a validated analytical model supports prototype design and process scaleup. The behavior of the paramagnetic particles as they interact with the magnetized matrix has been investigated through modeling studies in the past.[2-8] Areas investigated include the dynamic effect of particle trajectories in homogeneous materials, the influence of particle build-up on the matrix elements, and the effects of particles on the flow field. These studies, though requiring simplifying assumptions, identify appropriate independent variables that form the basis for this investigation.

The matrix materials currently used for HGMS are inhomogeneous and have a complex cross section. In addition, the paramagnetic particles are nonspherical and include a range of particle sizes. All of these factors preclude precise analytical treatment. If the particles are physically liberated from the host material and are not electrically charged, the principle forces governing their behavior are magnetic, viscous, and gravitational.

The performance of the magnetic separator is modeled using a static force balance on an individual paramagnetic particle in the immediate vicinity of a matrix element, as shown in Figure. 2. The model assumes that if the magnetic capture forces are greater than the competing viscous drag and gravity forces, the particle is captured and removed from the flow stream.

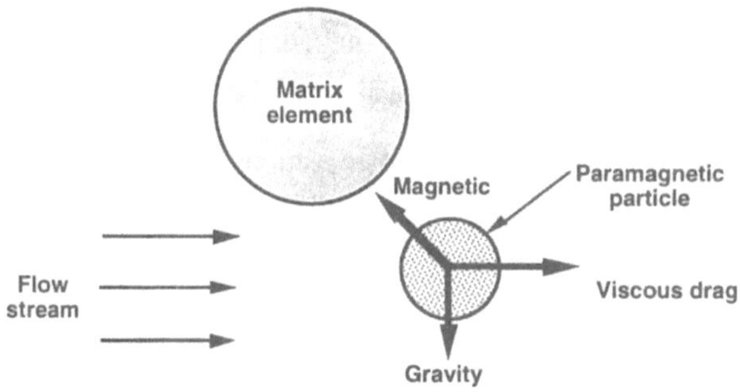


**Figure 2.** Force balance on a paramagnetic particle in the vicinity of a matrix element.

We propose a rate model for this process that depends on a separation coefficient, which we define in terms of a capture cross section and a potential function defined by the force balance on the particle. A series of tests was performed to evaluate separator performance as a function of the independent variables in the separation process. The results were used to determine the separation coefficient in the proposed rate model and were correlated using physically meaningful, dimensionless groups.

We assume that there exists for each matrix element a capture height, wherein each magnetic particle entering the window defined by the capture height and the matrix element diameter will be captured by the matrix element. We then define a capture cross section for

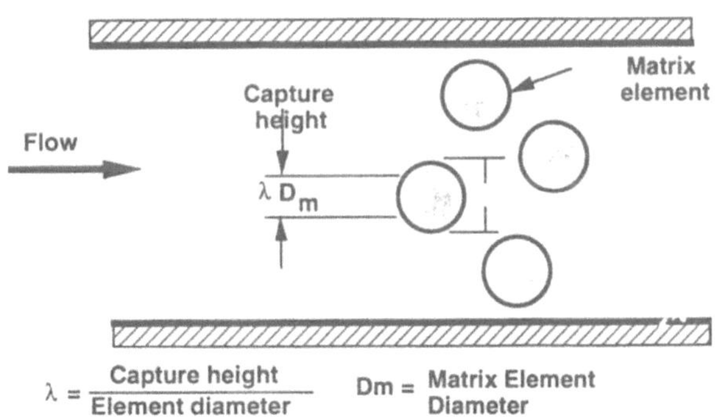


**Figure 3.** The matrix capture cross section defined as the ratio of capture height to matrix element diameter.

a matrix element to be the ratio of the capture height to the matrix element diameter, as shown in Figure 3. The separation coefficient is related to the capture cross section, and if the capture cross section is known, the separation efficiency can be calculated.

## HIGH GRADIENT MAGNETIC SEPARATION DATA CORRELATION

A comprehensive series of experiments was performed to determine the capture cross section as a function of the independent variables. Table 2 lists the variables investigated and their range. For convenience, the list has been divided into material characteristics and separator parameters. A total of 120 experiments were conducted; 76 involved water/contaminant mixtures and 44 involved water/soil/contaminant mixtures. (For more details see reference 9.) The capture cross section 1 for each test was based on the measured feed and effluent concentrations, the particle force balance, the matrix geometry, and the contaminant particle size.

The capture cross section is significantly less than 1.0, which implies that particles are swept around the matrix elements by the fluid flow even if the particles enter the region defined by the projected frontal area of the matrix element. In addition, the capture cross section is proportional to the magnetic force and the contaminant particle diameter and inversely proportional to the matrix element diameter and the matrix element spacing. These results are consistent with our observations. The cross section 1 is inversely proportional to the superficial velocity, which is the ratio of the separator length and the slurry residence time in the matrix. Intuitively, one expects to have to maximize both the separator length and the matrix residence time for optimum performance; however, only their ratio has to be considered, and this fact makes superficial velocity the more fundamental parameter.

**Table 2.** Variables in Separator Performance.

| MATERIAL CHARACTERISTICS | SEPARATOR PARAMETERS |
|---|---|
| Particle Size: Solids and Impurity | Matrix Element Size and Spacing |
| Impurity Concentration | Matrix Material/Orientation |
| Solids Concentration | Magnetic Field Strength |
| Solids Type | Matrix Residence Time/Flow Velocity |
| Magnetic Susceptibility | Matrix Volume/Temperature |
| | Surfactant, pH |

## RESULTS AND DISCUSSION

We have completed a comprehensive series of HGMS experiments with nonradioactive surrogates, as discussed above, and have progressed to tests with radioactive material. The results to date are very promising and are summarized below.

### Contaminated Soils

After completion of the work with surrogate systems, work on radioactive samples was started. One of the biggest contamination problems within the DOE Defense Complex is contaminated soils. To address the feasibility of using HGMS for these problems, preliminary tests using uranium and plutonium spiked systems were investigated. Much of the original radioactive work was performed using the 2.0 tesla electromagnet.

**Uranium**. Figure 4 shows the results of an HGMS experiment to extract $UO_3$ from water. The efficiency of separation was expected to be only approximately 70%, based on model calculations, due to the low magnetic susceptibility of $UO_3$ ($\chi vol = 41x10^{-6}$ (SI)). The results show much better separation than expected. The $UO_3$ had an equivalent Stoke's diameter of 2-5 μm and the experiment was conducted at 2.0 tesla at pHs of 7 and 12. One other variable that was investigated was the addition of magnetite to the $UO_3$ slurry. The use of magnetite as a plutonium and americium waste treatment process was reported by Kochen and Navratil in 1987.[10] The process works by adsorption of the actinide to the ferrite surface, followed by magnetic separation. Magnetite is ferromagnetic and is easily extracted using magnetic separation.

Our results show that with only one pass through the magnetic separator at 2.0 tesla and pH 7, 91.7% of the uranium can be removed from the flow stream with the addition of magnetite. Also encouraging was that even without the addition of magnetite, >85% of the uranium was removed on two passes through the separator at pH 7. These results were very encouraging as we anticipated moving toward experiments with authentic uranium contaminated soils.

**$PuO_2$ Spiked Soils.** Many sets of experiments have been performed on $PuO_2$-spiked clay and $SiO_2$ slurries. Variables that been investigated include pH, soil particle size, $PuO_2$ particle size, and field strength. In general, for 2-5 μm $PuO_2$ particles, >90% of the plutonium can be removed from the influent flow stream during an HGMS experiment. This number is calculated as the difference in the influent activity and the effluent activity (Ci-Co)/Ci)(100). Typically, final plutonium concentrations are between 20 and 80 pCi/g.

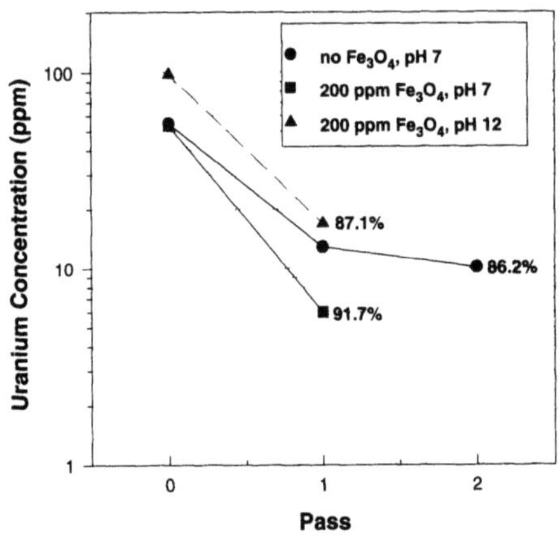


**Figure 4.** Extraction of $UO_3$ from Water. Variables: pH, $Fe_3O_4$; Field Strength = 2 tesla; $UO_3$ Particle Size = 2-5μm.

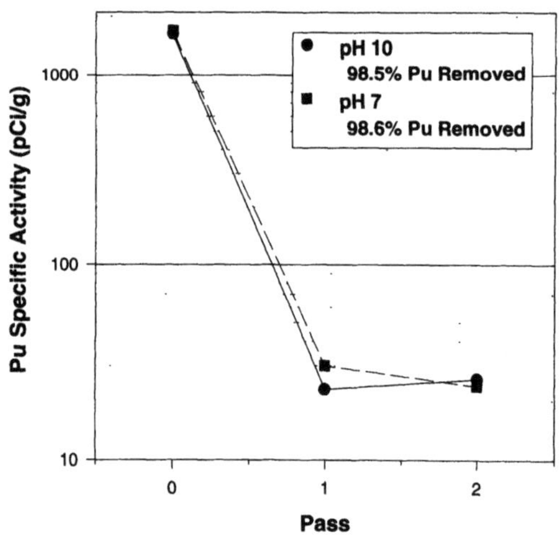


**Figure 5**. HGMS Extraction of $PuO_2$ from $SiO_2$: Effect of pH. Field Strength = 6.5 tesla; $PuO_2$ Size = 2-5mm; $SiO_2$ size = 10-45mm; 10% solids; 0.2% Hexametaphosphate.

Figure 5 shows the results of a typical HGMS experiment to extract $PuO_2$ from $SiO_2$. A 10 weight percent slurry was pumped through the separator at two different pH values. From the graph it is clear that there is little difference in the separation efficiency for removal of $PuO_2$ from $SiO_2$ at pH 7 and pH 10. About 98.5% of the plutonium activity is removed from the $PuO_2$-spiked mixture at both pH 7 and pH 10, with <2.5% of the bulk mass collected in the backflush. Similar results were seen for $PuO_2$-spiked clay slurries, however we observed *ca.* 10-15% better extraction of the $PuO_2$ at pH 10 than at pH 7.

The variable that had the most effect on the extraction of $PuO_2$ from $SiO_2$ was the solids particle size. When 10-45 μm $SiO_2$ was used the extraction efficiency was only about 85%, with 2.4% of the bulk mass collected in the backflush. With the <10 μm $SiO_2$ the extraction efficiency was >95% with 2.7% of the bulk mass collected in the backflush. This is probably due to the bigger $SiO_2$ particles knocking the smaller $PuO_2$ particles off of the matrix. This was not observed in the clay experiments because typically the clay has very few particles >20 μm in size.

**Authentic Soils.** At the Fernald Environmental Management Project (FEMP) near Cincinnati, Ohio, there are an estimated 2,000,000 to 4,000,000$m^3$ of soil contaminated with uranium. The source of the uranium contamination is from the deposition of airborne particulates from incinerator stacks, aqueous uranium wastes from chemical processing, and solid uranium product spills. The target clean-up for the soils is 30 ppm. Current technologies for remediation of the soil include 1) excavation and transportation of the soil to a repository, 2) immobilization of the uranium in place, 3) physical separation and removal of the highly contaminated soil fractions from the balance of the soil, and 4) soil washing. The incinerator areas, as mentioned above contain soil with uranium contamination in the form of particulates. This makes HGMS a strong candidate for partitioning of the uranium from the soil.

Figure 6 shows the results of an HGMS test run on an incinerator area soil from Fernald. The soil was dry sieved to <53μm and a 10 weight % slurry was run through the magnetic separator at 2.0 tesla and at three different pH values. The incinerator area soil contains about 500 ppm uranium. Spectroscopic characterization indicates that the uranium is ca. 75% U(VI) and 25% U(IV).

From the data it is clear that the separation did not work well at pH 10. However, pH 8 and pH 4 both show promising results. At pH 4 *ca.* 58% of the uranium was concentrated in 14% of the bulk mass. At pH 8 *ca.* 47% of the uranium was concentrated in only 3% of the bulk mass. Both of these results represent reasonable separation efficiencies for only

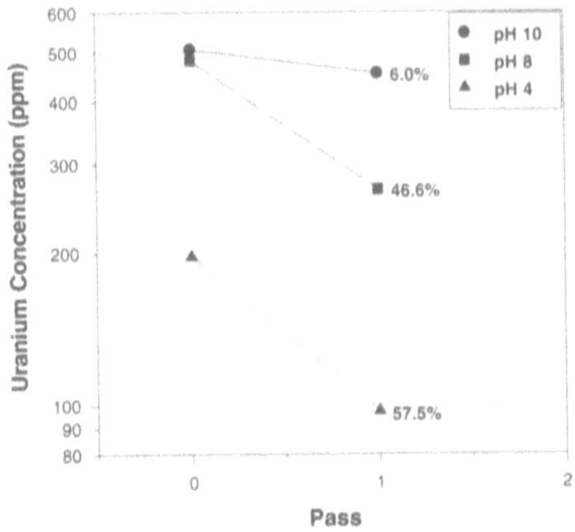


**Figure 6.** HGMS of Uranium from Fernald Soil. Field Strength = 2 tesla; 10% solids; 0.2% Hexametaphosphate.

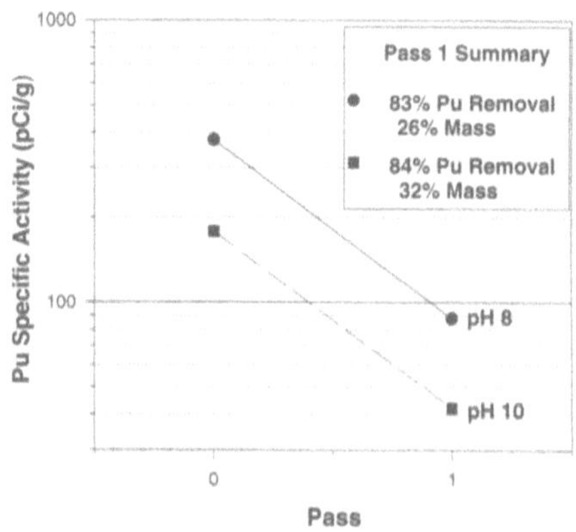


**Figure 7.** HMGS of plutonium from NTS Residue: LESAT Thickener Underflow Residue. Effect of pH. Field Strength: Pass 1 = 0.25 tesla; 10% solids; 0.2% Hexametaphosphate.

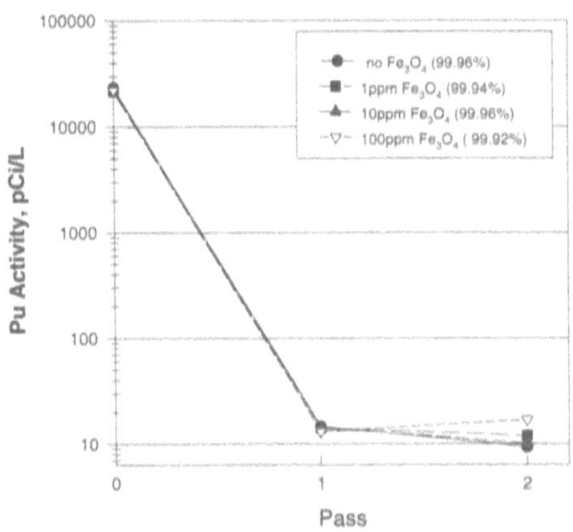


**Figure 8.** HGMS of LANL Waste Treatment Facility influent. Field Strength = 7.5 tesla; pH 8

one pass through the separator and at only 2.0 tesla field strength. However, in neither test run were we able to reach the 52 ppm uranium target. A second series of tests is planned with test variables of pH and field strength. Also planned are experiments involving reducing agents. If the U(VI), which has a low magnetic susceptibility, can be reduced to U(IV), ($\chi$ > $1000\times10^{-6}$(SI)) we have a much better chance at removing it by magnetic separation. Also planned are tests with magnetite seeding due to the success as had employing this technique with $UO_3$ in water.

The second DOE soil we have tested using HGMS is from the Nevada Test Site. The soil is a residue from Lockheed Environmental Systems & Technologies. The residue we received were the fines (<100μm) from the soil that had been processed through a centrifugal jig and a screw classifier. Again, the residue was sieved to <53μm and two passes of a 10 weight % slurry were passed through the magnetic separator at pH 8 and pH 10. The results from the first pass are shown in Figure 7. The first pass was performed at only 0.25 Tesla. At both pH values 83-84% of the plutonium was removed from the flow stream. However, *ca.* 30% of the soil mass was also caught in the separator.

On pass 2 at 0.5 tesla another 24% and 32% of the mass was collected at pH 8 and pH 10, respectively. These results indicate that the NTS residue is highly magnetic and the activity is probably tied up with the magnetic minerals. We are exploring other pretreatment options to try and liberate the plutonium from the soil. A second series of tests has been scheduled.

**Soil Results Summary**. We have conducted more than 100 HGMS experiments on paramagnetic surrogate materials in water and simulated soils. We have shown that >95% of the paramagnetic contaminant can be extracted from the surrogates. HGMS tests have begun on authentic contaminated soils. Preliminary tests have been performed on Fernald soil and NTS residue. Information from these tests is being used to determine test parameters and pretreatment steps. Future soils to be investigated include samples from Rocky Flats Plant, Johnston Atoll, Hanford, INEL, Mound and Los Alamos.

**Waste Water Treatment**

High Gradient Magnetic Separation (HGMS) is also being investigated for applications to waste water treatment. Los Alamos National Laboratory Waste Treatment Facility has an average annual influent of 20 million liters with an average gross alpha activity of 70nCi/L from isotopes of plutonium and americium.[11] The contaminants are in the form of particulates, colloids, polymerized colloids and dissolved ions. Another important characteristic of the waste is that most of the activity is associated with particles in the 1-20μm size range.

Current DOE regulations for the radioactive levels in the effluent allow 30pCi/L gross alpha for plutonium and americium combined. Current treatment at the waste facility results in an effluent activity of *ca.* 50pCi/L. Another major concern with the current treatment process is the large volume of low level radioactive sludge that is generated. Annually, 200 55-gal. drums of sludge are produced in the process. New technologies are being sought that can achieve the desired effluent activity levels while producing a minimum of secondary sludge waste. Because the contaminants are in the form of particulates or colloids in the 1-20μm size range, HGMS is a strong candidate to address the problems at the LANL Waste Treatment Facility. HGMS also has the advantage of producing a minimum of secondary waste.

One of the 3-inch bore, 8 tesla superconducting magnets was set up at the waste treatment facility to perform treatability tests on plant influents. The first test was performed on influent that had been in a holding tank for *ca.* six months. The pH was adjusted to pH 8. Three different concentrations of magnetite were added to the influent to aid in the separator performance and address the problem of the dissolved ions. The test was performed at a field strength of 7.5 tesla and the results are shown in Figure 8. In each run, over three orders of magnitude of reduction in plutonium activity were achieved. It appears that the addition of magnetite is not necessary to achieve these separation efficiencies. Activity levels below 10pCi/L were achieved, well below the 30pCi/L required by the DOE, with or without magnetite seeding.

Future HGMS tests are planned on raw waste treatment facility influent, (i.e. influent that has not been stored in a holding tank.) The question has been raised as to whether the influent is changing with time in the holding tank.

## CONCLUSIONS AND DEVELOPMENT PLANS

We have demonstrated that HGMS can be used to concentrate actinides in waste residues and contaminated soils. One of the major benefits of this technology is that it partitions the existing waste volume, creating a low volume, high contaminant concentration component and a high volume, low contaminant concentration component. If the separation efficiency is high, the high volume component can be discarded and the low volume component can be processed further or packaged for disposal. Concentration of the actinides from extraneous materials before further processing can result in more efficient recovery or treatment operations.

We plan to continue to utilize our magnetic separators for surrogate and authentic contaminated soils, residues and waste streams to determine operational parameters. The experimental results will continue to be used to develop a performance-based HGMS model. We hope to demonstrate the application of HGMS to the separation of actual residues, waste streams and contaminated soils and to develop HGMS for new applications. Results from the CRADA tests are being used to support LESAT's prototype separator design effort.

## REFERENCES


1. Avens, L.R., Worl, L.A., de Aguero, K.J., Padilla, D.D., Prenger, F.C., Stewart, W.F., Hill, D.D., Tolt, T.L. *Proceedings of the Symposium on Waste Management*, Tuscon, Arizona, Mar, 1993, p. 787; Worl, L.A., Avens, L.R., Aguero, K.J., Prenger, F.C., Stewart, W.F., Hill, D.D. *Proceedings of the Symposium on Waste Management,* Tuscon, Arizona, Mar, 1993, p. 1039.
2. Akoto, I.V. *IEEE Transactions on Magnetics* **1977**, *13*, 1486.
3. Friedlander, F.J., Takayasu, M., Rettig, J.B., Kentzer, C.P., *IEEE Transactions on Magnetics* **1978,** *14*, 1158.
4. Lawson, W.F., Simons, W.H., Treat, R.P., *J. Appl. Physics* **1977**, *48*, 3213.
5. Luborsky, F.E., Drummond, B.J. *IEEE Transactions on Magnetics* **1976**, *12*, 463.
6. Nesset, J.E., Finch, J.A., Industrial Appl. Mag. Sep., Conference Proceedings **1978,** 188.
7. Stekly, Z.J.J., Minervini, J.V. *IEEE Transactions on Magnetics* **1976**, *12*, 474.
8. Watson, J.H.P. *J. Appl. Physics* **1973**, *44*, 4209.
9. Prenger, F.C., Stewart, W.F., Hill, D.D., Avens, L.R., Worl, L.A., Schake, A.R., de Aguero, K.J., Padilla, D.D., Tolt, T.L. "High Gradient Magnetic Separation Applied To Environmental Remediation", Cryogenic Engineering Conference, Albuquerque, NM, July, 1993. DOE Report LANL LA-UR-93-2516.
10. Kochen, R.L., Navratil, J.Dl. *Lanthanide and Actinide Research* **1987**, 9, and references therein.
11. Covey, J.R., Midkiff, W.S., Cadena, F. *Industrial Waste Conference*, 47th Annual Purdue University, May, 1992.

**Appendix 1. List of Participants in the Symposium (Presenting Authors Underlined):**

Recent Advances in the Separation of Actinides by Liquid-Liquid Extraction, Extraction Chromatography, and Ion Exchange. E. P. Horwitz, R. Chiarizia, H. Diamond, M. Dietz, R. A. Leonard

Superlig®-Mediated Separations Relevant to Nuclear Waste Remediation. R. M. Izatt, J. S. Bradshaw, R. L. Breuning, B. J. Tarbet, K. E. Krakowiak

Removal of Plutonium and Americium from Hydrochloric Acid Waste Streams Using Extraction Chromatography. L. D. Schulte, J. R. FitzPatrick, B. T. Martinez, R. R. Salazar, B. S. Schake

The Sorption of Uranium(VI)/(IV) on Reillex™-HPQ Anion Exchange Resin from Nitric/Hydrochloric Acid Solutions. A. B. Pinkerton, K. R. Ashley, K. D. Abney, N. C. Schroeder

Selective Chelators for Actinide Complexation. V. J. Huber, N. Koshti, A. S. Gopalan, G. D. Jarvinen, P. H. Smith

New Water Soluble Phosphonate and Polycarboxylate Complexants for Enhanced f Element Separations. K. L. Nash, P. G. Rickert, E. P. Lessmann, M. D. Mendoza, J. F. Feil, J. C. Sullivan

EXAFS Studies of the Structure of Aqueous Pu(IV) Nitrato Complexes. D. K. Viers, P. Allen, C. A. Smith, S. Conradson

Separation of Americium from Europium by Solvent Extraction from Aqueous Phosphonate Media. D. D. Ensor, K. L. Nash

Determination of the Stability Constant for the Pu(IV)-Dinitrato Complex Using UV-Vis Spectrophotometry. C. A. Smith, D. K. Viers

Probing Metal Ion Interactions with Water Soluble Complexants: From Solvent Extraction to Pollution Prevention. J. V. Beitz, K. L. Nash

Design of Ligands for f Element Separations. R. T. Paine

Soft Donor Ligands for Separation of Trivalent Actinides and Lanthanides. G. D. Jarvinen, B. F. Smith, R. E. Barrans, Jr., M. Jones, R. R. Gibson

Extraction of f Elements with Soft Donor Adducts. G. R. Choppin, J. Yao, R. M. Wharf

Uranium Recovery from Contaminated Soil. D. J. Chaiko, R. Mensah-Biney

TRU Extraction by Bridge-Modified Carbamoylmethylphosphonates and Phosphine Oxides. B. M. Rapko

New Methods for Transplutonium Elements Isolation, Purification, and Separation from Rare Earth Elements and Selected Fission Products. B. F. Myasoedov

The Use of Selective Inorganic Ion Exchangers for Separation of Rare Earths. A. Clearfield, R. A. Cahill, S. B. Wright, P. C. Bellinghausen, B. Shpeizer

Magnetic Separation for Environmental Remediation. A. R. Schake, L. R. Avens, L. A. Worl, K. J. de Aguero, D. D. Padilla, F. C. Prenger, W. F. Stewart, D. D. Hill, T. L. Tolt

Plutonium and Americium Separation Using Organophosphorus Extractant Adsorbed onto Ferromagnetic Particles. L. Nuñez, G. F. Vandegrift

Factors Influencing the Efficiencies of Multistage Separations of Lanthanides. S. Muralidharan, G. Ma, H. Freiser

# INDEX


1-phenyl-3-methyl-4-benzoyl-5-pyrazolone (HPMBP), 177, 180-181
  lanthanide separation factors, 183
  mechanism of dissociation of Ln complex, 188-191

Actinide covalency, 31
Actinide/lanthanide solution chemistry, 126
  complexants for, general, 12
  soft-donor effects in, 126
Actinide processing from hydrochloric acid effluent streams, 199-208
Actinide wastes, 257
Americium
  (IV) in phosphotungstate solutions, 23
  (V) extraction, 24
  (VI) extraction, 25
  distribution onto MACS particles, 248, 250
  extraction by bridge-modified CMPO, 114-115
  extraction by CMPO and bisphosphine oxides, 12
  extraction by DMDHOPDA, 36
  extraction by tris(isobutyl)phosphine sulfide + HTTA, 31, 35
  extraction with HSTTA, 35
  fluoride complexes, 252
  phosphonoacetic acid complexes 143
Am/Eu separation factors
  comparison of dithiophosphinic acid extractants, 61
  with CMPO/diethylbenzene and phosphonoacetic acid, 151
  with Cyanex 301, 55, 57
  with DCHDTP, 50, 52
  with HDEHP/toluene and phosphonoacetic acid, 150
  with HDNNS/toluene and phosphonoacetic acid, 150
  with HDPhDTP, 59
  with HSTTA+TBP, 40, 41
Amberlite XAD-7, 200
Aqueous biphasic extraction, 225
  Arsenazo III as phase transfer agent, 226
  decontamination of soils by, 225
  silica partitioning in, 236-237
  use of polymeric flucculants in, 237-238
Aqueous biphasic extraction of uranium from soils, 225-239
Aqueous biphasic extraction with polyethylene glycols, 16-17
  $^{239}$Np isotope generator, 17
  use of Arsenazo III in, 16
Aqueous biphasic extraction... (*cont'd*)
  use of potassium phosphotungstate in, 17
Aqueous complexants
  α-hydroxyisobutyric acid, 4
  Arsenazo III, 16, 186-188, 226
  HEDPA, 19
  MMDPA, 133
  phosphonoacetic acid, 150-151
  PMDPA, 133
  THFTCA, 125, 128, 155-162
Aqueous phosphonate complexants, 128
  aminomethanediphosphonate complexes, 125
    MMDPA (morpholinomethanediphosphonic acid), lanthanide complexes, 133
    PMDPA (piperidinylmethanediphosphonic acid), lanthanide complexes, 133
    synthesis of, 129
    potentiometric titration of ligand and lanthanide complexes, 130, 132
    protonation constants, 133
  hydration of, from $Eu^{3+}$ fluorescence decay rate, 128, 132
  lanthanide complexes of, 131
    stability constants, 133
Aqueous processing of Pu residues, 200
Arsenazo III-lanthanide complex, 186-188

Bidentate neutral organophosphorus extractants, 12-14
Bifunctional organophosphorus compounds, 99
Binding constants, 82
Bis(2,4,4trimethylpentyl)phosphinic acid (Cyanex 272), 177, 180-181
  lanthanide separation factors, 183
  aqueous stability constants for lanthanide complexes, 184-186

Bis(2,4,4trimethylpentyl)phosphinic acid (*cont'd*)
mechanism of dissociation of lanthanide complexes, 191-193
Bismuth phosphate process, 1
Bridge-modified CMPO's, 99-121
synthesis of 102-103, 107-108
characterization of , 103-105
by chromatography, 108-109
by infrared spectroscopy, 110-111
by mass spectrometry, 109-110
by NMR spectroscopy, 111-113
distribution ratios, 105-106, 113-117
of Am(III), 114-115
of $HNO_3$ into, 113-114
of Pu(IV), 115-116
of U(VI), 116-117

Catecholates, 79
Centrifugal Partition Chromatography, 177, 178-179
band widths and chemical kinetics, 184-186
correlation of inefficiency with dissociation kinetics, 193-195
La/Pr/Sm/Gd/Dy/Tm separation with Cyanex 272, 184
Nd/Sm/Eu separation with Cyanex 272, 182-183
Pr/Eu separation with HPMBP, 181-183
Channel equivalent theoretical plate (CETP), 177, 179, 193-195
Chelating polymers for actinides, 91-93
binding of Fe(III), 93
binding of Nd(III), 93
binding of Th(IV), 93
Chelators for actinide specific bonding, 78
CMP ligands, 64
CMP's (bifunctional extractants), 6
CMPO, 100-101, 102, 103,199, 200, 201-202, 241, 241, 249, 252
CMPO's and bisphosphine oxides, 12-14
Am(III) extraction, 12
perchloric acid effect on extraction, 12-13
Pu(IV) extraction, 12
U(VI) extraction, 12
CMPO ligands, 64
CMPO's (bifunctional extractants), 6
bridge modified, 99-121
Comparison of dissociation kinetics of Ln-HPBMP and Ln-Cyanex 272 complexes, 189-190
Computer modeling, 80, 90
Covalent bonding of actinides, 43, 44
Crystal structure of Nd(III)-1-hydroxyethane-1,1-diphosphonic acid complex, 131
CYTROX (cyclam tetrahydroxamate), 90
CYTAC (cyclam tetraacetylacetonate), 91

DA(A)P (diamyl(amyl)phosphonate), 199
Determination of stability constants from distribution measurements, 145, 147-149
DHDECMP, 100, 102, 103
Dipyridine-N-oxide phosphonates, 72
Dispersion number, 230-231
Distribution coefficients
for Am(III), 202
for Pu(IV), 199, 201-202
for uranyl nitrate on Reillex™ HPQ, 213-214
Dithiophosphinic acid extractants, 44
bis(2,4,4-trimethylpentyl)dithiophosphinic acid (Cyanex 301), 45, 54-57
dicyclohexyldithiophosphinic acid (HDCHDTP), 45, 47-54
diphenyldithiophosphinic acid (HDPhDTP), 45, 46, 57-61
Dowex™ 1X8, 209, 210
uranyl nitrate sorption on, 209, 215

Effect of temperature on aqueous biphase formation, 230
Eu emission spectrum, 157-158
fluorescence lifetimes, 160, 162
$Eu^{3+}$ hydration by fluorescence, 160-161
in THFTCA complexes, 161-162
Environment remediation, 256
Extractants
2,4,6-tris[diarylphosphoryl]-1,3,5-triazine, 19-20
ß-diketones, 15
CMP's, 6
DHDECMP, 100, 102, 103
CMPO's, 6, 12-14, 64, 99-121, 199, 200, 201-202, 240, 241, 249, 251
Cyanex 272, 177, 180-184, 189-190
Cyanex 301, 45, 54-57
CYTROX (cyclam tetrahydroxamate), 90
CYTAC (cyclam tetraacetylacetonate), 91
DA(A)P, 199
dithiophosphinic acids
bis(2,4,4-trimethylpentyl)dithiophosphinic acid (Cyanex 301), 45, 54-57
dicyclohexyldithiophosphinic acid (HDCHDTP), 45, 47-54
diphenyldithiophosphinic acid (HDPhDTP), 45, 46, 57-61
HDEHP, 5, 7, 150
HSTTA, 32,33,35,39
HTTA, 31,35,38-40
pyrazolones
DMDHOPDA, 32, 33, 36-40
HPMBP, 177, 180-183
TBP, 40, 41, 199
Tetraphenylmethylenediphosphine oxide, 19

Extraction from alkaline media, 15-16
separation factors, 16
with alkylpyrocatechols, 15
with ß-diketones, 15
with quaternary ammonium bases, 15-16
Extraction chromatography, 199-208
loading capacities for EC materials, 203-204
Extraction of Pu, Am from HCl, 199, 200

f element coordination chemistry, 63-64

Fernald Environmental Management Project (FEMP)
soils, 265-267
uranium removal from, 266
Fernald (soil decontamination), 226
Ferromagnetic particles, 242-243, 246
Fluoride complexes of Am, Pu, 252
Fluorine containing diluents, 13-14
Fluopol 732, 14

Generic TRUEX model (GTM), 252

HDEHP (bis(2-(ethyl)hexyl)phosphoric acid), 5,7
High Gradient Magnetic Separation (HGMS),
257-259
analytical model, 260-261
contaminated soils in, 263
spiked with $PuO_2$, 263-264
field gradients, 258
plutonium removal by, 264
schematic diagram, 257
superconducting magnets, 259-260
uranium removal by, 263
variables in separation performance, 262
High-level waste treatment, 241, 251
PFP simulant waste stream, 252-253
composition of, 244
HTTA + DMDHOPDA extraction, 38-40
Humate behavior in aqueous biphasic treatment of
soils, 235-236, 237, 238
Hydration of $Eu^{3+}$ in the extractant phase, 38
Hydroxamic acids, 89
Synthesis, 89

Inorganic ion exchangers for separation of rare earths,
165-175
preparation of, 167
Intragroup separations of trivalent actinides and
lanthanides
by α–hydroxyisobutyric acid, 4
by HDEHP, 5
Ion exchangers
Dowex™ 1X8, 209, 210
Reillex™ HPQ anion exchange resin, 209-222
sodium nonatitanate, $Na_4Ti_9O_{20}•xH_2O$, 165,
167-170
$Zr(C_6H_4P_2O_6)_x(HPO_4)_{2-2x}•x\ H_2O$ (ZrMPP), 165,
167, 170-174
$Zr(C_{12}H_8P_2O_6)_x(HPO_4)_{2-2x}•x\ H_2O$ (ZrBPP), 165,
167, 174-175
Iron, 82

Karr column, 226, 227-228
contaminated soil treatment in, 231-235

Lanthanide cation radii, 166
Lanthanide/actinide separation, 2
2,4,6-tris[diarylphosphoryl]-1,3,5-triazine, 19-20
by anion exchange from LiCl, 3-4
by cation exchange from HCl, 3
by chloride anion exchange, 127
Lanthanide/actinide separation (*cont'd*)
by partition countercurrent chromatography, 17-20
by TALSPEAK, 127
HEDPA, used in, 19
need for, 2
Tetraphenylmethylenediphosphine oxide,
$NH_4SCN$, 19
Lanthanide/actinide solution chemistry, 2
Lanthanides, history of discovery, 1
Laser-induced fluorescence of $Eu^{3+}$, 154, 159
coordination numbers from, 160-162
hydration numbers from, 160-162
Ligand-metal charge transfer, 154, 158
Ligand preorganization, 128
Luminescence of $Eu^{3+}$, 38

MACS particles
coating procedure, 245
dissolution in $HNO_3$, 246, 250-251
distribution of $^{241}Am$ onto, 248, 250
temperature effect on, 252-252
distribution of $^{238}Pu$ onto, 248, 250
extraction process using, 245
magnetite distribution in 247, 249
microscopy and electron dispersive
spectroscopy, 245, 246-247
removal of plutonium, 253-254
stripping procedure, 253-254
transmission electron micrographs, 249
Magnetic separation, 257
Magnetic separation for environment
remediation, 257-269
Magnetically Assisted Chemical Separation
(MACS), 241-255
Mathematical modeling of sorption
equilibria, 217-218.
Membrane separation of actinides, 25-28
transfer kinetics of actinides through SLM's, 25-26
Metallochromic indicator method, 186-188
Molecular modeling, 132, 134
Multistage separation efficiency of
lanthanides, 177-197

Nevada Test Site soil, 267
N,N'-dimethyl-N,N'-dihexyl-3-oxapentanediamide
(DMDHOPDA), 32,33,36-38
Am extraction by, 36
Nuclear waste treatment, 154

Paramagnetic species, 257
Phase diagrams and tie lines, 229-231
PEG-1500/$Na_2CO_3$, 229-230
Phosphonoacetic acid, 142-153
Am(III) complexes, 143
stability constants, 149-150
Eu(III) complexes, 143
stability constants, 149-150
potentiometric titration, 145
protonation constants, 146

Plutonium
removal by HGMS, 259
removal by MACS particles, 253-254
(III) retention by extraction chromatographic materials, 205
fluoride complexes, 252
extraction by extraction chromatographic materials from HCl, 199, 200
extraction by CMPO, 12
extraction by bridge-modified CMPO's, 115-116
Polyethylene glycol(s), 225, 227
Polyhydroxamate Chelators, 77-95
Production of transplutonium elements, 11
PUREX process, 2
Pu(III) retention by extraction chromatographic materials, 205
Pu(IV) oxidation state purity, 105-106
Pu and Am separation by magnetic particles, 241-255
Pyridine-N-oxide ligands-2,6-phosphonate esters, 64-73
Bi(III) complex, 68-69
Pr(III) complexes, 67-69
Tb(III) complex, 68
uranyl complex, 66
Y(III) complex, 69
Pyridine-N-oxides-2,6-methylenephosphonates, 71
Nd(III) complex, 71
species calculation plot, 71

Radioactive waste remediation, 78
Radioactive waste treatment, 199
REDOX process, 2
Reillex™ HPQ anion exchange resin, 209-222
elution of uranyl from, 219-220
breakthrough volumes, 209,-210, 212, 220-221
technetium sorption, 210, 215
uranium sorption from nitrate media, 209, 212
effect of nitrate salts on sorption of, 214, 216
uranium sorption from chloride, 210, 220-221
RE-Spec resin, 200, 202, 204

Seaborg, Glenn T., 1
Secondary waste treatment, 238
Selective removal of TRU's from waste, 101
Separation factors
amine extraction with phosphotungstate for separation of Bk(IV) from various metals, 26
Am/Eu,
HDEHP/toluene and phosphonoacetic acid, 150
HDNNS/toluene and phosphonoacetic acid, 150
CMPO/diethylbenzene and phosphonoacetic acid, 151
with HSTTA+TBP, 40, 41
with organophosphorus complexants, 143-152
for extraction from alkaline media, 16
lanthanide,
with Cyanex 272, 183

Separation factors
lanthanide (*cont'd*)
with HPMBP, 177, 180-181
with ZrBPP, 173
with ZrMPP, 173
trivalent actinides/lanthanides, 31-41
Separation methods for transplutonium elements based on oxidation states, 21-25
Am(IV) in phosphotungstate solution, 23
Am(V) Extraction , 24
Am(VI) extraction with electrochemical oxidation, 25
Bk(IV) extraction with amines, 25-26
Separation of Am/Eu
with CMPO/diethylbenzene and phosphonoacetic acid, 151
with Cyanex 301, 55, 57
with DCHDTP, 50, 52
with HDEHP/toluene and phosphonoacetic acid, 150
with HDNNS/toluene and phosphonoacetic acid, 150
with HDPhDTP, 59
with HSTTA+TBP, 40, 41
with organophosphorus complexants, 143-152
Separation of trivalent actinides/lanthanides, 31-41
Separation processes
anion exchange, 3-4, 127, 209-222
bismuth phosphate, 1
cation exchange, 3, 165, 167-175
chromatography, 17-20, 108-109, 177-179, 199-208
HGMS, 257-269
MACS, 241-255
PUREX, 2
REDOX, 2
TALSPEAK, 5, 8
TRUEX, 6, 100, 242, 253
Siderophores, 79
Sodium nonatitanate, $Na_4Ti_9O_{20}\cdot xH_2O$, 165, 167-170
lanthanide $K_d$'s into, 169
Soft-donor atoms, 43, 44
Sorption kinetics, 201, 202, 207
Speciation, 154
Spectrophotometric analysis for iron complexes, 86-88
Synergistic extraction, 43

TALSPEAK process, 5, 8
TBP (tributylphosphate), 199
Tetrahydrofuran-2,3,4,5-tetracarboxylic acid (THFTCA), 125, 128, 155-162
as a stripping agent, 153
in the Combined Process, 135
comparison of lanthanide complexes with other ligands, 139
crystal structure of the monohydrate, 134
distribution of into Combined Process solvent, 130, 135-137
lanthanide complex stability constants, 138

Tetrahydrofuran-2,3,4,5-tetracarboxylic acid (THFTCA) (*cont'd*)
photochemistry of, 154-155
photodestruction of, 155
potentiometric titration of free ligand and lanthanide complex, 130
purification of, 129
species distribution of lanthanide complexes, 138
ultraviolet photolysis, 157-158, 162-163
Tetrahydroxamates, 80-87
Fe(III) complex stability constants, 84
preparation of, 81
protonation constants, 83
Pu(IV) complex stability constants, 84
Th(IV) complex stability constants, 84
Thenoyltrifluoroacetone extraction, 36-37, 39
Thiothenoyltrifluoroacetone (HSTTA), 32,33, 35,39
Am/Eu extraction, 35
Thorium, 82
Transuranium elements, history of discovery, 1
Tris(isobutyl)phosphine sulfide + HTTA, 31,35
Am/Eu extraction, 34-35
Triton X-100 micelle, 183, 186, 188, 191, 193, 195-196
Trivalent actinide/lanthanide separation, 43-62
TRU-Spec resin, 200, 203-205, 206
TRUEX process, 6, 100, 241

Uranium
distribution onto bridge-modified CMPO's, 116-117
extraction by CMPO's and bisphosphine oxides, 12-14
extraction from soils by aqueous biphasic extraction, 225-240
phosphate, 226
removal by HGMS, 263
removal from Fernald Environmental Management Project (FEMP) soils, 265-267
silicates, 226
sorption onto Reillex™ HPQ anion exchange resin
from chloride, 210, 220-221
from nitrate, 209, 212
trioxide, 226
Uranium contaminated soils, 226, 227
Uranium phosphate, 226
Uranium silicates, 226
Uranium trioxide, 226

Water-soluble polymers, 94
polyethyleneimine, 94-96

$Zr(C_6H_4P_2O_6)_x(HPO_4)_{2-2x}{\cdot}x\ H_2O$ (ZrMPP), 165, 167, 170-174
idealized structure, 170
lanthanide $K_d$'s into, 173
separation factors for lanthanides, 173
thermogravimetric analysis of, 169

$Zr(C_{12}H_8P_2O_6)_x(HPO_4)_{2-2x}{\cdot}x\ H_2O$ (ZrBPP), 165, 167, 174-175
idealized structure, 174
lanthanide $K_d$'s into, 173
separation factors for lanthanides, 173
thermogravimetric analysis of, 172

MIX
Papier aus verantwortungsvollen Quellen
Paper from responsible sources
FSC® C105338

If you have any concerns about our products,
you can contact us on
**ProductSafety@springernature.com**

In case Publisher is established outside the EU,
the EU authorized representative is:
**Springer Nature Customer Service Center GmbH**
**Europaplatz 3, 69115 Heidelberg, Germany**

Printed by Libri Plureos GmbH
in Hamburg, Germany